Praise for Howard Gardner's *Creating Minds*

"A humanistic spirit pervades [*Creating Minds*]. . . . Gardner isn't trying to reduce creativity to maze-running. . . . He is tentative rather than dogmatic, attuned to exceptions and complexities as well as commonalities."
—*The Houston Chronicle*

"Gardner may well have uncovered some fundamental aspects of the creative personality and of the process of creativity. His discussion will inevitably open up more of this fascinating territory." —*New Scientist*

"[Gardner's] books are lucid, cross-disciplinary examinations of heady topics: *Creating Minds* . . . and *Leading Minds* . . . are rarities, being academic studies that are as readable as they are compelling." —*The Boston Globe*

"Mention Howard Gardner's name to a growing cadre of educators and the response verges on the reverence teenagers lavish on a rock star. . . . [*Creating Minds*] is sure to get attention not only for Gardner's typology of intelligence but also because of his guru-like status." —*Newsweek*

"One of the notable characteristics of creativity that Howard Gardner emphasizes in this new study is the special amalgam of the childlike and the adult: Creative personalities, he argues, often display features such as innocence and freshness, as well as selfishness and retaliation. . . . Their demanding personalities and devotion to their own creative breakthroughs (which tend, Gardner argues, to take place at 10-year intervals) also make creative people very hard to live with. But the creative process depends upon the support of caring individuals."
—*Washington Post*

"Few things inspire more wonder than the power of genius. . . . Gardner derives his view of genius from his earlier, groundbreaking research on the specialized nature of intelligence. . . . From this perspective, he questions whether creative minds of the caliber of Freud's or Einstein's will ever come to dominate the 21st century. These earlier geniuses made their mark by challenging the well-established thinking of the day. But today, Gardner says, there is really no such thing as establishment thought." —*US News and World Report*

"[Gardner's] enthusiasm and long experience show. *Creating Minds* is a stimulating work that fulfills the author's wish to write a book of the sort he himself likes to read: 'a jargon-free one with only the most essential visual aids'. . . . Gardner's writing style is remarkable in other regards too. He is a fluent writer, at great ease with the English language—and so confident of his ideas that he is not afraid to express them clearly. . . . Everyone who is interested in understanding and fostering creativity—and maybe that should be all of us—should read this rich, enthusiastic book to learn more about creativity, about seven fascinating creative minds—and maybe about the creative potential of ourselves and those in our care." —*Times-Picayune* (New Orleans)

"Rejecting the idea that creativity can be measured on a single linear scale, [Gardner] argues instead that many forces are at work to drive the creative individual, including such unexpected motivations as competition, ego, vanity and fear of death. This is a nice thought for those of us who feel that if pushed enough we could all write the next great symphony."

—*The Dallas Morning News*

"[This] groundbreaking work on brain functioning by Harvard researcher Howard Gardner has shed further light on the vitality and centrality of imagination and its close intellectual relatives." —*Pittsburgh Post Gazette*

"[A] boldly ambitious study. . . . Each of the seven creative geniuses whom Gardner incisively limns transcended interpretive frames or conventions that became entrenched during the 19th century; each forged a new 'system of meaning'; and each, in Gardner's view, struck a 'Faustian bargain,' sacrificing a rounded personal life for the sake of an all-consuming mission. . . . This highly stimulating synthesis illuminates the creation of the modern age."

—*Publishers Weekly*

"A delightful look at creativity . . . rich, readable, and thought-provoking."

—*Vision*

"Gardner has uncovered other intelligences we had failed to notice because we had no tools sensitive enough to measure them." —*BusinessWorld*

"It takes chutzpah to come up with a scheme for analyzing creativity—especially in subjects already exhaustively examined. But for psychologist and MacArthur fellow Gardner (Harvard Graduate School of Education), it amounts to a natural progression from his earlier dissections of intelligence."

—*Kirkus Reviews*

"Illuminating and entertaining, *Creating Minds* provides an unforgettable synthesis of the ideas that have shaped contemporary culture. . . . As the guide in this tour of the theater of the mind, Gardner is at his best: insightful, civilized, and precise. I can't think of a more stimulating book about creativity."

—Mihaly Csikszentmihalyi, author of *Creativity*

"*Creating Minds* is both informative and a wonderful read."

—Robert Ornstein, author of *Roots of the Self*

Creating Minds

OTHER BOOKS BY HOWARD GARDNER

The Quest for Mind (1973)

Arts and Human Development (1973)

The Shattered Mind (1975)

Developmental Psychology (1978)

Artful Scribbles (1980)

Art, Mind, and Brain (1982)

Frames of Mind (1983)

The Mind's New Science (1985)

To Open Minds (1989)

The Unschooled Mind (1991)

Multiple Intelligences: The Theory in Practice (1993)

Leading Minds (with Emma Laskin) (1995)

Extraordinary Minds (1997)

The Disciplined Mind (1999)

Intelligence Reframed (1999)

Good Work (with Mihaly Csikszentmihalyi and William Damon) (2001)

Changing Minds (2004)

Multiple Intelligences: New Horizons (2006)

The Development and Education of the Mind (2006)

Five Minds for the Future (2007)

Truth, Beauty, and Goodness Reframed (2011)

Creating Minds

An ANATOMY of CREATIVITY

SEEN THROUGH *the* LIVES *of* FREUD, EINSTEIN,

PICASSO, STRAVINSKY, ELIOT, GRAHAM, *and* GANDHI

HOWARD GARDNER

BASIC BOOKS
A Member of the Perseus Books Group
New York

Published by Basic Books,
A Member of the Perseus Books Group

Book design by Linda Mark
Text set in 10.75 point AGaramond by the Perseus Books Group

The Library of Congress has cataloged the hardcover edition as follows:

Gardner, Howard.
    Creating minds / Howard Gardner.
        p. cm.
Includes bibliographical references and index.
    ISBN-10: 0-465-01454-2 (paper)
    ISBN-13: 978-0-465-01454-5 (paper)
    ISBN: 0-465-01455-0 (cloth)
    1. Creative ability—Case studies. 2. Gifted persons—Biography.
I. Title.
BF408.G33 1993
153.3'5'0922—dc20
[B]92–56172
CIP

2011 paperback edition ISBN: 978-0-465-02774-3
E-book ISBN: 978-0-465-02786-6

10 9 8 7 6 5 4 3 2 1

For Andrew
Benjamin
Jay
Kerith

CONTENTS

Contents

ACKNOWLEDGMENTS

NUMEROUS COLLEAGUES GENEROUSLY read and commented on sections of this book. I should like to thank Mihaly Csikszentmihalyi, William Damon, Rupen Das, Iris Fanger, Ina Hahn, Gerald Holton, Arthur Miller, Ricardo Nemirovsky, Robert Ornstein, David Perkins, Dean Keith Simonton, and Ellen Winner. For information about Martha Graham I owe a special debt to Jane Sherman. With thanks and with sadness, I acknowledge the invaluable help of my friend Stephen Albert, a great composer and a wonderful critic, who died in December 1992.

At Basic Books, Susan Arellano, Martin Kessler, and Jo Ann Miller provided valuable editorial suggestions. Several colleagues at Project Zero, including Karen Donner Chalfen, Lela Collins, Samantha Kelly, and Mindy Kornhaber, helped with the preparation of the manuscript; Emma Laskin played an indispensable role during the last year in helping me with every facet of the book, from chasing down a footnote to clarifying an obscure thought. For their help in the final phases of publication, I would like also to thank Melanie Kirschner, Michael Mueller, and Sharon Sharp. Cynthia Dunne and Joan Greenfield are responsible for the book's attractive design.

My work over the past twenty-five years has been possible only because of the generosity of many foundations. If over that period in the United States there has been any progress in our understanding of the most fundamental issues about human nature, it is due in significant measure to the wisdom and flexibility of these funders.

For permission to quote materials or use images, I wish to thank the following:

- Artists Rights Society, N.Y., and SPADEM, Paris, for permission to reprint the artistic works of Picasso.

- Eidgenössische Technische Hochschule, Zürich, Switzerland, for permission to reprint the photograph of Albert Einstein.

- Faber and Faber, Ltd., London, and Harcourt Brace Jovanovich, Inc., Orlando, Fla., for permission to reprint excerpts from *The Letters of T. S. Eliot: Volume 1 1898–1922*, copyright © 1988 by Valerie Eliot. Reprinted by permission of the publishers.

Acknowledgments

- Harcourt Brace Jovanovich, Inc., Orlando, Fla., for the permission to reprint excerpts from "The Lovesong of J. Alfred Prufrock" from *Collected Poems, 1909–1962,* copyright © 1936 by Harcourt Brace Jovanovich, Inc., and 1964, 1963 by T. S. Eliot. Reprinted by permission of the publisher.

- The Houghton Library, Harvard University, Cambridge, Mass., for permission to reprint the photograph of T. S. Eliot.

- François Meyer, the estate of André Meyer, Artephot-Ziolo, and the Bibliothèque National, Paris, for permission to reprint pages from *Igor Stravinsky: The Rite of Spring, Sketches 1911–1913,* published by Boosey and Hawkes, 1969.

- Czeslaw Milosz Royalties, Inc., and the Ecco Press for permission to reprint "Youth," from *Provinces* by Czeslaw Milosz, copyright © 1992 by Czeslaw Milosz Royalties, Inc., first printed in 1992 by the Ecco Press.

- Lloyd Morgan and the Willard and Barbara Morgan Archives, Dobbs Ferry, N.Y., for permission to reprint photographs of Martha Graham by Barbara Morgan.

- Museé Picasso, Paris, for permission to reprint the photograph "Picasso sur la place Ravignon."

PREFACE TO THE 2011 EDITION

THOUGH PARENTS SHOULD NOT have a favorite child, authors ought to be permitted to have a favorite book. I've written well over twenty books, but in many ways *Creating Minds* is my favorite. Preparation of the book was a labor of love. I reveled as I dove into the rich repositories of information about the seven master creators whom I was describing: examining primary sources, reading biographies and notebooks of Sigmund Freud, watching the films of dancer Martha Graham shot in the 1930s and 1940s, poring over the drafts of T. S. Eliot's *The Waste Land*, listening again and again to Igor Stravinsky's pathbreaking compositions, looking at sketches for Pablo Picasso's boldest canvases, attempting to piece together the many fragments of autobiography left by Mahatma Gandhi, and puzzling over Albert Einstein's most important scientific papers. It was like being enrolled in seven elective college or graduate school courses.

As a studious youngster growing up in Scranton, Pennsylvania, in the 1950s, I loved to read. What captured my interest most were biographies and histories, drawn from many lands, but focused particularly on Western Europe, from which my family came, and the United States, our new home. I had scarcely heard of psychology when I entered college, and so it was natural for me to declare myself a history major. But only when I encountered the psychohistorical and psychobiographical writings of Erik Erikson did I find an intellectual home. And so I shifted my studies to the social sciences and found myself increasingly drawn to the psychology of human development.

A conflict within me between an interest in the emotional side of human experience and a curiosity about its more cognitive dimensions was resolved—at least temporarily—in favor of cognition when I began to read the works of the Swiss psychologist Jean Piaget at the close of my college career. I read Piaget intensively during a postgraduate year in England. During that time of leisure, I also became far better acquainted with the ideas and art forms of the modern era: the music of Igor Stravinsky; the paintings of the cubists; the writings of T. S. Eliot; and the astonishing outpouring of scientific, artistic, and political creativity that had taken place in the principal European countries in the first decades of the twentieth century. Although I decided to pursue graduate studies in developmental psychology, I had already become keenly fascinated with the society that had produced such sparkling works while at the same time plunging into two devastating world wars and then engaging in a dogged cold war.

Interest in history and biography took a back seat for a while as I mastered the methods and techniques of experimental developmental psychology. I am grateful for this systematic training. However, soon after my graduate studies began, I felt keenly the lack of interest among my teachers and peers in the puzzles of artistic creation. My own background had included intensive work in music; I had spent innumerable evenings during my postgraduate year exploring the arts of the modern era; and yet I searched in vain for any reference to these facets of life in my professors' lectures and in the assigned readings. I was therefore primed when I learned of a new research enterprise at Harvard called Project Zero, which was focusing specifically on the nature of artistic knowledge and education.

Under the aegis of Project Zero, I have for more than forty years studied human development in normal and gifted children, as well as the breakdown of human capacities and gifts under conditions of brain damage. The project's animating interest has been the nature of human symbolization, with particular reference to those forms of symbolizing that are key to the arts. Put more concretely, my colleagues and I have probed how youngsters become musicians or poets or painters, why most of them do not, and how these and other artistic capacities develop or atrophy or are nurtured within our own and others' cultures.

By a curious twist, the words *art* and *creativity* have become closely linked in our society. It is for this reason, I suppose, that during the recent decades I was generally considered to be studying "creativity." There is no necessary association: People can be creative in any sphere of life, and the arts can be the scene of bathos or boredom, as well as of beauty, beatitude, or bedlam. Nonetheless, because of the quirk, I was regularly invited to conferences on creativity; regularly interviewed by journalists interested in creativity; and, in general, assimilated inappropriately to membership in the "creativity research mafia." I did not mind this slight case of mistaken identity, to be sure, given my lifelong interests in the achievements of certain extraordinary human beings.

Although I had written a good deal about creativity, particularly in the arts, I had not initially thought about doing a comparative biographical study of this type. The impetus came following the publication in 1983 of the book for which I am best known: *Frames of Mind: The Theory of Multiple Intelligences* (3rd ed., 2011). In that book, drawing on examples from both ordinary and extraordinary individuals, I described seven relatively autonomous forms of human intelligence. Once I had pluralized intelligence, many people asked me whether I also believed that there were several kinds of creativity. Though I had intuitions about this question, it occurred to me that it would be fascinating to study creative individuals who, by hypothesis, stood out in the various intelligences and to see what I might discover about the nature of their several creativities.

And so, having made that initial decision, I then had the challenge of selecting individuals of undeniable creativity who seemed to stand out in terms of a particular intelligence. I flirted with the idea of selecting subjects from the full span of human history but rejected that tack. So that my subjects would be at least roughly comparable, I elected to pick individuals who had lived in the latter part of the nineteenth and the first part of the twentieth century. The one other criterion, important for many biographers, was that I had at least a fundamental sympathy with the subjects. I did not want to devote myself to the study of the juvenilia of a writer, or the childhood sketches of a painter, whose mature work I did not like.

Though the details of the subjects' lives were fascinating in themselves, I saw this work as fundamentally social scientific rather than humanistic. That is, I was looking for concepts and generalizations that might illuminate the study of creativity more broadly. As I detail in Chapter 2, I wanted to begin to build a bridge between the detailed psychological studies of individual creators, by scholars such as Howard Gruber, and the quantitative, historiometric studies of scores of creators, by scholars such as Dean Keith Simonton. And so the chief substantive chapters, dedicated to my seven masters, are bookended by more general reflections about how to study creative processes and what findings have emerged from this study.

Toward the conclusion of the book, I interrogated myself about several issues: Did I select the right persons and the proper domains, did I have the right measures of creativity, would my conclusions have obtained with reference to other persons and to other historical eras? The republication of this book, twenty years after it was initially drafted, gives me the opportunity to revisit these and other questions that have occurred to me in the interim.

I have had few second thoughts about the eras or the individuals studied. A century ago, Europe and, to a lesser extent, Russia and America hosted individuals of remarkable creativity. I might have studied James Joyce or Marcel Proust rather than T. S. Eliot, and I had similar choices in the other performance domains, but the list has held up well. Closer to our own era, I might have studied a film director such as Ingmar Bergman or a scholar such as Noam Chomsky, but I don't think that the conclusions would have been substantially altered.

But there are other issues where conclusions might have been different. In my original sample, individuals had been born on the periphery of cultural centers and had moved to such places as London or Vienna. Had I studied the philosopher Ludwig Wittgenstein instead, I would have detected an opposite pattern.

Coming from a powerful and wealthy family, Wittgenstein was born and grew up at the center of culture in Vienna: no need for him to move to a major metropolis to encounter other young people of enormous potential! But Wittgenstein found Vienna to be oppressive: He moved first to Cambridge, England, then to Norway, and finally to the United States. Perhaps genius needs to gain distance from wherever it first resided.

My sample also may have led to the unwarranted conclusion that creators are necessarily difficult persons, particularly in their later years. I would have difficulty making this argument with reference to Charles Darwin who, by all reports, was a humane family member and a generous scholar. But in addition to living in an earlier era, Darwin was distinguished in other ways. Like Wittgenstein, he came from wealth and never had to work for a living. Like Wittgenstein, Darwin was born in the center of things—in England—and achieved distance through his famous five-year trip around the world on the *Beagle*. Perhaps most important, Darwin had or feigned illness throughout his adult life; first his wife Emma protected him from unwarranted intrusions, and then his colleague Thomas Henry Huxley (nicknamed "Darwin's bulldog") barnstormed Britain, defending Darwin's controversial claims. Perhaps Darwin had the protection that genius needs and did not have to erect barriers on his own.

My point here is not to argue about each of my initial conclusions but rather to illuminate the nature of the bridge that I was trying to build, from case studies to broad generalizations. One must begin with patterns observed in the original sample. When an apparent exception arises—such as the absence of a move from the periphery to the center of culture—one needs to see whether the basic point can be rescued by a broader reformulation—gaining distance from one's customary locale. And one has to be open to the possibility that the generalization was an accident, based on the particular sample chosen or the specific time period focused upon. Examples such as that of Darwin force one to rethink earlier conclusions.

As it happens, just a few years after finishing *Creating Minds*, I had the opportunity to carry out a further study. Editor and agent John Brockman asked me if I wanted to write a short book with the topic wide open. At the time, I had become friendly with D. Carleton Gajdusek, an outstanding Nobel laureate in biology and a fascinating, larger-than-life personality. In thinking about Carleton's remarkable career, I planned to conceptualize my portrait around four different roles that he had assumed: The Master (who has climbed to the top of an already existing field), the Maker (who devises a new area of study or practice), the Influencer (who changes the behaviors of others), and the Introspector (who thinks deeply about him- or herself).

As I was nearing the end of my research, and preparing to write this short book, Gajdusek was arrested and subsequently convicted of pedophilia. (After serving a short sentence, he left the United States and died of natural causes a decade later.)

Faced with the question of whether to write a book about a convicted felon, I decided that I could not do so. Instead, I chose to write brief accounts of four individuals, each of whom exemplified one of the four Gajdusek-inspired roles.

Two of the roles were well filled by personalities from this book: Freud (a Maker) was an individual who had founded psychoanalysis—a new field of study *and* practice; Gandhi (an Influencer) was an individual who had influenced the thoughts and behaviors of many thousands, perhaps millions, of his fellow human beings.

The book, called *Extraordinary Minds*, gave me an opportunity to test the scheme of the present book in two ways. For the Master, I decided to write about Wolfgang Amadeus Mozart, certainly a master of classical music and a person who had lived more than a century before my seven subjects. Even that relatively short period of time revealed many differences between the Europe at the crest of the Enlightenment and the same pursuits at the crest of the modern era. Far more so than those who came later, Mozart was the practitioner of a craft, who wrote basically on commission, and who had to worry throughout his life about basics of health and money.

For the Introspector, I chose to write about Virginia Woolf. Her achievements as a writer of fiction and of essays spoke for themselves. But I also had felt—and had been criticized for—the dearth of women in my original sample. The inclusion of Woolf proved equally instructive. Though she came from an illustrious family and lived amid a highly literate and intellectual family, she did not have any formal education. As she famously observed, it was difficult to become a writer unless one had three hundred guineas (slightly more than 300 pounds) and a room of one's own. Also, as is well known, Woolf suffered from severe mental illness and eventually committed suicide. The inclusion of Woolf forced me to consider the very different challenges facing a talented woman, a century ago, and the devastating effects of depression in an era when treatments were woefully inadequate.

Researching and writing *Extraordinary Minds* helped me to consider two other issues. The first was the lessons that the rest of us can learn from individuals who are highly creative. I culled three: (1) Creative individuals spend a considerable amount of time reflecting on what they are trying to accomplish, whether or not they are achieving success (and, if not, what they might do differently). (2) Creative individuals leverage their strengths. They determine their strongest area and build their achievements around these potent intelligences. They do not worry about what they do not do as well; they can always get help from others and perhaps barter their areas of strength with those who have complementary skills. (3) Creative individuals frame their experiences. Such people are highly ambitious, and they do not always succeed, by any means. But when they fail, they do not waste much time lamenting; blaming; or, at the extreme, quitting. Instead, regarding the failure as a learning experience, they try to build upon its lessons in their future endeavors. Framing is most succinctly captured in aphorism by French economist and visionary Jean Monnet: "I regard every defeat as an opportunity."

The other issue, touched upon in *Creating Minds*, concerns the role of pathology in creative genius. Of course, the imputed relationship between the wound and the bow has long been a staple of studies of those who achieve. And in *Creating Minds* I noted that my creators each had had periods of mental fragility and that they ranged from a distanced relationship to other persons (Einstein) to an inclination toward frank sadism (Picasso). While the two new persons did not directly challenge my earlier conclusions, my immersion in the life of Gajdusek catalyzed much reflection on the relationship between his great gifts and his predilection for pushing the envelope, both in terms of his scientific work and his relations to others, in this case, particularly young boys.

I continue to ponder whether the link between monumental achievement on the one hand and a tendency to behave according to one's own rules on the other is a cardinal feature of creative genius. At present, I am reluctant to state such a strong conclusion. There are certainly individuals such as Darwin or composer Johannes Brahms or writer Thomas Mann who did not stand out in terms of their defiance of convention. And yet, I feel confident in declaring that the character trait of thinking outside the box with reference to one's own work life often, if not inevitably, spills over into other sectors of life.

My focus on the modern era was quite explicit: In fact, my working title for this book was "The Creators of the Modern Era"—a pun that appealed more to me than to my publishers! Not only did my seven creators reflect the era in which they were brought up, but as would be the case with any remarkable septet, they helped to create the art, the science, and even the politics (Einstein and the atomic bomb, Gandhi and the disappearance of colonization) of the middle of the twentieth century. But even twenty years ago, I was aware that this era was at an end, and that we had embarked on an era that was postmodern: both in the literal sense, of succeeding the modern era, and in the rhetorical sense, an era exhibiting its own epistemology and aesthetics.

A tad presciently, I include in *Creating Minds* a short discussion of how creativity might differ in the era that succeeded modernism.

Briefly, the postmodern era is a time when any claim of ultimate truth or morality is shunned, where genres are blurred and readily mixed, and when seriousness is challenged and irony is favored. And had I been more prescient, I would have anticipated the dominance of the digital media: global communication, the collapse of time and space, instant access to knowledge and to personal messages, and powerful interpersonal networks.

Even the short span of some decades is significant enough to raise the question of whether creativity, circa 2010, differs qualitatively from creativity in

1910. (Woolf famously quipped, "On or about December 1910, human character changed.") I believe that in a number of ways, the kind of solitary individual, the lonely creativity of earlier eras, is far rarer.

To begin with, anything that becomes known in one part of the world is readily available throughout the world. And so, whether we are talking about a new genre of painting or a new line of scientific work, all interested persons can have access to it right away, and their subsequent activities may thus be affected.

Second, the potential for, and in some cases the necessity for, collaboration is patent. One hundred years ago, science was largely an individual matter; fifty years ago, science was carried out by small teams, or even pairs, most famously by the two men who deciphered the genetic code, James Watson and Francis Crick. Nowadays, we are in the era of Big Science, where dozens or even hundreds of scholars collaborate on a single project. An experiment at the Hadron collider can involve three thousand scientists! And while far less prevalent, more of artistic work is collaborative—across genres and disciplines, and even with teams of creators. Consider this testimony from Carla Peterson: "Rather than creating a unique movement language à la Martha Graham or Merce Cunningham . . . [choreographers today] are focusing on conceptual issues, drawing on collaborators, appropriating, sampling, referencing, and dialoging with other artists' works, notions of authorship, dissolving of genres, the rethinking of dance's relationship with movement, and with audiences are all in play" (quoted in Gardner 2011, p. 73). And with the instant availability of work in one's own genre, the prospect of borrowing or even stealing works of others is prevalent.

At least in developed countries, the relationship between "pure creativity" and the actual or potential marketability of works has also been altered. At the time of this writing, and despite the financial meltdown that began in 2008, a far larger proportion of work, particularly in the sciences, is paid for by commercial entities and is oriented toward potential marketability. Both in the physical and the natural sciences, the relation between "pure" science on the one hand and technology on the other has become increasingly blurred. And in the famous examples of Bill Gates, Steve Jobs, Mark Zuckerberg, and other creators of the digital revolution, we see traditional conceptual creativity wedded to skill in the realms of technology and commerce.

In the arts, the role of commerce is complex. Potentials for huge amounts of money—for the artists and for those tied to them commercially—are greater than ever before. At the same time, because of the power of advertising and consumerism, artworks fall increasingly in two camps: the esoteric, now available via technologies such as the web, and those works that have mass appeal and entail vast sums of money. It is increasingly difficult for works that fall between these poles to see the light of day.

Finally, and happily, the advent of the digital media is a boon for education around the world. Not only is it possible for anyone with access to a computer to sample the world's achievements but also those who want to produce have an unprecedented opportunity to acquire skills, either alone at their personal computer or smart communication device or via live or robotic tutors. The ten years it used to take to master a domain can happen much more rapidly.

Could creative work of the highest caliber emerge from an individual who resides far from creative centers and who works almost entirely in remote fashion? At present, I think that it is unlikely; creative people seem to want and need the "offline" contact with others as much as they ever did. Metropolises and megatropolises seem as seductive as ever. However, I have no trouble envisioning a future where the proverbial isolated creator—call her Isabella—seated at her computer could come to the forefront of her field though she'd never met any of the leaders firsthand. Indeed, Isaac Newton virtually accomplished this feat nearly a half a millennium ago. It is even possible that the generalizations that have emerged from my own studies would turn out to be truer for Isabella than for those who were engaged in regular face-to-face contacts with others in her field.

To the extent that creativity in our time has a different shade than creativity in earlier eras, what happens to the findings and analytic frameworks that predominated when I was a student and a younger scholar? Truth to tell, the study of creativity remains a marginal topic in psychology and related fields, and quantum advances are few and far between. So far as I have been able to ascertain, in the past two decades no powerful new approaches have threatened dominant concepts, frameworks, or paradigms.

That said, we may be on the cusp of important breakthroughs in two areas. Turning first to the realm of computers, intelligent systems have advanced enormously in recent decades. Not only has this advance blurred the lines between art, science, and technology, but increasingly, creative individuals work so intensely with computational devices that the role of each is harder to ascertain. I am not predicting the onset of the singularity, where brain and machine or body and silicon merge. Rather, I am suggesting that our judgments of creativity will no longer be restricted to outputs of individual humans or human beings in the future. Of course, we should remember that, at least to this point, the programming is done by human beings, and so are the rules governing the display and evaluation of creative products. But all of this could change, and perhaps more quickly than most can imagine.

The other realm poised for breakthrough is that of the biological understanding of creativity. I do not mean that we will discover the genetics of creativity; I believe that creativity is an emergent of individuals "at promise" when they live in a specified society, with certain values and opportunities. But I do believe that we will discover a good deal about what happens in the brains, as well as the minds, of creative individuals. Perhaps we will discover the extent to which those brains and those minds may have been different from the beginning or, more likely, how they learn and how they make use of what they have learned. And these lessons, in turn, may help us to encourage creativity in a larger portion of the populations, in domains new as well as old.

Of increasing interest to me as I approach the age of seventy, we will learn more about the psychology and biology of creativity in the later years of life. It has long been maintained, with both anecdotal and empirical evidence, that creative breakthroughs—and especially the most dramatic ones—occur in the early decades of adulthood. I don't expect this to change soon; indeed, given the digital explosion, the mean age of breakthroughs may even drop further—look out, current record-holders Mozart and Picasso!

And yet, given the greater health and the increased longevity of our population, and the ease of "keeping up," we may well encounter creativity across the decades and perhaps even new and cherished forms of creativity during the later years of life. After all, just considering the seven individuals in this book—all reasonably long-lived—although their seminal breakthroughs occurred during their first decades, they continued to produce work of significance into their 70s and perhaps even beyond. We will need to determine whether the apparent decline in creativity is a direct function of the aging process, in which case it will be difficult to change, or whether factors of health, access to information, and motivation are key. In the latter case, the examples of Verdi, Titian, Picasso, and—in our own era—such individuals as biologist E. O. Wilson, linguist Noam Chomsky, composer Elliott Carter, painter Jasper Johns, or choreographer Merce Cunningham will be more common.

As noted earlier, my study focused on seven individuals who, broadly speaking, were the products of Western civilization and education. To be sure, it was Mahatma Gandhi who, asked about Western civilization, famously quipped, "It's a good idea, it should be tried." Yet Gandhi freely admitted that he was a product of a British education in a British colony, and his achievements would have been inconceivable without his lengthy immersion in the life and achievement of the West. And I would go so far as to say that the picture of individual revolutionary creativity sketched here is a distinctly Western one, born in Athens and Rome; confirmed in the Enlightenment Era of Europe; and then marshaled for various ends in the America of Hollywood, Silicon Valley, and Wall Street.

American—indeed Western—hegemony are at an end. This century will either be dominated by models from the East—China, India, and Japan—or be even more multipolar, with beachheads spanning the Southern as well as the Northern Hemispheres. Any serious study of creativity in the future must avoid parochialism. It must look both at the approaches to creative work in major non-Western civilizations and at the achievements, perhaps measured by a new set of criteria. Such a study will need to examine individual versus group creativity, revolutionary versus evolutionary creativity, creativity in new as opposed to standard domains, and the ways in which societal fields (institutions, gatekeepers, teachers) steer the promotion and evaluation of creative efforts. The tools and insights of computational studies and brain studies will be key. Realistically speaking, I will not be able to undertake such a study, nor is it likely that I'll be able to read about it. But if the study of creativity is to advance, it will have to become a global undertaking.

REFERENCES

Cited in the Preface

Gardner, H. 1983/2011. *Frames of Mind: The Theory of Multiple Intelligences*. New York: Basic Books.

Gardner, H. 1997. *Extraordinary Minds: Portraits of Four Exceptional Individuals and an Examination of Our Own Extraordinariness*. New York: Basic Books.

Works Published Since Creating Minds

On Creativity

Amabile, T. 1996. *Creativity in Context*. Boulder: Westview Press.

John-Steiner, V. 2006. *Creative Collaboration*. New York: Oxford Univ. Press.

Kaufman, J., and R. J. Sternberg, eds. 2010. *The Cambridge Handbook of Creativity*. New York: Cambridge Univ. Press.

Runco, M. 2006. *Creativity: Theories and Themes*. New York: Academic Press.

Sawyer, Keith. 2006. *Explaining Creativity: The Science of Human Innovation*. New York: Oxford Univ. Press.

Simonton, D. K. 1999. *Origins of Genius: Darwinian Perspectives on Creativity*. New York: Oxford Univ. Press.

Works about Creators

Albert Einstein

Galison, P. 2004. *Einstein's Clocks, Poincare's Maps: Empires of Time*. New York: Norton.

Isaacson, W. 2008. *Einstein: His Life and Universe*. New York: Simon & Schuster.

T. S. Eliot

Eliot, V., and H. Houghton. 2009. *The Letters of T. S. Eliot, Volume 2: 1923–1925*. London: Faber and Faber.

Gordon, Lyndall. 1999. *T. S. Eliot: An Imperfect Life*. New York: Norton. (This work discusses Eliot's anti-Semitism and misogyny.)

Hargrove, Nancy. 2010. *T. S. Eliot's Parisian Year*. Gainesville: Univ. of Florida Press.

Julius, Anthony. 2003. *T. S. Eliot, Anti-Semitism and Literary Form*. London: Thames and Hudson.

Sigmund Freud

Breger, L. 2001. *Freud: Darkness in the Midst of Vision*. New York: Wiley.

Gay, P. 1998. *Freud: A Life for Our Time*. New York: Norton.

Kramer, P. 2006. *Freud: Inventor of the Modern Mind*. New York: Eminent Lives.

Mahatma Gandhi

Andrews, C. F., and Arun Gandhi. 2003. *Mahatma Gandhi: His Life and Ideas*. Woodstock, VT: Skylight Paths. (In this book Gandhi's grandson updates a book by an associate of Gandhi's, looking at his ideas and their impact today.)

Lelyveld, J. 2011. *Great Soul: Mahatma Gandhi and His Struggle with India*. New York: Knopf.

von Tunzelmann, A. 2008. *Indian Summer: The Secret History of the End of an Empire*. New York: Picador.

Martha Graham

Horosko, M. 2002. *Martha Graham: The Evolution of Her Dance Theory and Training*. Gainesville: Univ. of Florida Press. (This work contains interviews, syllabi, and techniques.)

Tracy, R. 1997. *Goddess: Martha Graham's Dancers Remember*. Limelight Editions. (Thirty dancers remember Martha Graham.)

Pablo Picasso

Flam, J. 2003. *Matisse and Picasso: The Story of Their Rivalry and Friendship*. New York: Basic Books.

Richardson, John. 1991. *A Life of Picasso: The Prodigy, 1881–1906*. New York: Knopf.

———. 2007. *A Life of Picasso: The Cubist Rebel, 1907–1916*. New York: Knopf.

———. 2010. *A Life of Picasso: The Triumphant Years, 1917–1932*. New York: Knopf.

Staller, N. 2001. *A Sum of Destructions: Picasso's Cultures and the Creation of Cubism*. New Haven: Yale Univ. Press.

Igor Stravinsky

Taruskin, R. 1996. *Stravinsky and the Russian Traditions*. Berkeley: Univ. of California Press.

Walsh, S. 2002. *Igor Stravinsky: A Creative Spring: Russia and France*. Berkeley: Univ. of California Press.

Walsh, S. 2002. *Igor Stravinsky: France and America: The Second Exile*. Berkeley: Univ. of California Press.

YOUTH

Your unhappy and silly youth.
Your arrival from the provinces in the city.
Misted-over windowpanes of streetcars,
Restless misery of the crowd.
Your dread when you entered a place too expensive.
But everything was too expensive. Too high.
Those people must have noticed your crude manners,
Your outmoded clothes, and your awkwardness.

There were none who would stand by you and say,

You are a handsome boy,
You are strong and healthy,
Your misfortunes are imaginary.

You would not have envied a tenor in an overcoat of camel hair
Had you guessed his fear and known how he would die.

She, the red-haired, because of whom you suffer tortures,
So beautiful she seems to you, is a doll in fire.
You don't understand what she screams with her lips of a clown.

The shapes of hats, the cut of robes, faces in the mirrors,
You will remember all that unclearly, as something from long ago,
Or as what remains from a dream.

The house you approach trembling,
The apartment that dazzles you—
Look, on this spot the cranes clear the rubble.

In your turn you will have, possess, secure,
Able to be proud at last, when there is no reason.

Your wishes will be fulfilled, you will gape then
At the essence of time, woven of smoke and mist,

An iridescent fabric of lives that last one day,
Which rises and falls like an unchanging sea.

Books you have read will be of use no more.
You searched for an answer but lived without answer.

You will walk in the streets of southern cities,
Restored to your beginnings, seeing again in rapture
The whiteness of a garden after the first night of snow.

—Czeslaw Milosz

(Translated, from the Polish, by the author and Robert Hass)

PART I

INTRODUCTION

1

CHANCE ENCOUNTERS
IN WARTIME ZURICH

TOM STOPPARD'S COMEDY *Travesties*, first performed in 1974, is set in Zurich during the First World War. The plot ostensibly revolves around the efforts of Henry Carr, a minor official in the British consulate and an amateur actor, to stage Oscar Wilde's turn-of-the-century farce *The Importance of Being Earnest*. But the glitter of *Travesties* comes from the portraits of historical personalities who happen to be living in Switzerland at the time, and whose activities are being recalled many years later by an aging, forgetful, and self-congratulatory Carr.

Though Stoppard does not hesitate to mix fact and fancy, it is indeed true that many individuals of historical moment congregated in war-spared Zurich during what was universally called the Great War. Carr reminisces: "Zurich during the war. Refugees, spies, exiles, painters, and poets, writers, radicals of all kind."* The Stoppard play centers on three of these figures: a little-known Irish writer, James Joyce; an obscure Russian revolutionary, V. I. Lenin; and a half-crazed Rumanian artist-intellectual, Tristan Tzara, who is fashioning dadaism, an aesthetic brand of nihilism. These exiles go about their business—respectively, writing the great novel; planning the Russian Revolution; and redefining art, politics, and life.

While Wilde's play, redolent of an earlier and less turbulent era, is being rehearsed, the protagonists banter about themes of the modern era. Lenin declares: "Literature must become party literature. . . . As for me, I'm a barbarian. Expressionism, futurism, cubism . . . I don't understand them and I get no pleasure from them." Tzara puts forth his view: "Doing the things by which is meant Art is no longer considered the proper concern of the artist. . . . Nowadays, an artist is someone who makes art mean the things he does." Joyce puts him in his place: "You are an overexcited little man, with a need for self-expression far beyond the

---

* The sources of quotations and other materials are listed in endnotes keyed to the pages of this book.

scope of your natural gifts. This is not discreditable. Neither does it make you an artist. An artist is the magician put among men to gratify—capriciously—their urge for immortality."

As members of the audience, we can view *Travesties* not only as an amusing farce about makeshift theater but also as a backstage glimpse at a trio of individuals in the course of creating what would come to be regarded as the modern era. It is as if the principal figures of the Enlightenment had all been choirboys during the 1740s, or the leading American transcendentalists had been college classmates in the 1820s.

Though casting a glance three-quarters of a century backward, *Travesties* is very much a play of our era. Its rapid shifts across conflicting interpretive frames, political credos, and aesthetic codes constitute homage to the contributions of pivotal "modern" figures like Joyce, Lenin, and Tzara. The arguments placed in the mouths of the characters represent a conversation that has continued throughout the twentieth century. Above all, the conceit that these culturally diverse individuals, armed with their radically different agendas, could have congregated in the same European city and, at least in theory, have come to know each other, is plausible only in a world that is no longer a collection of localities—André Malraux's museum without walls, Marshall McLuhan's global village.

SEVEN CREATIVE THINKERS

Remaining anchored in the Great War, between 1914 and 1918, we can readily locate a comparable set of historical figures—potential characters in an expanded *Travesties*—whose impact on our time has been compelling.

- The neurologist-turned-psychologist Sigmund Freud (1856–1939) was living in Vienna, continuing to see patients, observing with satisfaction the rising influence of his psychoanalytic movement, glancing nervously at his Zurich-based rival Carl Jung, and worrying about not only the fate of his son on the battlefront but also, more generally, the survival of an inherently destructive human society.

- The theoretical physicist Albert Einstein (1879–1955) had just moved from Zurich to Berlin, where he served as a distinguished professor of physics at the university and also as the director of a new physics institute. A pacifist, he declared his opposition to the war being waged by his countrymen. Having separated from his wife, he promised her the proceeds of the Nobel Prize, which he was confident he would soon receive for his pathbreaking reconceptualizations of time, space, and light.

- Pablo Picasso, the Spanish-born painter (1881–1973) who had moved to Paris at the beginning of the century, was gradually shedding the cubist style of painting, which he and Georges Braque had constructed together in the early 1900s. The war years had seen the death of his beloved mistress, Eva; traveling to Rome to devise scenery for the Ballets Russes, he met and fell in love with the Russian ballerina Olga Koklova.

- Igor Stravinsky (1882–1971), the Russian-born composer, had created for the Ballets Russes a set of spectacular ballets, including the highly controversial *Le sacre du printemps* (1913). When the First World War broke out, he elected to remain in Western Europe and was in fact headquartered in Switzerland for most of the war. There he worked on two of his most innovative works, *Histoire du soldat* and *Les noces*.

- T. S. Eliot (1888–1965), the St. Louis–born poet, moved to Europe in the early 1900s; defying his family's wishes, he decided to remain there when the First World War broke out. With startling speed, he became an important literary figure in England. Eliot published his first important poem, "The Love Song of J. Alfred Prufrock," at the start of the war and then worked over the next years on his pathbreaking *The Waste Land* (1922).

- Martha Graham (1894–1991), the dancer born near Pittsburgh, had moved with her family to Southern California. Ignoring her parents' wishes, she began during the war years to study with dance pioneers Ruth St. Denis and Ted Shawn. In the early 1920s, she traveled to Europe and also across much of America. Breaking away from the Denishawn troupe, she formed her own company and soon fashioned a distinctly modern form of dance.

- Mahatma Gandhi (1869–1948), the Indian political and spiritual leader, had just returned to his native land after two decades abroad in England and in South Africa. Despite his opposition to British rule, he elected to support the efforts of the Allied powers during the Great War. He continued to develop innovative methods of peaceful resistance; at the end of the war he launched in India a nonviolent political revolution with worldwide reverberations.

Any short list of individuals who gave birth and form to the modern era must be notable for its absences: Why T. S. Eliot rather than Marcel Proust or Virginia Woolf? Why Mahatma Gandhi rather than Mao Zedong or Martin Luther King Jr.? Why Martha Graham rather than Isadora Duncan or George

Balanchine? Equally, any set of domains will call attention to those that have been bypassed: why dance rather than athletics, why statesmanship rather than business, why physics rather than biology? Even the date of focus is not immune from argument: A case can be (and has been) made that the modern era ought properly to begin with the political revolutions of 1776, 1789, or 1848 or with the ideas or events of 1500 or 1815; that the true origin of the modern aesthetic lies in the fin de siècle paintings of Paul Cézanne, the music of Gustav Mahler, or the poetry of Stéphane Mallarmé; or that scientific breakthroughs of a fundamental nature are more properly associated with the quantum mechanics of the late 1920s, the deciphering of the genetic code at mid-century, or the recent insights of chaos theory.

Less controversial, however, is my assertion that the seven individuals I have named and the domains that they represent constitute *a representative and fair sample* drawn from the larger pool of individuals whose discoveries gave rise to one or another version of the modern era. Any study that ignored all of them would be suspect; any study that includes the present ensemble is at least on the right track. More crucially, if we can understand the creative breakthroughs achieved by Freud, Einstein, Picasso, Stravinsky, Eliot, Graham, and Gandhi, we will surely understand some facets of human creation construed more broadly. I claim as well that a grasp of the underpinnings of their creations should help to elucidate the "modern era"—an era whose fundamental ensemble of ideas has animated the twentieth century, an era that is fast receding from our "postmodern" perspective. Had these seven figures all inhabited the Zurich of Tom Stoppard's imagination—and it is not physically impossible that they could all have found themselves at the same table in a café on the *Bahnhofstrasse* in the summer of 1916—*Travesties* could serve as a tract on, of, and for our times.

THE PURPOSES OF THIS BOOK

In writing about seven creative "modern masters" or "masters of the modern era," I have in mind three principal purposes. First, I seek to enter into the worlds that each of the seven figures occupied during the period under investigation—roughly speaking, the half century from 1885 to 1935. In so doing, I hope to illuminate the nature of their own particular, often peculiar, intellectual capacities, personality configurations, social arrangements, and creative agendas, struggles, and accomplishments.

Illumination of seven disparate creative breakthroughs carried out by seven singular characters is no small assignment. Were I working in the humanistic tradition, I would likely focus on one of these individuals and try to understand his or her contributions as fully as possible. Comparisons would play only a minor role in my presentation. But because I approach this assignment as a social scientist, my focus takes the form of a *search for patterns*—for revealing similarities and for instructive differences.

As my second goal, I seek conclusions about the nature of the Creative Enterprise writ large. I believe that if we can better understand the breakthroughs achieved by individuals deliberately drawn from diverse domains, we should be able to tease out principles that govern creative human activity, wherever it arises. I shall argue that creative breakthroughs in one realm cannot be collapsed uncritically with breakthroughs in other realms; Einstein's thought processes and scientific achievements differ from those of Freud, and even more so from those of Eliot or Gandhi. A single variety of creativity is a myth. Yet, I shall also supply evidence that certain personality configurations and needs characterize creative individuals in the twentieth century, and that numerous other commonalities color our ways of conceiving, articulating, and responding to ideas.

And finally, I seek conclusions about the sparkling, if often troubled, handful of decades that I term "the modern era." While creative individuals can be drawn from disparate periods of history and from diverse cultures, there is an advantage in selecting a group of individuals who were roughly contemporaneous (Freud was born in 1856, Martha Graham in 1894, and the remaining figures in between these framing dates) and who were influenced by the civilization of Western Europe, broadly construed. Such a selection allows me to comment not only on the particular achievements of a group of talented persons but also on the times that formed them, and that they in turn helped define.

I argue that the arts, crafts, scientific understandings, and intellectual syntheses that were regnant in the nineteenth century were no longer viewed as adequate; and that, in response to the perceived inadequacies, these seven creators forged a new agenda, which has accordingly been worked through—and perhaps exhausted—in this, "their" century. The character of that reformulation entails, paradoxically, a return to the basic elements of each domain: the simplest forms, sounds, images, puzzles—a purification process that involves a strange yet productive amalgam of the most elemental impulses with the most sophisticated understandings. I argue, further, that each creative breakthrough entails an intersection of the childlike and the mature; the peculiar genius of the modern in the twentieth century has been its incorporation of the sensibility of the very young child.

The heart of such a study is, necessarily, the intensive probing entailed in our seven case studies. But before embarking on that assignment in part II, I must undertake several tasks. In the remainder of this chapter, I comment on the particular creative breakthroughs focused on in this book, thus providing an informal introduction to my major themes. I also review briefly some of the appeals and some of the pitfalls of focusing on a particular historical era. Then, in chapter 2, I introduce my own approach to studies of creativity, locating it within the perspective of other recent efforts by social scientists.

ORGANIZING THEMES

Although I cannot summarize the contents of this book in a single phrase or a simple set of elements, I can ease the entry into my more complex framework by introducing a set of key distinctions. To begin, this framework has three core elements: a creating *human being*, an *object* or *project* on which that individual is working, and the *other individuals* who inhabit the world of the creative individual. The superstructure needed to account for creative activity is based on these three core elements and on the relationships among them, specifically:

1. *The relationship between the child and the master.* In a developmental study, it is natural to look for continuities, as well as disjunctions, between the world of the talented, but still unformed, child and the realm of the confident master. Equally important in a study of creativity is sensitivity to the innovator's ways of drawing on the worldview of the young child.

2. *The relationship between an individual and the work in which he or she is engaged.* Every individual works in one or more domains or disciplines, in which he or she uses the current symbolic systems or contrives new ones. Throughout these pages, I am concerned with individual ways of mastering, then laboring in, and ultimately revising the nature of such domains.

3. *The relationship between an individual and other persons in his or her world.* Though creative individuals are often thought of as working in isolation, the role of other individuals is crucial throughout their development. In this study I examine the roles of family and teachers during the formative years, as well as the roles of crucial supportive individuals during the times in which a creative breakthrough seems imminent.

As a provisional representation of these factors, I propose the following:

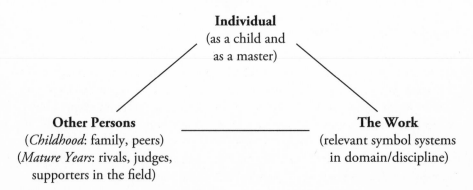

Individual
(as a child and
as a master)

Other Persons
(*Childhood*: family, peers)
(*Mature Years*: rivals, judges,
supporters in the field)

The Work
(relevant symbol systems
in domain/discipline)

By introducing these elements at the outset, I wish to stress that all creative activity grows, first, out of the relationships between an individual and the objective world of work and, second, out of the ties between an individual and other human beings. Later I detail some of the dynamic interactions among these three "nodes" in the triangle of creativity. Now, however, as a perhaps welcome respite from the terminological thicket, let me indicate how these themes—each to be highlighted in a separate chapter—will be realized in the case studies.

From the World to the Self—And Back Again

While our seven masters all exerted a profound effect on the domains in which they worked, Freud has the strongest claim to the actual creation of a new domain—that amalgam of psychological theory and therapeutic practice called psychoanalysis. As a youth, Freud displayed a protean ability to absorb information from a gamut of disciplines and a galaxy of other individuals. And, indeed, Freud came to the discovery of psychoanalysis through a synthesis of various scientific perspectives and clinical approaches that he had mastered. Still, at least as important was the solitary activity entailed in his own nightly analysis of himself—the first psychoanalysis.

Few investigators of any era have had as strong a conviction that they were solitary explorers of virgin territory as did Freud, but even in this case Freud received strong, and perhaps indispensable, support from a single other individual—his valued friend Wilhelm Fliess. Once he had enunciated his basic theory, Freud could risk a break with the eccentric Fliess. But shortly thereafter, he began drawing into his fold an ever-expanding circle of individuals with whom he shared his psychoanalytic understanding and who eventually became the principal vehicles for its future developmental course. The trajectory from the solitary investigator to a dialogue among confidants to interaction with the many members of the newly emerging discipline constitutes the framework for my first case study.

The Child and the Master

For most readers the actual conceptualizations of Einstein—while known superficially from capsule summaries in the media—constitute a formidable intellectual challenge. After all, Einstein was making technical contributions to physics, that most advanced of the sciences. Yet Einstein was able to effect a breakthrough precisely because he did *not* simply accept as given the paradigms and agendas of the physics of his time. Instead, he insisted on going back to first principles: in setting for himself the most fundamental problems and in looking for the most comprehensive yet simplifying explanatory axioms.

In so doing, Einstein was, in a way, returning to the conceptual world of childhood: the search for basic understandings unhampered by conventional delineations of a question. Indeed, the very puzzles that he first pursued as a youth—the behavior of the point of the compass, the "thought experiment" of riding on a light beam—later fueled his most innovative scientific work. My treatment of Einstein accordingly stresses the continuing dialectic between the common experiential agendas of childhood and the complex challenges of a finely articulated intellectual domain.

Prodigiousness and Beyond

All of our modern masters showed formidable gifts in childhood, but none other approached the spectacular level of skill displayed by the youthful Picasso. A gifted draftsman in the first decade of his life, he was by late adolescence painting with as much finesse as any other artist of his time—and laying the groundwork for seventy-five more years of productivity. Picasso provides an opportunity to consider the contributions of prodigiousness to early dazzling attainments and its transmutation into a form that permits the achievement of more lasting contributions—the finest example in our time of the "Mozart enigma."

The Politics of Music

Picasso's and Stravinsky's names are often coupled—and appropriately so since these two individuals were almost exact contemporaries who knew, respected, and learned from each other. Each had launched a fundamental reorientation of his domain by the time he was thirty years old, and each pursued a lengthy subsequent creative life, during which he introduced further innovations while also revisiting, always in a distinctive manner, the major artistic milestones of earlier eras.

With Arnold Schönberg and Béla Bartók, Stravinsky dominates classical music in the twentieth century. His creative breakthrough stimulates a consideration of what it is like to initiate a fundamental transformation of a traditional domain while also reorienting neighboring domains, such as the dance and theater. Some creative figures—for instance, theoretical physicists—can work in relative isolation, but not a musical composer. Because nearly all of Stravinsky's work was collaborative, an examination of his creative activity casts light on the political factors that permeate the planning, staging, and critical review of artistic performances.

The Marginal Master

Moving readily from one culture to another is a distinctly modern phenomenon, and our creative masters found it necessary—as well as inviting—to immerse themselves in diverse cultural settings. Their gravitation to cosmopolitan settings like Paris or Zurich is hardly a coincidence. More so than others, Eliot affords an opportunity to consider the marginality of the modern creative figure—caught between cultures, "inhabiting" diverse time periods, experiencing painful personal anxieties and disjunctions on the border of mental disturbance. And, because Eliot was born into a decidedly nonmarginal family, he also exemplifies the extent to which creative individuals may strive to *make* themselves ever more marginal.

Representative of poets—individuals whose creative heights are typically reached by the thirties—Eliot also yields insight into the formation of subsequent productive identities: in his case, as a critic, playwright, and editor. His life provides an opportunity to consider which options remain open to an aging creative individual.

A Creative American Woman

The youngest of our modern masters and the only one still alive at the time of this book's conception, Graham stood out from her contemporaries in two instructive respects. First, she was quintessentially American. She drew her inspiration from her homeland—her New England heritage, the Appalachian surroundings of her youth, and the spaces and populations of the broad plains—as well as from the traditions of Western Europe and the Orient. Second, as a woman, she faced obstacles stemming from prevalent attitudes and expectations in a male-dominated creative world.

Transcending the limits placed on women in earlier eras, Graham created her own artistic forms, her own institution, her own legacy. Perhaps more so than other creative figures—and in a manner reminiscent of the biologist Barbara McClintock, the anthropologist Margaret Mead, the artist Georgia O'Keeffe, the writer Virginia Woolf, and other pioneering twentieth-century women— Martha Graham had to create her own paragons, her own role models. Not surprisingly, she ended up inspiring many female artists, making it easier for them to find or create an audience for their distinctive mode of expression.

A Person Who Affects Others' Lives

In my study Einstein and Freud function as individuals of science, embodying what I call the logical-mathematical intelligence (in Einstein's case) and the intrapersonal intelligence (in Freud's case). Four other figures are identified with artistic breakthroughs, each representing a different intellectual strength: Picasso as a visual-spatial master, Stravinsky as a musical innovator, Eliot as a manipulator of language, and Graham as a fashioner of bodily-kinesthetic intelligence.

Less likely to arise in discussions of creativity is the realm of human relations— as it can be seen at work in politics, religion, teaching, commerce, and the clinical professions. One reason is parochial: Artists and scientists, more than other professionals, have often become involved in discussions of creativity. The other reason is more substantive: Creative breakthroughs in the human realm tend to occur gradually, over centuries rather than decades, and so are less readily identified with a specific individual at a particular historical moment.

In my view the only figure of recent times who warrants comparison with the great *interpersonal innovators* of earlier times—Christ, Buddha, Mohammed, Confucius, Socrates—is the Indian statesman and religious leader Mahatma Gandhi. Following extensive analysis, as well as carefully fashioned experiments in which he was personally engaged, Gandhi fashioned a novel, nonviolent approach to human conflict: *satyagraha* sought the attainment of valued political goals without enervating confrontations, demeaning submissions, or recourse to violence. In my study Gandhi represents a person whose ideas and, even more dramatically, whose courageous personal example directly affected the behaviors of millions of people. Moreover, Gandhi exerted his impact in ways more constructive than those adopted by twentieth-century totalitarian leaders and potentially more significant than those associated with commercialism and the mass media.

As this montage indicates, my approach to the study of creativity begins in focused biography—in an intensive examination of the periods in the life of a creative individual when a breakthrough was conceptualized, realized, and reacted to by knowledgeable individuals and relevant institutions. I seek to transcend a concatenation of specific biographies by searching for common properties and illuminating distinctions across a small set of instructive cases. In terms of the core elements described earlier, the studies of Einstein and Picasso focus on the relationship between the child and master; the studies of Freud, Stravinsky, and Gandhi, on the relationship between the creator and other individuals; and the studies of Eliot and Graham, on the marginal position of creators with respect to the domains and fields in which they work. A concern with the dialectic between creator and work permeates the studies. Each includes a focus on the changing relationship of the individual to the domain of work, as well as an examination of how that individual formulated and promulgated new symbol systems in the domain. Finally, to provide a perspective on an important period in recent human history, I deliberately focus on individuals who were roughly contemporaneous.

THE STUDY OF CONTEMPORARIES

While many researchers deem the study of creativity to be a difficult challenge, and some consequently elect not to pursue it, few question the legitimacy of such an undertaking. When it comes to studying a particular historical era and to drawing general conclusions therefrom, however, more ticklish questions arise.

As noted, my decision to investigate a number of masters active between, roughly, 1885 and 1935 grew out of a complex of circumstances. Initially, I had wanted to study individuals who represented the range of human intelligences in which I had become interested. It was important that there exist sufficient information about these individuals so that their creative processes and interim products could be examined. Alas, we do not have enough demythologized information about Bach or Aquinas, let alone Confucius or Moses, to analyze their creativity with much confidence.

It seemed wise to select individuals who had lived at a time when record keeping was widespread and the documents of that period remained for inspection. Also, enough time needed to have elapsed for a solidifying of judgments about the quality and significance of each person's breakthrough.

The decision to study individuals who lived and worked in the first part of the twentieth century seemed a reasonable response to this nexus of issues. I could then assemble individuals who were different from one another, in respect to chosen domains, yet at least somewhat similar in terms of the milieus in which they lived. As it happens, five of our principal figures lived in Western Europe, and the other two—Gandhi and Graham—were each influenced decisively by the civilization of Europe. In a sense, then, these individuals serve as "controls" for one another—individuals inhabiting the same general life and cultural space, yet ones who have chosen (or who were chosen) to work in distinct realms of experience.

Of course, each of these individuals possessed a full range of intelligences and drew on those in his or her work. Yet, it seems fair to maintain that each highlighted a different human intelligence, and that each one's creative breakthrough represented the sophisticated use of the symbols, images, and operations associated with a particular intelligence operating in a particular discipline or domain.

THE ILLUMINATION OF AN ERA

Transcending discussions of individuals with their particular intelligences and personalities, can one say something substantive about an era? Certainly this Hegelian notion has been bruited about quite a lot. In its relatively pure form, the claim is that history has its own dynamic, with specific issues and ideas necessarily coming to the fore at a given time, and then giving way in crisp fashion to another set of issues at another time. Perhaps, indeed, even the specific issues have been preordained, or, if not so, the need for some kind of a reaction to prior events dictates the particular form of an era.

I have no commitment to the view that there exists some kind of zeitgeist, some spirit of the time that expresses itself through particular individuals who happen to be present in its wake and who thereby serve (perhaps unwittingly) as its vehicles. I see history as contingent: No spirit has determined in advance what will happen. In fact, it is often accidents—such as a stray bullet or an erupting volcano—that cause the most dramatic historical upheavals.

But belief in an underlying organizational framework is not restricted to those of a Hegelian disposition. In recent years, the innovative French scholar Michel Foucault argued that historical eras are characterized by certain underlying (and typically unconscious) assumptions about the nature of knowledge. Assuming such a structuralist stance vis-à-vis the seventeenth century, Foucault discerned the same taxonomic assumptions about knowledge operating in such diverse fields as biological classification, economic exchanges, and linguistics. Though these "frames" do not operate in lockstep, they tend to appear and disappear at about the same time.

Suppose one could demonstrate that a number of actors living at the same moment in fact epitomized the same forces or achieved comparable accomplishments, or that a number of disciplines exhibited the same conceptualization or classificatory scheme. Such a demonstration would by no means prove the operation of some sort of an overarching spirit. It is far more prudent to assume that the very fact that one individual was working in a certain way influenced others, either directly or indirectly. When a number of individuals live in the same era and actually know of one another's work, such mutual influence may well become the rule.

In the present study, it is important that individuals like Picasso and Stravinsky knew each other and worked together. Eliot and Stravinsky were friends in later life. Freud and Einstein had a casual acquaintance and engaged in a memorably pointed correspondence about war. Certain ideas associated with these creators became such common coin that anyone working during the era would have encountered them. Thus, Picasso's cubist paintings, Eliot's portrait of *The Waste Land*, Freud's ideas about unconscious motivation, and Einstein's incorporation of the observer into the space-time complex were all widely known within a decade after having been formulated. Many reasons account for common themes and co-occurrences across disciplinary domains; accordingly, we need not posit occult forces at work. Indeed, it would be odd if highly creative individuals did *not* somehow take others' novel conceptualizations into account in their own work.

THE MODERN ERA

Many, perhaps most, eras of human history seem to have proceeded without a memorable label attached to them. Yet, as observers, we often focus on eras that seem to have been characterized by a pervasive mood or ethos. For example, the rediscovery of ancient texts paralleled the growth in art, science, and "civilization" of the European Renaissance; and the highlighting of the ideas of rationality, progress, secularism, perfectibility, and liberty marked the Enlightenment.

Other eras, in contrast, are defined by their relative lacks, such as the Dark Ages in Western Europe, or the Period of the Warring States in feudal China.

At least with respect to the domains of knowledge and culture, the era of concern here has also acquired a set of labels: "modernism," "modernity," or, in my terms, "the modern era." These labels are typically seen as positive ones, though they are not without disquieting connotations. They have already developed a mythic status in potted accounts of the twentieth century.

According to the "standard historical story" (which, like most standard stories, has been revised of late), a period of cultural quiescence and conservatism followed the revolutions of the late eighteenth century and the upheavals of the Napoleonic wars. Bourgeois civilization, with its tight-laced moral code, increasingly determined standards of behavior and thought. Science and art evolved gradually and without dramatic breakthroughs or reversals. Still, by the end of the nineteenth century, these entrenched norms were being widely challenged; themes of decadence were particularly noticeable in the arts (consider Oscar Wilde's writings and his life) but apparent as well in politics (the decline of liberalism) and humanistic writings (the nihilism of Nietzsche). In the sciences, the Newtonian mechanical world order and the rational view of human behavior were seen as insufficient at best and perhaps fundamentally flawed.

Fatal blows to the nineteenth-century consensus were struck in rapid succession around the turn of the century, with our seven modern creators playing major roles in the onslaught. First, in a series of powerful revelations around 1900, Freud punctured the veneers of middle-class morality and human rationality, discerning a complex of unconscious motivations and strivings, often of a sexual or aggressive nature. Just a few years later, Einstein challenged long-accepted assumptions about the absolute status of time and space, replacing a stable, "objective" Newtonian world with an observer-determined, relativistic one.

Following sharply upon these scientific reorientations, accepted canons of artistic practice had all been undermined by the end of the second decade of the twentieth century. In the visual arts, Picasso, Braque, and their contemporaries demonstrated that faithful representation was not of the essence in the arts, created a genre in which aspects of form were dominant, and laid the foundation for a purely abstract art. Stravinsky and Schönberg struck with equal fervor at the assumptions of a single tonality and a simple rhythmic base; Stravinsky embraced primitive but complex rhythmic pulsations and polytonality, while Schönberg created his own cerebral, twelve-tone approach to composition. Similar revolts against classical verse and narrative forms were led by English authors like Eliot, Joyce, and Woolf, as well as their counterparts in other European lands, and against classical balletic forms by such innovators as Duncan, St. Denis, and Graham, and, ultimately as well, by modern ballet's Balanchine.

It is no accident that many cultural historians have fixed their attention on Vienna in the period from 1890 to 1920. If there was a single birthplace of the modern sensibility, it can most defensibly be located in the fading milieu of the Hapsburg empire. But examinations of other cities—Paris, Budapest, Prague, Berlin, St. Petersburg—tell similar stories, and there would have been a modern era, even had Vienna unaccountably sunk into the Danube a century before. Prevalent throughout Western and Central Europe in the late 1800s and early 1900s were declining institutions and disappearing shared understandings, on the one hand, and feverish creative impulses, often disturbingly unchanneled and sometimes disconcertingly wild, on the other.

An application of the label "modern" to the political realm proves more problematic. The momentous decline of the Pax Europa, the formation of nation-states in Italy and Germany, the launching of the first wars that could be called global conflicts, and the rise and ultimate defeat of fascism do not lend themselves to any simple characterization. Nor am I entirely convinced by Modris Eksteins's intriguing argument that intimations (if not provocations) of ultimate military conflagrations can be discerned in the groundbreaking artworks at the beginning of the century. Rather than realizing new conceptions of life and death, the wars of the twentieth century in many ways simply rehearsed ancient human frailties in sharper tones.

If anything, innovative political activities and novel political forms appeared not in the established nations of Western Europe but rather in the developing nations—the founding of the first Communist state in the Soviet Union, the successful accomplishment of a peasant revolution in China, and the relatively nonviolent transition to independence in India. It can be argued that the creative geniuses of the twentieth century in the political realm are Lenin, Mao Zedong, and Gandhi, rather than Benito Mussolini, Adolf Hitler, Winston Churchill, Charles de Gaulle, or even Jean Monnet, the architect of the European Market. If a bond exists among the creators of the modern era, it may need to extend across the Ural Mountains to the far side of the Euroasian landmass.

We know little about the roots and the unfolding of the creative impulses in specific individuals, yet we know even less about how to locate, name, and characterize historical eras. In the present study, an effort to apply social- and cognitive-scientific insights to the phenomena of creativity, I cannot hope to resolve vexing epistemological issues in intellectual and cultural history. And yet, in pursuing this inquiry into the lives, minds, and worlds of seven remarkable individuals, I have been tempted to determine whether in fact there may be a larger story as well.

And I have concluded that there is: This "story of the modern era" chronicles the dissolution of conventions, practices, and interpretive frames that grew up over the centuries and became entrenched throughout Europe (and the many regions affected by Europe) during the nineteenth century. Once these conventions came to be strongly challenged in certain artistic and scientific domains, the chances that they would be questioned elsewhere were greatly enhanced, for two related reasons: first, because the very knowledge that there *could* be a new painting raised the likelihood of a new dance or poetry or politics; second, because for the first time in human history, events in one part of the world could be known virtually instantly all over the world. Cubism was displayed in the New York Sixty-Ninth Regiment Armory less than a half dozen years after it had been invented; Einstein's general theory of relativity was tested in a solar eclipse halfway around the world after an even briefer interval of time; Gandhi's hunger strikes were politically effective only because the telegraph reported them immediately throughout India and all over the world.

That conventions will be challenged is one thing, and, indeed, a characteristic of all eras of revolution. The *nature* of the challenge is another. I find a noteworthy similarity in challenges that occurred across the domains under investigation. The similarity inheres in the search for *the most elementary, the most elemental forms* within a domain; a wrestling with the kinds of issues and concepts that traditionally occupy the young child; and an attempt to capture—for the record, so to speak—the death of one kind of civilization and the birth of a new, as-yet-undefined one. Such a revolution may occur but once in a century, and perhaps even only once in a millennium. I elaborate on this major change in the epilogue.

To sum up then: While the characterizing of historical eras is fraught with risk, the special characteristics of the period around 1900 warrant such an effort. Not only were the principal creative figures exposed to common forces and events, but they often were actually cognizant of and influenced by one another's activities. A study of each one's efforts, while illuminating in itself, gains significance when considered in light of the parallel events and insights occurring in the lives of the cocreators of the modern era.

So far, I have introduced our seven creative figures, presented major themes that are realized in their creative efforts, and discussed some perils and promises of a study that purports to portray a historical era. In what follows, the creative breakthroughs achieved by the seven individuals form my major area of focus. Before embarking on the individual case studies in part II, however, I need to locate this effort within the broader landscape of earlier studies of creative individuals, works, and processes.

2

APPROACHES TO CREATIVITY

IN A SURPRISINGLY faithful way, the history of behavioral scientists' attempts to study human creativity parallels the history of their attempts to investigate human intelligence. Like *intelligence*, the term *creativity* has been applied over the years as an honorific label to a wide range of individuals, situations, and products. Such lay use of the terms *creative, creativity,* or *creating* may have sufficed on the streets; but as happened with the term *intelligence*, the variant forms of *creativity* have seemed in need of more precise formulation.

THE STUDY OF CREATIVITY SHADOWS
THE STUDY OF INTELLIGENCE

Thanks to the revolution in psychological measurement (or psychometrics), associated particularly with the work of Alfred Binet in Paris and Lewis Terman in California, the concept of "intelligence" and its putative measure "IQ" were operationalized early in the twentieth century—as it happens, at the birth of the modern era, as I have defined it. Every individual was thought to possess a certain amount of intelligence, possibly as his or her birthright, possibly as a result of nurture; the kinds of brief verbal and numerical items that populate IQ tests were thought sufficient to indicate an individual's intelligence. Many intelligence tests were devised, but they tended to incorporate the same kinds of items and to correlate highly with one another; if one is psychometrically "bright" on a Stanford-Binet measure of intelligence, one is likely to stand out equivalently on the measures devised by David Wechsler and by other leaders of the intelligence intelligentsia.

It was not surprising—and was perhaps overdue—when, at mid-century, a leading psychologist, Joy P. Guilford, called for a scientific focus on creativity. As a psychometrician, Guilford had in mind a program that would parallel the apparently successful mission undertaken earlier in the century with reference to intelligence. Arguing that creativity is by no means equivalent to intelligence, Guilford asserted the need for an arsenal of measures designating *which* individuals had the potential to be creative.

The key idea in the psychologist's conception of creativity has been *divergent thinking*. By standard measures intelligent people are thought of as convergers—people who, given some data or a puzzle, can figure out the correct (or at any rate, the conventional) response. In contrast, when given a stimulus or a puzzle, creative people tend to come up with many different associations, at least some of which are idiosyncratic and possibly unique. Prototypical items on a creativity test ask for as many uses as possible for a brick, a range of titles for a story, or a slew of possible interpretations of an abstract line drawing: A psychometrically creative individual can habitually issue a spectrum of divergent responses to such an item, at least some of which are rarely encountered in the responses of others.

After considerable debate and experimentation in the decades following Guilford's challenge, psychologists reached three conclusions. First, creativity is not the same as intelligence. While these two traits are correlated, an individual may be far more creative than he or she is intelligent, or far more intelligent than creative. Moreover, when talented individuals are examined, it is clear that psychometric creativity is independent of psychometric intelligence, once a threshold IQ of 120 has been reached.

The other two conclusions pertain to the classical issues surrounding all testing. Creativity tests *are* reliable. That is, if an individual takes the same creativity test more than once, he or she is likely to get a similar score. Moreover, correlations in a person's measured creativity score are robust even across creativity tests (of course, creativity tests, like intelligence measures, are typically considered valid if their results correlate with other measures presumed to reflect the construct in question).

The remaining conclusion is, in my view, devastating for the enterprise of measuring creativity using paper-and-pencil tests. Despite a few suggestive findings, it has not been possible to demonstrate that creativity tests are *valid*. That is, high scores on a creativity test do not signal that one is necessarily creative in one's actual vocation or avocation, nor is there convincing evidence that individuals deemed creative by their discipline or culture necessarily exhibit the kinds of divergent-thinking skills that are the hallmark of creativity tests.

Even more so than intelligence tests, then, tests of creativity have failed to satisfy the expectations they were designed to meet. Except for certain targeted research purposes, creativity tests (and the thinking that underlines them) have made little difference in the broader research and educational communities. They have, however, triggered some constructive reactions among cognitively oriented researchers.

COGNITIVE APPROACHES TO CREATIVITY

Many commentators have criticized creativity tests for the seemingly banal view of human creativity they embody. One alternative tack has been to devise more demanding test items—ones that seem to require genuine insight or mental leaps rather than cocktail-hour glibness. Researchers in the tradition of Gestalt psychology have favored items like the "tumor problem": In this classic puzzle, the solution for dealing with a pernicious tumor without destroying the vital surrounding tissue is to direct sublethal dosages of radiation from several vantage points. In another favorite, the "three-line" problem, the solver is challenged to connect nine dots arranged in a three-by-three matrix without lifting the pencil. The creative move here is to extend the line beyond the confines of the target configuration. Such problems begin to refute the charge of banality; but they tend to favor individuals who are already familiar with the domain in question (e.g., X-ray technology, geometrical puzzles), and they have little demonstrable relation to creativity outside of the testing environment. Both also reward individuals who happen to excel in visual problem solving, while penalizing those more comfortable with numbers or words.

A second reaction has been characteristic of cognitive science (particularly that branch called artificial intelligence). Researchers from this investigative tradition disparage the superficiality of psychometric creativity items as well as the lack of clarity about the mental processes allegedly used to solve these items. Instead, such cognitive researchers call for a computer-based investigation of full-scale scientific problem solving, a process requiring creative thought processes for attainment of an original solution.

In a prototypical instance, researchers have devised a computer program called BACON. When supplied with raw (unprocessed) data—for example, about the varying pressures on a gas and the volume that the gas accordingly occupies—the program computes an underlying principle—in this case, the same inverse ratio between pressure and volume that Robert Boyle discovered in the seventeenth century and that has come to be known as Boyle's law. Computer programs of this sort have been able to rediscover many scientific laws through induction and generalization.

At the very least, these computer simulations constitute demonstrations or existence proofs—illustrations that a computing entity can, when furnished with the relevant data, ferret out a scientific law. However, it is by no means evident that BACON and human scientists use identical or equivalent processes. As Mihaly Csikszentmihalyi has pointed out, the computer program must begin with the problem and the data that are supplied in the particular form favored by the cognitive scientist; and it must use the algorithms it has been programmed to employ. In contrast, the human problem solver must select the problem to be investigated; determine which of an infinite array of potential data are relevant to a solution of the problem; and ascertain which kinds of analyses to perform on that data in reaching for a solution, painstakingly inventing new means of analyses when necessary.

While the particular claims seem to me to be overwrought, the general approach taken by the cognitivists constitutes a definite step forward. Cognitive researchers, among them Margaret Boden, David Perkins, and Robert Sternberg, have described the ways in which creative individuals identify problem and solution "spaces" that appear promising; search within these spaces for approaches appropriate to the problem at hand and for leads that may pay off; evaluate alternative solutions to problems; deploy resources of energy and time to advance their program of investigation in an efficient manner; and determine when to probe further and when to cut losses and move on, and more generally, reflect on their own creating processes. Some cognitive researchers have shown these principles at work in specific domains, such as jazz improvisation or imaginative writing. In all, the cognitivists have identified ways to examine creative work at the appropriate level of complexity.

A third arrow of criticism aimed at the psychometric approach calls for a focus on *unambiguous* instances of creative processes, as embodied in the behavior and thinking of productive artists, scientists, and other workers. In general, those sympathetic to this tradition have favored careful case studies of individuals like Charles Darwin (as carried out by the psychologist Howard Gruber), Antoine Lavoisier (as carried out by the historian of science Frederic Holmes), or Pablo Picasso (as carried out by the Gestalt psychologist of art Rudolf Arnheim). Such studies differ from the investigations usually carried out by humanistically oriented biographers in their focus on the development of networks of ideas, their use of concepts and models drawn from the cognitive sciences, and their search for principles that may extend beyond the particular individual under investigation.

The most elaborated instance of this work has come from Howard Gruber and his students over the past few decades. Gruber's work is characterized by careful attention to the ways in which generative ideas, and sets of ideas, evolve and deepen over significant periods of time. The Gruber team has uncovered a number of principles that seem to characterize the work of major scientists, like Charles Darwin or Jean Piaget (the latter was Gruber's own teacher). Such individuals engage in a wide and broadly interconnected network of enterprises; exhibit a sense of purpose or will that permeates their entire network, giving direction to their daily and their yearly activities; favor the creation and exploitation of images of wide scope (such as the branching tree of evolution); and display a close and continuing affective tie to the elements, problems, or phenomena that are being studied. Gruber speaks of an "evolving systems" approach to the study of creativity: That is, one monitors simultaneously the organization of knowledge in a domain, the purpose (s) pursued by the creator, and the affective experiences he or she undergoes. While these systems are only "loosely coupled," their interaction over time helps one understand the ebb and flow of creative activity over the course of a productive human life.

In many ways, my approach is faithful to Gruber's tradition. I favor the use of individual case studies, the adoption of a developmental perspective, the monitoring of several different systems, and the examination of the ways in which they interact. I diverge from the Gruber tradition by using a deliberately broad comparative focus on creativity across diverse domains; examining instances of creativity drawn deliberately from a specific historical-cultural era; and focusing on how individual breakthrough, domain of practice, and reactions of the surrounding community are dynamically related.

APPROACHES IN TERMS OF PERSONALITY AND MOTIVATION

Up to this point, my discussion of creativity has drawn primarily on two approaches within psychology: the venerable testing, or psychometric, tradition and the more recent cognitive perspective. For many years, however, there has existed a complementary approach to creativity within psychology, one associated with the noncognitive aspects of the individual—in particular, with facets of personality and motivation.

Paralleling the psychometric approach, researchers using one paradigm have examined the personality traits of individuals deemed creative by their community. Typically, individuals participating in these studies are asked to select apt descriptions of themselves and also to respond to ambiguous stimuli (such as inkblots or silhouettes) in ways thought to "evoke" or "project" their underlying personality structure.

In a representative study conducted by the Berkeley Institute of Personality Assessment, "creative architects," as distinguished from their less creative peers, exhibited a greater incidence of such personality traits as independence, self-confidence, unconventionality, alertness, ready access to unconscious processes, ambition, and commitment to work. However, it is not clear whether people who already exhibit these characteristics become creative or whether, as a result of acknowledged creativity, people come to exhibit such positively tinged traits. Also, individuals who work closely with those deemed creative seem to exhibit a similar profile of traits.

Psychoanalytic Perspectives

It is not surprising that Freud, arguably the most important psychologist of his era, also contributed to an understanding of creativity—and this, despite his oft-quoted laments that "before creativity, the psychoanalyst must lay down his arms" and that "the nature of artistic attainment is psychoanalytically inaccessible to us." To begin with, Freud's illustration of the centrality of unconscious processes underscored the point that creative activity is not a direct reflection of deliberate intention; much of its impetus and significance remain hidden from the individual creator and, quite possibly, from those in his or her community as well.

Having demonstrated the importance of sexuality in motivating human behavior in general, Freud called attention to the sexual factors that undergird a creative life. In Freud's view, creative individuals are inclined (or compelled) to sublimate much of their libidinal energy into "secondary" pursuits, such as writing, drawing, composing, or investigating scientific puzzles. He would have found many data of interest in the seven cases presented here.

Freud's convictions about the importance of infantile development also colored his view of creative activity. Freud was impressed by the parallels between the child at play, the adult daydreamer, and the creative artist. As he once phrased it:

> Might we not say that every child at play behaves like a creative writer, in that he creates a world of his own, or, rather, rearranges the things of his world in a new way which pleases him? . . . The creative writer does the same as the child at play. He creates a world of phantasy which he takes very seriously—that is, which he invests with large amounts of emotion—while separating it sharply from reality.

Freud's view of the creative life, particularly that of the artist, has attracted considerable attention—and much criticism. The evidence on which Freud drew his conclusions is considered shaky, particularly in instances where the subject is long dead (for example, Leonardo, Shakespeare) and has left little reliable autobiographical material. And, while Freud's characterizations may apply to some creative individuals, they apply to noncreative individuals as well; hence, they cannot distinguish the effective artist or scientist from the ineffective or banal one. Yet, despite such criticisms, Freud's work remains appropriately influential in the study of creativity, including the present inquiry. Like other revolutionary figures, Freud helped frame the terms within which the personality and motivation of creative individuals have subsequently been described.

Behaviorist Perspective

While the psychoanalytic tradition shares little else with the American behaviorist school, representatives of both perspectives agree that individuals engage in creative activity largely because of the material rewards they secure. In Freud's account, artists seek power and money and, unable to secure these directly, find a haven in creative activities; or they attain indirectly from their creative work some of the libidinal and Oedipal pleasures they crave. In Skinner's behavioral terms, people engage in creative activity because of a previous history of rewards, or "positive reinforcements." Recently, however, a number of psychologists have put forth a rather different picture of the factors that motivate creative activity.

Intrinsic Motivation

In a series of illuminating experimental demonstrations, social psychologist Teresa Amabile has called attention to the importance of "intrinsic motivation." Contrary to what is predicted by classical psychological accounts, Amabile has shown that creative solutions to problems occur more often when individuals engage in an activity for its sheer pleasure than when they do so for possible external rewards. Indeed, knowledge that one will be judged on some criterion of "creativeness" or "originality" tends to narrow the scope of what one can produce (leading to products that are then judged as relatively conventional); in contrast, the absence of an evaluation seems to liberate creativity.

Embracing a different vocabulary, Mihaly Csikszentmihalyi has described a highly sought-after affective state called the flow state or flow experience. In such intrinsically motivating experiences, which can occur in any domain of activity, people report themselves as fully engaged with and absorbed by the object of their attention. In one sense, those "in flow" are not conscious of the experience at the moment; on reflection, however, such people feel that they have been fully alive, totally realized, and involved in a "peak experience." Individuals who regularly engage in creative activities often report that they seek such states; the prospect of such "periods of flow" can be so intense that individuals will exert considerable practice and effort, and even tolerate physical or psychological pain, in pursuit thereof. Committed writers may claim that they hate the time spent chained to their desks, but the thought that they would *not* have the opportunity to attain occasional periods of flow while writing proves devastating.

During an individual's immersion in a domain, the locus of flow experiences shifts: What was once too challenging becomes attainable and even pleasurable, while what has long since become attainable no longer proves engaging. Thus, the journeyman musical performer gains flow from the accurate performance of familiar pieces in the repertoire; the youthful master wishes to tackle the most challenging pieces, ones most difficult to execute in a technical sense; the seasoned master may develop highly personal interpretations of familiar pieces, or, alternatively, return to those deceptively simple pieces that may actually prove difficult to execute convincingly and powerfully. Such an analysis helps explain why creative individuals continue to engage in the area of their expertise despite its frustrations, and why so many of them continue to raise the ante, posing ever-greater challenges for themselves, even at the risk of sacrificing the customary rewards.

THE HISTORIOMETRIC APPROACH

Studies in the cognitive tradition, on the one hand, and the personality and motivation tradition, on the other, have constituted the large majority of social-scientific investigations of creativity in recent years. One additional perspective, less well known perhaps, also deserves mention in this survey: the historiometric approach associated particularly with the work of the psychologist Dean Keith Simonton.

Unlike the approaches reviewed so far, Simonton's is inherently a *methodology* for investigation and hence can be applied equally to issues of cognition, personality, motivation, or creative works themselves. Simonton formulates (or operationalizes) classical puzzles concerning creativity as clearly as possible; he then seeks quantitative data that can help resolve those issues. The topics considered range from the personality traits of creative individuals to the circumstances of their training to the properties of their most highly esteemed works. In contrast to Gruber, Simonton uses a quantitative approach and deals with as large a database as possible; in contrast to Amabile, Simonton eschews experimental intervention and relies instead on the historical record.

In a typical approach, historiometric investigators like Simonton review large bodies of data to determine the decade of life in which creative individuals are most productive. Such studies have led to the findings that maximal productivity typically occurs between ages thirty-five and thirty-nine, but that profiles differ appreciably across disparate domains of knowledge: Thus, poets and mathematicians reach an apogee in their twenties or thirties, while historians or philosophers may peak decades later.

In another, different line of work, Simonton demonstrates that the most highly esteemed creators not only are more productive in general, but that they produce more "bad" works that have long been ignored as well as more "good" works that are esteemed by posterity. Using such an approach, Simonton and other historiometricians have managed to provide at least provisional answers to a host of questions long debated by experts on creativity. Of course, the investigations depend on the particular ways in which the historiometrician chooses to frame the problem and on the quality of the available historical data. The method provides few fresh insights about particular creative breakthroughs or particular creators, but it proves invaluable for assessments of individuals in a broader context.

In my own view, the work by Gruber and his associates on individual case studies, along with the work by Simonton and his associates in the historiometric tradition, are among the most exciting recent lines of investigation in the area of creativity. Not surprisingly, they are also the most pertinent to my exploration of creative individuals. While my method and sympathies are closer to Gruber's evolving systems approach, I crave as well the precise and copious background information that the historiometric school can provide. In my judgment, a comprehensive science of creativity must somehow succeed in spanning the gap between these approaches. This book may be regarded as one effort to proceed at least the first step away from findings rooted in individual cases to generalizations that can elucidate creativity within and across domains. Accordingly, in part III, I compare the case studies on a series of dimensions.

MY APPROACH TO CREATIVITY

Even to begin to encompass creativity, one must take into account a huge number of factors and their multifarious interactions. In this book, I seek to provide a readable account of my conclusions while also presenting enough technical information for interested readers to evaluate and build on my methods, data, and findings. In the remainder of this chapter, and in chapter 10, I focus on these research issues. It is possible to appreciate the case studies and the conclusions without immersing oneself in these matters of detail; armed with the brief apparatus introduced in chapter 1, readers have the option of jumping immediately to part II. But I have arranged my presentation in such a way that the nonspecialist can follow the gist of my methods.

My approach consists of four separate components. There is no hard-and-fast line separating these components, but I find it useful to conceive of them as discrete contributors to the study.

1. *Organizing Themes:* These are the most general themes that guided my inquiry and that gave rise to the principles around which the individual case studies have been formulated.

2. *Organizing Framework:* My study presupposes an interdisciplinary analytic framework, worked out in conjunction with a number of valued colleagues.

3. *Issues for Empirical Investigation:* Growing out of the framework are a host of issues and questions, which case studies ought, at least in principle, to be able to clarify.

4. *Emerging Themes:* Two themes not originally part of my research agenda emerged with increasing clarity as I pursued the individual case studies. Because these themes were not anticipated, they in some sense constitute my discoveries in this study.

In the four following sections, I provide further details on each of these components. For the sake of convenience, I have labeled each component, and its respective subcomponents, with separate letters and numbers; these are reproduced in table 2.1 and discussed again in the final chapters.

Let me now, in turn, touch on each of the components, providing sufficient background so that its role in the case studies can be properly appreciated.

TABLE 2.1. FOUR PRINCIPAL COMPONENTS IN THE STUDY OF CREATIVITY

I. Organizing Themes
   A. Relation Between the Child and the Adult Creator
   B. Relation Between the Creator and Others
   C. Relation Between the Creator and His or Her Work

II. Organizing Framework
   A. Developmental Perspective
      1. Life-Course Perspective
      2. Creation of a Work
   B. Interactive Perspective: Interaction Among Individuals, Domains, and Fields
      1. Definition
      2. Multidisciplinary Framework
      3. "Where Is Creativity?"
   C. Fruitful Asynchrony

III. Issues for Empirical Investigation
   A. Individual Level
      1. Cognitive Issues
      2. Personality and Motivational Issues
      3. Social-Psychological Issues
      4. Life Patterns
   B. Domain Level
      1. Nature of Symbol Systems
      2. Kind of Activity
      3. Status of Paradigm
   C. Field Level
      1. Relation to Mentors, Rivals, and Followers
      2. Level of Political Controversy
      3. Hierarchical Organization

IV. Emerging Themes
   A. Cognitive and Affective Support at the Time of the Breakthrough
   B. The Creator's Faustian Bargain

ORGANIZING THEMES REVISITED

The organizing themes, introduced in chapter 1, are the most intuitive of the components. As such, they provide an accessible way for us to discuss each of the seven creators of the modern era. Any of the themes could be employed with reference to any of the creators, but I have chosen to discuss each creator in terms of a theme that is particularly appropriate to his or her life circumstances.

The organizing themes can be grouped broadly into three categories that may be ordered in any way. The first concerns *the relationship between the child and the adult creator*. This theme reflects my belief that important dimensions of adult creativity have their roots in the childhood of the creator. In the study of Einstein, this theme provides a way of examining the connection between the kinds of questions a gifted child ponders, and the nature of training and thinking required for the adult practitioner to answer such questions. In the study of Picasso, I turn my attention to the relationship between the productivity associated with youthful prodigiousness, on the one hand, and with mature mastery, on the other.

The second organizing theme probes *the relationship between the creator and other individuals*. These other individuals include those who are closest to the creator (family members, confidants), as well as those who are involved in his or her education (as teachers or mentors) or subsequent career (as colleagues, rivals, or followers). In three of the remaining case studies I treat the relationship between the creator and other people directly; in two cases, more abstractly.

With Freud, I examine the relationship between Freud and many other individuals during his youth; the gradual narrowing of the ensemble until Freud is virtually alone; and then, following the principal discoveries of psychoanalysis, the opening up again to a wider circle of associates. With Gandhi, I highlight the ways he affected the behaviors of others, while with Stravinsky, I portray the vexing political pressures that surround an individual who chooses to work in a collaborative domain.

In two of the cases I explore a more abstract kind of relationship to the world of others. With Eliot and Graham, I discuss two kinds of marginality: a marginality of choice, which Eliot sought; and an enforced marginality, which Graham experienced because of her gender and her nationality.

The third organizing theme focuses on the *relationship between the creator and work in a domain*. Early in life, the creator generally discovers an area or object of interest that is consuming. At first the creator seeks to master work in that domain in the manner of others working within the culture; increasingly, however, the very relationship to the domain becomes problematic. The individual then, willingly or unwillingly, feels constrained to try inventing a new symbol system—a system of meaning—that is adequate to the chosen problems or themes and that can eventually make sense to others as well. In each chapter I examine in detail the ways in which a creator forges a new system of meaning in a distinctive domain; it turns out that surprising commonalities hold across the domains as well.

ORGANIZING FRAMEWORK

In effect, this trio of organizing themes embodies the principal features of the framework that has guided this study. They serve as an informal way of introducing enduring concerns with human development (as in the relation between child and master); the development of work (as happens when the creator begins to deviate from common practices in the domain); and the relations and tensions among individual talent, the domain of work, and the field of judges. I turn now directly, and somewhat more formally, to an introduction of the principal features of this framework.

A Developmental Perspective

To a developmental psychologist, the study of creativity is necessarily anchored in the study of human development. Both the evolution of specific creative works and the more general trajectory of growth of mastery in a domain require consideration in the light of principles governing development.

The Life-Course Perspective. By virtue of species membership, all normal children undergo a lengthy period of exploration of their environment, a period during which they have the opportunity to discover the principles that govern the physical world, the social world, and their own personal world. Not only does this discovery of universals become the background against which further learnings and discoveries necessarily take place, but the very *processes of discovering* themselves become models for later exploratory behaviors, including efforts to probe phenomena never before conceptualized.

The quality of these early years is crucial. If, in early life, children have the opportunity to discover much about their world and to do so in a comfortable, exploring way, they will accumulate invaluable "capital of creativity," on which they can draw in later life. If, on the other hand, children are restrained from such discovering activities, pushed in only one direction, or burdened with the view that there is only one correct answer or that correct answers must be meted out only by those in authority, then the chances that they will ever cast out on their own are significantly reduced.

Many creative individuals do point with some distress to the restrictiveness of their early childhood; and in the pages that follow, I describe parents who were quite strict. (Sometimes, as a reaction, creative individuals bend too far in the opposite direction in rearing their own children.) But even those who suffered a strict regimen somehow managed to retain the spark of curiosity, possibly because they were strong and rebellious personalities, but even more likely because they encountered at least one role model who did not simply toe the line but rather encouraged a more adventurous stance toward life.

What may distinguish creative individuals is their ways of productively using the insights, feelings, and experiences of childhood. For some purposes, it may prove adaptive to erase memories of childhood. But when it comes to the forging of new understandings and the creation of new worlds, childhood can be a very powerful ally. Indeed, I contend that the creator is an individual who manages a most formidable challenge: to wed the most advanced understandings achieved in a domain with the kinds of problems, questions, issues, and sensibilities that most characterized his or her life as a wonder-filled child. It is in this sense that the adult creator draws repeatedly on the capital of childhood. In different eras, different periods of childhood will be drawn upon; it seems that the special burden of the modern era is to mine the early years of childhood.

As the educational psychologists Benjamin Bloom and Lauren Sosniak have documented with respect to talented adults, one can usually identify a situation or even a moment when these young individuals first fell in love with a specific material, situation, or person—one that continues to hold attraction for them. Following the philosopher Alfred North Whitehead, they speak of an initial romance; borrowing a term from David Feldman, I allude to a crystallizing experience.

No matter how potent such an intoxication, at least ten years of steady work at a discipline or craft seem required before that métier has been mastered. The capacity to take a creative turn requires just such mastery, and accordingly, significant breakthroughs can rarely be documented before a decade of sustained activity has been accomplished. Even Mozart, arguably the exception that proves the rule, had been composing for at least a decade before he could regularly produce works that are considered worthy of inclusion in the repertory. With the seven creators in question, at least a decade—and in some instances, more time—elapsed before innovative achievement had coalesced. And, as typical, another decade passed before a second major innovation was forged.

Yet it would be unwarranted to contend that one first follows die craft for ten years and only then strikes out on one's own. My own analysis suggests the reverse pattern. Individuals who ultimately make creative breakthroughs tend from their earliest days to be explorers, innovators, tinkerers. Never satisfied simply to follow the pack, they can usually be found experimenting in their chosen métier, and elsewhere as well. Young musical performers, for example, often reveal their gift for composing by a constant effort to "rewrite a piece" according to their own preferred specification, budding scientists do not brook received wisdom, but rather demand to see for themselves. Often this adventurousness is interpreted as insubordination, though the more fortunate tinkerers receive from teachers or peers some encouragement for their experimentation.

At any rate, after a period of skill development, with or without overt challenge to authority, the future innovator clearly shows a readiness to cast off in new directions. A specific personality configuration must be at work here, since so many people who attain a comparable level of competence are satisfied simply to remain at that level or to make minor adjustments, rather than to strike out audaciously on their own. Sometimes events intervene, for example, when a crisis in the discipline alerts a whole generation of young workers to the need for rethinking. Even here, however, a certain doggedness is required. For example, many young researchers knew that a struggle was under way to decipher the structure of DNA, but it took the special gifts and pertinacity of James Watson and Francis Crick—as well as strokes of luck—to crack the code.

Creation of a Work. At this point in the general trajectory of development, I examine human creativity most intensively: Einstein as he engages in thought experiments about light, Graham as she searches for a distinctly American form of bodily expression, and Gandhi as he experiments with various stances between human beings in an effort to resolve bitter conflict without violence. I look at the individual's construction (or constructing) of the domain in which he or she works, the location of problem areas or uncertainties in this domain, and the casting out for new leads or perspectives that more adequately address a felt lack or a promising new direction.

Consistent with a cognitive perspective, I attempt to re-create the mental model, the representational map, that each individual formed about the chosen creative task. At first accepting the common language or symbol system of the domain, each creator finds soon enough that it proves inadequate in one or more respects. He or she will probably try minor changes at first, because no one finds it that inviting or facile to alter the entire legacy of a domain, one that may have been built up painstakingly over decades or even centuries.

Yet, characteristically, the creator finds further change necessary—whether because the creative individual is dissatisfied with an ad hoc solution or because the particular problem can be solved only by a fundamental reorientation or because of some other factor(s) depends on the particular circumstances. But in any event, a seemingly local solution needs to be abandoned in favor of a far more extensive reorientation or reconceptualization.

These are the times that try the mettle of the creator. No longer do the conventional symbol systems suffice; the creator must begin, at first largely in isolation, to work out a new, more adequate form of symbolic expression, one equal to the problem or product in all of its complexity. Often initial efforts do not work out satisfactorily, and the creator must return to the drawing boards (sometimes literally!). In this pursuit there are no guarantees or even reliable guides; the creator must trust his or her own intuition and must be braced for repeated and unrequited failures.

In each of the case studies, I carefully examine the moments of creative breakthrough. Cognitive work at this time differs in terms of not only the particular intelligences that are mobilized but also the kinds of creative activity in which the individual is involved. Put succinctly, it is a different matter to solve a mathematical problem or define a psychological construct than it is to stage an effective performance or to influence the behavior of millions of one's countrymen. The readily invoked terms *problems* and *solutions* prove far more suited for standard scientific work than for creation in the arts or in the social sphere.

Let me summarize the developmental features that recur in the seven analyses: (1) a concern with the universals of childhood as well as the particularities of specific childhoods; (2) an examination of initial interest and its conversion into sustained mastery of a domain; (3) the discovery or creation of novel or discrepant elements at some point after mastery has been obtained; (4) the ways in which the creator deals with the initial novelty and embarks on a *program* of exploration; (5) the supportive or inhibitory roles played by other individuals during the period of isolation; (6) the ways in which a new symbol system, language, or mode of expression is gradually worked out; (7) the initial reactions of the relevant critics and the ways in which these reactions are transformed over a significant period of time; and (8) the events surrounding a second, more comprehensive innovation that often occurs during middle life.

Interactive Perspective: Interaction among Individuals, Domains, and Fields

Over the past several years, in conjunction with colleagues, and especially with Mihaly Csikszentmihalyi and David Feldman, I have evolved the "interactive perspective" on creativity, which informs this book. While not overly complex, the perspective is multifaceted and requires some background and elaboration. I have begun to introduce the framework in the guise of the three intuitive organizing themes. I propose here to introduce the perspective more formally in three phases: via a definition, a multidisciplinary research perspective, and the reformulation of a familiar question. Following this introduction, I explain how the perspective informs the case studies that constitute the heart of this investigation.

Definition. Let me begin, then, by offering a definition of the creative individual, which I have found useful in my own work: The creative individual is a person who regularly solves problems, fashions products, or defines new questions in a domain in a way that is initially considered novel but that ultimately becomes accepted in a particular cultural setting.

Parts of this definition (such as the notion that creativity involves problem solving and that it connotes both initial novelty *and* ultimate acceptance) would be accepted by nearly every psychologically oriented researcher of creativity. Less standard (and therefore more revealing) are four other features:

1. My statement that a person must be creative *in a domain*, rather than across all domains, directly challenges the conceptualization of an all-purpose

creative trait that underlies tests of creativity. I am focusing on the particular domains or disciplines within which an individual works, and the ways those domains may be refashioned as a result of a creative breakthrough.

2. My claim that creative individuals *regularly* exhibit their creativity calls into question the possibility of having a once-in-a-lifetime burst of creativity. Indeed, as Gruber has so well illustrated, creative individuals wish to be creative, and they organize their lives so as to heighten the likelihood that they will achieve a series of creative breakthroughs. In general, only the creative individual who dies at a young age is a likely candidate for one-shot creativity.

3. By insisting that creativity can involve the *fashioning of products* or the *devising of new questions* as well as the solution of problems, I challenge psychometric and computer simulation approaches, which prove far better at resolving extant problems than at forging new products or at defining new problems. Of course, much creative work does involve the solution of problems already recognized as such. But at the higher reaches, creativity is far more often characterized by the fashioning of a new kind of product, or by the discovery of an unknown or neglected set of issues or themes that call for fresh exploration.

4. I assert that creative activities are only known as such when they have been *accepted in a particular culture.* No time limit is assumed here; a product may be recognized as creative immediately—or not for a century or even for a millennium. But the crucial (if controversial) point here is that nothing is, or is not, creative *in and of itself.* Creativity is inherently a communal or cultural judgment. The most one can say about an entity before it has been evaluated by the community is that it (or he or she) is "potentially creative." And evaluation must be undertaken by a relevant portion of one's community or one's culture: No other arbiters are available.

Multidisciplinary Framework. Clearly, the bulk of work in the area of creativity has been carried out by researchers trained in psychology and related individual-centered disciplines. Yet it has become increasingly clear that creativity is precisely the kind of phenomenon or concept that does *not* lend itself to investigation completely within a single discipline. As Peter Medawar, the Nobel Prize–winning immunologist, once declared:

The analysis of creativity in all its forms is beyond the competence of any one accepted discipline. It requires a consortium of talents: Psychologists, biologists, philosophers, computer scientists, artists, and poets would all expect to have their say. That "creativity is beyond analysis" is a romantic illusion we must now outgrow.

I believe that, ultimately, the understanding of creativity will entail explorations at four different levels of analysis:

1. *The Subpersonal.* As yet, little is known about the genetics and the neurobiology of creative individuals. We know neither whether creative individuals have distinctive genetic constitutions, nor whether there is anything remarkable about the structure or functioning of their nervous systems. Yet, any scientific study of creativity will ultimately need to address these biologically oriented questions, and I expect that such study will soon be undertaken.

2. *The Personal.* Those trained in the psychological tradition will continue to provide major input into our understanding of creative individuals, processes, and products. As in the past, and in my own review of earlier psychological research, there will be two major lines of investigation. One will focus on the cognitive processes that characterize creative individuals; a complementary tradition will focus on the personality, motivational, social, and affective aspects of creators.

3. *The Impersonal.* Inherent in my view of creativity is a conviction that an individual cannot be creative in the abstract; as Feldman has insisted, all of us exhibit whatever creativity we have via specific *domains or disciplines.* Thus, any creative individual makes his or her contributions in particular domains, which can themselves be described in terms of the current level of knowledge and practice. Einstein's achievements must be appreciated with reference to the physics of 1900, just as Gandhi's recipe for human interaction must be seen in light of earlier modes of interaction between occupying and indigenous populations. The impersonal study is carried out by historians, philosophers, students in artificial intelligence, and, most especially, experts drawn from the domain itself. Because this perspective represents an attempt to capture the nature of knowledge per se, I see it as primarily epistemological in nature.

4. *The Multipersonal.* Surrounding any potentially creative individual or product is a host of other individuals and institutions sanctioned to evaluate the appropriateness and quality of the contribution at hand. I adopt Csikszentmihalyi's term "field" to describe this congeries of forces, the study of which is fundamentally sociological. Such a multipersonal perspective

examines the ways that members of the field—judges, editors, agents, media professionals, encyclopedia writers, and other evaluators—make initial, provisional assessments, as well as the processes by which, aided by the perspective of time, they render more authoritative judgments. Sometimes, as with physics, the field consists of a small cohort of trained experts; but in areas like popular entertainment, the field may number in the millions.

The full-blown study of creativity can best proceed—as I try to do here—through examination of creative phenomena from the multiple perspectives of the neurobiologist, the psychologist, the domain expert, and the sociologically oriented student of the field. Yet, because of my own training and because of the preponderance of psychological studies, it is probably inevitable that I will place heaviest emphasis on personal factors, drawing on biological, epistemological, and sociological perspectives to enhance the picture I am fashioning. There is a sense—for which I do not apologize—in which this study of creativity reflects the "great man/great woman" view of creativity.

Where Is Creativity? In a difficult and complex area of study like creativity, important conceptual advances are not easy to come by. It was therefore a significant moment when Csikszentmihalyi suggested that the conventional question, "What is creativity?" be replaced by the provocative inquiry, "Where is creativity?"

Csikszentmihalyi identifies three *elements* or *nodes* that are central in any consideration of creativity: (1) the individual person or talent; (2) the domain or discipline in which that individual is working; and (3) the surrounding field that renders judgments about the quality of individuals and products. (These three nodes correspond, roughly, to the core elements introduced in chapter 1 and to the second, third, and fourth disciplinary perspectives just outlined.) In Csikszentmihalyi's persuasive account, creativity does not inhere in any single node, nor, indeed, in any pair of nodes. Rather, creativity is best viewed as a dialectical or interactive process, in which all three of these elements participate:

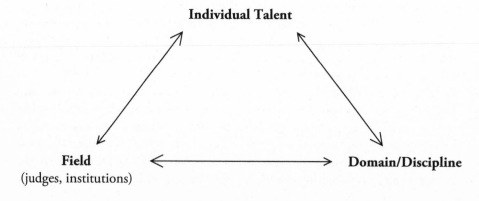

Individual Talent

Field
(judges, institutions)

Domain/Discipline

We can now revisit the figure introduced in chapter 1 and consider its operation in dynamic form. One begins with a set of individuals of varying abilities, talents, and proclivities, each engaged in work in a particular domain. At any historical moment, that domain features its own rules, structures, and practices, within which the individuals are socialized and according to which they are expected to operate. Such individuals address their work to the field, which in turn examines the various products that come to its attention. Of the many individuals and works that undergo scrutiny by the field, only a few are deemed worthy of sustained attention and evaluation. And of those works that are appreciated at a given historical moment, only a small subset are ever deemed to be *creative*—highly novel, yet appropriate for the domain. The works (and the workers) so judged come to occupy the most important spot in the dialectic: They actually cause a refashioning of the domain. The next generation of students, or talents, now works in a domain that is different, courtesy of the achievements of highly creative individuals. And in this manner the dialectic of creativity continues.

To help concretize this scheme, assume that there are a thousand budding painters at work in Paris, each with his or her peculiar strength and style. All of these individuals attempt some mastery of the domain of painting, as it now exists; and all address their work to the field—the set of critics, art school departments, gallery owners, agents, and the like. Of these thousand individuals, a few will be selected as worthy of attention by the field; and at least today, sheer novelty of the work is likely to constitute a significant factor in their selection. Of this smaller circle of talented individuals, one or two at most will paint in a manner that becomes so highly valued that their efforts will ultimately exert some effect on the domain—on the structure of knowledge and practice to be mastered by the next generation of painters. Thus, creativity lies not in the head (or hand) of the artist or in the domain of practices or in the set of judges: Rather, the phenomenon of creativity can only—or, at any rate, more fully—be understood as a function of interactions among these three nodes. I have sought to capture the complexity of this dialectic interaction through the multidirectional arrows in the figure.

Painting, however, may seem an idiosyncratic domain, perhaps one where the field assumes undue importance. What of a contrasting domain, such as mathematics, where monetary considerations are presumably unimportant and where less fickle standards can accordingly be invoked? I submit that the processes at work here are parallel. Substitute for our thousand visual artists an equal number of young mathematicians, say, topologists. Each of these students must master the domain as currently practiced. Those who wish to advance must then address their proofs and discoveries to the field—in this case, a set of journal editors, professors, prize committees, and supportive or jealous peers. Only a few of the young topologists will stand out in terms of professorships and publications; and of these, even fewer will sufficiently affect the domain in which they work so that the next generation of youthful topologists must master a somewhat altered domain.

Biographies of significant figures are studded with instances (sometimes bizarre) where an ultimately esteemed work was either ignored, misinterpreted, or completely scorned by the field. One might go so far as to maintain that initial rejection is the likely fate of any truly innovative work. But it is also possible to mount the opposite case. For example, with respect to our modern masters, following their initial, usually lonely struggles, most of them became well known and esteemed within a decade—a remarkably short period of time, given the perspective of history.

Such a statement is likely to rekindle a question that has doubtless been on the minds of many readers. Surely, one might say, there are many other individuals whose work is just as original, just as creative, just as notable, but who for one reason or another have had the misfortune of being ignored. And surely one can come up with examples from history—Gregor Mendel in biology, Vincent Van Gogh in painting, Emily Dickinson in poetry, or J. S. Bach in music—where an individual was not prized during his or her lifetime and sometimes for decades thereafter. Are we not concerned here with success and celebrity, rather than with sheer, unadorned creativity?

In addressing this question, it is important to indicate what has *not* been claimed: I am not claiming that there are necessarily different biological or psychological processes at work in the individual who is ultimately deemed creative as compared to the one who is not so judged. Emily Dickinson may have looked no different to her peers in Amherst, or to a neuroanatomist probing her brain, than would her hypothetical untalented twin sister, Amelie, who also felt that she herself was an estimable poet. Nor am I claiming that those who are esteemed are necessarily any greater in any absolute sense than those who are not.

My claim is simply the following: In the absence of a judgment by a competent field, one simply cannot determine *whether* an individual merits the epithet "creative." We can confidently state that Freud and Einstein were creative because there is more than a century of supporting opinion (and relevant controversy) on which to draw. We can with some confidence conclude that their close friends Wilhelm Fliess and Michele (Michelangelo) Besso were not, except perhaps within that little-recognized domain of intellectual midwifery. As for the dozens of other physicists or psychologists who may have thought of themselves as creative but who have not as yet been judged by the field, we must simply render the proverbial Scottish verdict "Not proved."

Fruitful Asynchrony

In the case studies in part II, I discuss the modern masters in terms of their personal talents, the nature of the domain in which they worked, and the operation of the relevant field of individuals and institutions. In addition to the intrinsic importance of these factors, I explore the utility of one final organizing theme. Specifically, I claim that *there exist certain kinds of asynchrony within or across these nodes* and that these may well enhance the likelihood of creativity.

Where there is *pure synchrony*, all three of the nodes mesh perfectly. One might claim that, in the case of a universally acclaimed prodigy, the prodigy's talents mesh perfectly with the current structure of the domain and the current tastes of the field. Creativity, however, does not result from such perfect meshes. In using the term *asynchrony*, I refer to a lack of fit, an unusual pattern, or an irregularity within the creativity triangle. Asynchrony *within a node* occurs when there is an unusual pattern at one of the three nodes. For instance, there may be an unusual profile of intelligences within an individual (as when the young Picasso displayed precocious spatial intelligence but very meager scholastic intelligences); a domain that is experiencing a large amount of tension (as when different schools of music were vying for hegemony in Stravinsky's time); or a field that is just beginning to shift in a new direction (as occurred when certain enterprising critics emerged around the time that modern dance was taking form).

Asynchrony *across nodes* is equally important. For example, the talent profile of an individual may be unusual for a domain (as when Freud's acute personal intelligences proved atypical in a scientist). Or an individual may find himself or herself in tension with a field as currently constituted (as when Einstein could not get a job after completing his degree). Or there may be tension between a domain and a field (as when classical music was moving sharply in an atonal direction, while the audiences and critics continued to favor tonal music).

Naturally, some asynchrony will mark any productivity, whether highly creative or not. My claim is based on two other propositions: First, there can be cases of asynchrony that are too modest or too pronounced; neither proves productive for creativity. An intermediate amount of tension or asynchrony, here termed *fruitful asynchrony*, is desirable. Second, the more instances of fruitful asynchrony that surround a case, the more likely that genuinely creative work will emerge. However, an excess of asynchrony may prove nonproductive: What is desirable is to have substantial asynchrony, without being overwhelmed by it.

As with the developmental perspective and the creativity triangle, I do not submit the hypothesis of fruitful asynchrony as a claim to be tested empirically. Rather, it constitutes an integral part of the framework I have brought to this set of case studies. The value of this framework will be determined by whether the phenomena of creativity have been elucidated. Growing out of the framework, however, are a host of issues that can be illuminated empirically, and to these I now turn.

ISSUES FOR EMPIRICAL INVESTIGATION

In the following case studies I focus on a number of issues that fall, roughly speaking, under the organizational rubric described thus far. I do not undertake to discuss and evaluate every issue systematically with respect to every subject, as in rigorous application of the historiometric approach. Rather, I hold these issues in the background until they appear pertinent in a particular case. In part III, I revisit these issues, in light of the relevant case studies, and present my tentative conclusions. When appropriate, I "score" a dimension and present a rough-and-ready assessment of its fate; in other cases, I offer only an impressionistic account.

Individual Level

Among the issues to be examined are the following at the level of the individual:

1. I begin by addressing cognitive issues—the nature of the intellectual strengths and weaknesses (the particular intelligences) displayed by specific creators and evidence of prodigious behavior in early life.

2. With respect to personality and motivation, I explore the extent to which these creators conform to the traditional view of the creative personality. I focus on the nature of relations to other individuals, the extent of self-promotion, and the kinds of childlike features that seem preserved in these creative masters. I also touch on the individuals' ways of expressing emotions and the degree of tension they had to sustain in their lives.

3. Turning next to social-psychological factors, I examine the nature of the relationship between the child and his or her parents, the attitude toward discipline and permissiveness within the home, and the extent of marginality that characterized each individual's relationship to the society and toward other persons in the chosen domain.

4. Finally, with respect to life patterns, I look for evidence of peaks and valleys in the productivity of the creators—particularly a test of the ten-year rule of productivity, that tendency to make major breakthroughs at ten-year intervals. Also, I consider what it means to be productive in different domains and at different points in the life cycle.

Domain Level

At the level of the domain I take the following steps:

1. I consider the nature of the symbol systems with which the creators worked.

2. I describe the individuals' kinds of creative practices in terms of five distinct kinds of activities. These activities are also touched on in the interludes.

3. Finally, I consider the status of the paradigms, or principal approaches, as they exist in the domains wherein the creators are working. Included is a consideration of the susceptibility of the paradigm to continued innovation throughout the life of the creator.

Field Level

At the level of the field my approach is as follows:

1. I begin with an examination of the relation of the creators to the mentors, rivals, and followers in the field.

2. I then treat the extent and nature of political controversy within the domain.

3. In conclusion, I touch on the extent to which hierarchical organization dominates the functioning of the field.

To repeat: It is beyond the scope of this work to arrive at a decisive, quantitative answer to these questions. Instead, they should be regarded as issues that guided my empirical inquiry, issues that ultimately ought to be resolved by a combination of case study and historiometric investigation.

EMERGING THEMES

Each of these empirical issues grew organically out of the framework and thus constituted part of my overt agenda when this inquiry began. Two other issues, however, were not part of the initial inventory, and their emergence constitutes a discovery for me. Because these emerging themes became an important part of the story of creativity that has ultimately emerged, I need to mention them briefly here.

Support at the Time of Breakthrough

The first issue surfaced during examinations of the period during which a creator made his or her most important breakthrough. I knew that at least some creators had close confidants during this time. But what emerged from the study was more dramatic: Not only did the creators all have some kind of significant support system at that time, but this support system appeared to have a number of defining components.

First, the creator required both affective support from someone with whom he or she felt comfortable and cognitive support from someone who could understand the nature of the breakthrough. In some situations, the same person could supply both needs, while on other occasions, such double duty was unsuccessful or impossible.

The relationship between the creator and "the other" can be usefully compared with two other kinds of relationship: the relationship between the caretaker and the child, in early life, and the relationship between a youngster and his or her peers, in the course of growing up. In some respects, the individual who is attempting to convey a new symbol system resembles the caretaker who is introducing a youngster to his or her language and culture; and in some respects, an individual developing such a system resembles a youngster interacting with a sympathetic peer. In any event, as a psychologist interested in the *individual* creator, I was surprised by this discovery of the intensive social and affective forces that surround creative breakthroughs.

The Creator's Faustian Bargain

The second discovery covers a longer time span, sometimes encompassing much of the creator's adult life. My study reveals that, in one way or another, each of the creators became embedded in some kind of a bargain, deal, or Faustian arrangement, executed as a means of ensuring the preservation of his or her unusual gifts. In general, the creators were so caught up in the pursuit of their work mission that they sacrificed all, especially the possibility of a rounded personal existence. The nature of this arrangement differs: In some cases (Freud, Eliot, Gandhi), it involves the decision to undertake an ascetic existence; in some cases, it involves a self-imposed isolation from other individuals (Einstein, Graham); in Picasso's case, as a consequence of a bargain that was rejected, it involves an outrageous exploitation of other individuals; and in the case of Stravinsky, it involves a constant combative relationship with others, even at the cost of fairness. What pervades these unusual arrangements is the conviction that unless this bargain has been compulsively adhered to, the talent may be compromised or even irretrievably lost. And, indeed, at times when the bargain is relaxed, there may well be negative consequences for the individual's creative output.

I have now introduced the full armamentarium (no less a word will do!) undergirding this study: a trio of broad themes that guided me as I embarked on the original study; a developmental, interactive, and synchronous organizational framework that has directed the particular investigations; a set of empirical issues that the study was designed to illuminate; and a pair of themes that emerged quite unexpectedly during the study and that may be considered discoveries in themselves.

In part II, I turn attention to those individuals who, to my mind, made signal creative breakthroughs at the start of the twentieth century. These individuals have been chosen because of the indisputable importance of their work; they have been chosen as well because each exemplifies a particular intellectual strength, talent, or intelligence as realized in a domain of their culture. As far as I am concerned, there is no order of priority among intelligences, nor is there order with respect to the question of whose work is more important, more innovative, or more creative than that of others. I have therefore elected to present these individuals roughly in the order of the historical moments of their most important breakthroughs. It is perhaps apposite to begin with Freud, because he is the one individual whose tools have themselves helped enhance our understanding of the creative mind. Throughout part II, the organizing framework just introduced will remain largely in the background; but in the interludes and in part III, I return explicitly to it, as I attempt to summarize what I have learned from this study.

PART II

THE CREATORS
OF THE MODERN ERA

3

SIGMUND FREUD: ALONE WITH THE WORLD

Freud, ca. 1891

BEGINNING IN 1902, somewhere between five and twenty Viennese men began to gather regularly on Wednesday evenings in the home of a physician-turned-psychologist named Sigmund Freud. Included within these ranks over the next few years were several young physicians, among them Wilhelm Stekel, Paul Federn, and Alfred Adler. Other group members, drawn from various corners of society, included the musicologist Max Graf, the music critic David Bach, the publisher Hugo Heller, the army surgeon Major General Edwin Hollering, and a former glassblower Otto Rank. Guests included the psychiatrists Ernest Jones from London, Carl Jung from Zurich, and Sandor Ferenczi from Budapest. After enjoying coffee and cigars, these men listened to papers prepared by one of the members or, occasionally, by a guest; then, adhering to a rigidly observed procedure, each offered commentary on the paper. In general, the last to speak was Sigmund Freud.

THE FIRST DISCIPLES

To an eavesdropper, Freud's Wednesday evening fellowship would have seemed an eccentric lot, engaged in a decidedly exotic activity. Several of the members were mavericks; most were loners; nearly all were Jews, accustomed to a degree of ostracism in Vienna. Most papers described unusual medical or psychiatric conditions, such as hysteria, obsession, or paranoia: Often the etiology of the condition was sexual and, on occasion, the presenters vividly described their own dreams or sexual practices. Critiques of one another's papers were often harsh, sometimes needlessly so. But until Freud had spoken, it was not clear what the "official verdict" on a paper would be.

Convening of the Wednesday Psychological Society marked an important turning point in Freud's life. Following a reasonably promising, but hardly triumphant, career as a medical researcher and a physician in private practice, Freud had withdrawn from professional circles. With little support—and sometimes considerable opposition—from his peers, he had developed the basic tenets of psychoanalytic theory and practice. Now, following the publication of *The Interpretation of Dreams* (1900), which he believed to be his most important work, Freud sensed for the first time that others might take his ideas seriously and even help introduce them to a wider world.

Perhaps, as his long-term associate Ernest Jones maintained, the particular collection of men who gathered at Nineteen Berggasse were mostly of the second rank: "I was not highly impressed with the assembly. It seemed an unworthy accompaniment to Freud's genius, but in the Vienna of those days, so full of prejudice against him, it was hard to secure a pupil with a reputation to lose, so he had to take what he could get." Indeed, they represented an almost irresistible target for the Viennese satirist Karl Krauss, and one can envision a latter-day Tom Stoppard recreating in *Travesties* format this collection of marginal neurotics. But from the central core of the Wednesday Psychological Society emanated the formidable Vienna Psychoanalytic Society and, ultimately, the International Psychoanalytic Society. Like faithful foot soldiers, these men, their associates, and their successors helped spread their commanding officer's revolutionary ideas throughout the intellectual world.

BACKGROUND AND EARLY CHILDHOOD

From one perspective, Freud came to the world with little in his favor. He was born in Freiberg, Moravia, a town of five thousand inhabitants situated 150 miles northeast of Vienna. His family was Jewish, and the Jews had not been treated well in the Austro-Hungarian empire. Kindly, well-intentioned, and for the most part optimistic, his father, Jakob, proved a ne'er-do-well in the mercantile world who never quite lived up to the expectations of his wife, Amalie. The young Sigismund (his given name, which he retained until early adulthood) at first lived in tiny and uncomfortable quarters; later, as Jakob's circumstances improved somewhat, the family—which now included seven children—moved to larger housing. Sigismund was surrounded by a bewildering family constellation: a father who had apparently been married twice before and was twice as old as his mother, two grown brothers who were as old as Freud's mother, a nephew who was a year younger than Sigismund, and a niece of roughly the same age.

Thanks to Freud's own writings, we now appreciate the crucial contributions of the early years to one's ultimate life course. In most respects, despite the trying conditions I have just described, Freud was blessed. As a firstborn, he received and maintained special attention from his mother, who lived until Freud was over seventy. He also had a doting nurse, who seems to have reinforced the message that Freud was somehow special. While the Jews had been subjected to marked prejudice in earlier generations, Freud grew up at a time when anti-Semitism was at least temporarily on the wane.

Perhaps most important, Freud was a very talented child, and those around him responded to his gifts. Indeed, among our seven creators, he was probably the one with the greatest academic strengths. By any definition he was extremely intelligent. In his own words: "At the Gymnasium I was at the top of my class for seven years: I enjoyed special privileges there, and was required to pass scarcely any examination"; he graduated summa cum laude. Family members organized much of their daily regime around the talented boy's needs: He was given his own room and his own bookcases; he did not have to dine with the rest of his family but was provided with his own eating chamber; and when his sister's piano practicing annoyed him, the piano was removed from the house.

While bookish, Freud seems to have had a reasonably rounded childhood. He enjoyed being out of doors and became a good walker, swimmer, and skater. He had his share of friends, extending well beyond the large family circle. Like many other young males of the time (and of other times), he identified with the life of the soldier. Particularly attached to Hannibal, the great Carthaginian general, Freud precisely plotted out Hannibal's battles and remarked that had he himself not been Jewish (and therefore prohibited from military leadership), he would have pursued the career of a military officer. In contrast to his fascination with the military, Freud had little attachment to formal religion. But he felt himself to be strongly Jewish, was well informed on biblical and other Jewish lore, and bridled at any anti-Semitic talk or behavior.

As a talented Jewish lad in the increasingly liberated Austro-Hungarian capital, Freud was clearly headed for the professions. His father, smitten with Freud's talents ("My Sigmund's [sic] little toe is cleverer than my head" he once remarked), gave Freud free reign in his choice of a career. For an ambitious Jewish boy, military or political careers were unlikely, but that still left the law, science, other academic disciplines, or medicine. Freud was headed for a career in law until he heard a recitation of Johann Wolfgang von Goethe's essay "On Nature." This grand paean to the world of creation, with Nature depicted as a nurturant mother figure, catalyzed Freud to study medicine and to become a natural scientist.

FREUD'S UNIVERSAL GIFTS

In his autobiographical study, Freud said: "[I] felt no particular partiality for the position and activity of a physician in those early years. . . . Rather I was moved by a sort of greed for knowledge." This comment is an understatement. In the eight years that passed between graduation from the gymnasium and receipt of the medical degree, Freud immersed himself spectacularly in the world of knowledge. He read extremely widely: the Bible, ancient classics, William Shakespeare in German and in English, Miguel de Cervantes, Molière, Gotthold Lessing, Johann Wolfgang von Goethe, and Friedrich von Schiller. He mastered English and French and also taught himself Spanish so that he could read Cervantes in the original. Fond of art and the theater, he attended many exhibitions and plays and commented penetratingly on what he had observed. Succumbing for a while to philosophy, he joined a society in which he read the major philosophers, translated John Stuart Mill into German, and took courses for three years with Franz Brentano, a respected philosopher at the University of Vienna with a special interest in psychological issues. And, not neglecting the area of science, he mastered the writings of Darwin as well as scientific texts by the most important scientist of the period, Hermann von Helmholtz.

A vivid sense of that exploring mind permeates Freud's letters of this period to his close friends Emil Fluss and Eduard Silberstein, and a bit later, to his fiancée, Martha Bernays. Freud comes across as a lively, enthusiastic, witty, sometimes sardonic, and highly ambitious but occasionally self-deprecatory young man. Equipped with a developed imagination, he is able to invent scenes, characters, institutions, and flights of poetic fancy; and vivid characterizations and dramatizations issue readily from his pen in several languages. He shifts from literature to art, from science to philosophy, from the personal to the professional, the political, and the worldly. Already he is the teacher—reporting what he has learned, asking pointed questions of his correspondents, and seeking to synthesize knowledge. One infers that his correspondents got the better end of the epistolary bargain.

Freud indicated in his autobiography that he was moved by a sort of curiosity directed "more towards human concerns than towards natural objects." What comes across most strikingly in his letters is a fascination with, and a surprisingly sophisticated understanding of, the foible-filled world of human beings. Freud spun out detailed, hilarious, touching paragraphs, pages, even short-story-length passages about family, friends, and strangers. In evocative prose he described an encounter with a stern professor, the dreams of an ambitious Jewish physician who marries the boss's daughter, and the suicide of a brilliant but troubled friend; he offered advice about how to deal with tense interpersonal situations in the family; and he did not spare himself, dissecting his own emotions, ambitions, and conflicts.

With the glare of hindsight, it is difficult to read these letters without being struck by the ambitious goals that Freud set for himself. From the days of his youth he seemed convinced that he would achieve something important during his time on earth. If anything, the question he had to confront was not *whether* he would achieve but, rather, in *which* of the possible spheres such attainment would be wrought. Here, again, he represents an extreme of explicit ambition and self-confidence in my sample of creators. His letters contain much talk about his future biography, his fame, and the pitfalls that might accompany the attainment of his laurels; the merits and demerits of a presentation, a lecture, or a paper on which he was working; the options of taking a more direct or a more circuitous route to greatness; and whether a particular discovery might harbor the key to fame. This "achieving" leitmotif is saved from obsession or megalomania by a pervading humor and sense of fatalism, as well as by the generous conviction that others in his circle had such options open to them as well.

Almost as notable is Freud's preoccupation with riddles and puzzles. Freud obviously loved to arrive at a paradox of some sort and then to ponder thereupon until a solution emerged. Like the contributors to the Talmud, he perenially posed the whys and wherefores with respect to every cranny of existence: why he and his fiancée are being tested by years of separation; how to live happily with little; what the reasons for suicide are; why one labors for months in the laboratory with but a slender hope of making a discovery; whether women can both manage a home and participate in professional life.

On occasion Freud voiced regret that he did not have a more retentive memory or a better brain. His contemporaries probably had little sympathy with this complaint (nor, for that matter, have his successors). While Freud did not excel in mathematics or the physical sciences, and while his sensitivity to music was limited, he ably and comprehensively tackled humanistic and scientific studies. Nor, in contrast with many other brilliant scholars, was he cut off from the world of human intercourse. The young Freud had the capacity for friendship and was, from most reports, an engaging companion, a dazzling lecturer, and a loyal family member. In my terms, Freud was superbly endowed in the linguistic and the personal intelligences—comfortable and competent in dealing in the realm of words and the realm of human beings. One feels that by the time he had completed his studies, Freud had the world arrayed in front of him: As an individual on the brink of the modern era, living in one of the centers of world civilization, well connected to influential mentors, Freud had virtually unlimited options.

A "FIRST" CAREER IN NEUROLOGY

Studying medicine, Freud wrote to his fiancée: "Strange creatures are billetted in my brain . . . here bacteria live, sometimes turning green, sometimes blue, here are the remedies for cholera, all of which make good reading but are probably useless." Indeed, much of this textbook knowledge turned out to be extraneous because Freud never undertook a practice in general medicine. Instead, he elected to pursue a research career in neuroanatomy.

Freud went to work in the laboratory of Ernst Bruecke, a major scientist who, under the influence of the even more illustrious Helmholtz, embraced the position that organisms were physical entities that must be understood in the same way as the rest of the material world. Spurning any consideration of vitalism or intentionality, Bruecke and his associates adopted a completely materialistic and reductionist tack: The answers to all natural questions must come from the careful study of cells and their connections, and from the specification of the chemical and physical forces that control their biological interactions. Freud was thrust into this ideologically charged atmosphere and given delimited tasks: to study the histology of a peculiar kind of large cell in *Petromyzon*, a primitive genus of fish; to determine the fine structure of the nerve cells of the crayfish; and to investigate the gonadal structure of the eel. In the process, he was granted his first taste of scientific invention: He devised a method for staining nervous tissue with gold chloride.

In addition to the scientific knowledge and technical skills he obtained in Bruecke's institute, Freud was touched personally. Indeed, along with the neurologist Jean-Martin Charcot and the physician Josef Breuer, Bruecke was one of the three unquestioned father figures for Freud. Bruecke's personal traits impressed the youthful Freud: This first scientific mentor was demanding, conscientious, unfailingly fair, a natural leader. He asked much of his students and did not tolerate the weak ones; but to those who could follow his formidable scientific and personal example, he was warm and supportive. In the end, Freud embraced Bruecke's faith in materialistic explanations and shared his disdain for the mystical, but Bruecke would likely have been aghast at the kinds of questions Freud ultimately tackled under the banner of a physically oriented science.

Freud's list of publications in his twenties and early thirties constitutes a respectable output in the area of neuroanatomy. The writing reveals Freud's eye for detail and his ability to ferret out at a microscopic level the principles governing the organization of the nervous system. Freud came close to discovering that the neuron*—the individual cell plus its extension—was the basic functioning unit of the nervous system, but the ultimate prize went to Wilhelm von Waldeyer in 1884. It seems fair to say that Freud was still awaiting that flash of insight that would separate him from the pack. That this race for primacy was on Freud's own mind is well conveyed in a letter where he declared:

> . . . Not so sure, though, whether I shall publish a paper, the subject being very rewarding and highly popular. Somebody might easily precede me in publication. . . . From next month I shall start experimenting with salivary secretions in dogs.

For some time Freud thought that he might have realized a decisive breakthrough as a result of his experiences with cocaine. Experimenting with what was then a little-known substance, Freud had found that a twentieth of a gram converted his foul moods into cheerfulness; since the drug acted as a gastric anesthetic, Freud thought it might also prevent vomiting. Buoyed by these discoveries, Freud gave cocaine to a close friend who was in pain and concluded that it operated as a "magical drug." This euphoria led in turn to Freud's ill-considered decision to pass the drug on to other friends and colleagues, as well to his fiancée and his sisters. Freud wrote a monograph about cocaine in which, for perhaps the only time in his scientific writings, he expressed personal enthusiasm, speaking, for example, of "the most gorgeous excitement" induced by the substance.

But the cocaine episode ended disastrously. First, as Freud's friend was to discover, cocaine turned out to be highly addictive; moreover, it only produced the desirable side effects temporarily in depressed individuals. To sharpen the psychic pain, the one unambiguously positive discovery about cocaine—that it

---

*Earlier writings employ the term *neurone*.

served as an effective anesthetic for eye surgery—was made not by Freud but by a close colleague, Carl Koller. Once more, Freud had sensed the excitement of discovery, but not its rewards.

CHARCOT AND THE MOVEMENT TO PSYCHIATRY

The year following the unproductive cocaine incident was much happier. In the wake of that political struggling so characteristic of his career, Freud was awarded a fellowship that took him to Paris from October 1885 to February 1886. This brief visit to a city of unsurpassed excitement changed his life. He worked at the Salpetriere Hospital in the clinic operated by Jean-Martin Charcot. There he was introduced to a range of fascinating neuroses and especially to hysteria—a condition where individuals, usually women, displayed strange behaviors, such as the paralysis of a limb, psychic blindness, or seizures, all in the absence of an obvious medical condition. Charcot presented these cases in the course of dramatic ward rounds; he analyzed the symptoms in detail; he laid out the rules that governed such attacks and classified the hysterias systematically into different groups. In Freud's words, Charcot "succeeded in proving the presence of regularity and law, where the inadequate or half-hearted clinical observations of other people saw only malingering or a puzzling lack of conformity to rule."

Charcot's particular interests lay in classifying hysteria and in documenting its hereditary bases. But his demonstration that nonhysterical patients can, under hypnosis, mimic hysterical systems suggested to Freud the possibility that at least some physical symptoms might be due exclusively to psychological or mental causes. Noting that these patients often failed to recall experiences undergone when under hypnosis, Freud was also struck for the first time by the operation of powerful unconscious processes. Freud wrote home with unbridled enthusiasm:

> I do nothing here except allow myself to be wound up by Charcot in the mornings, and in the afternoon I have time to unwind and to write letters in between. . . . Charcot, who is one of the greatest of physicians and a man whose common sense borders on genius, is simply wrecking all my aims and opinion. I sometimes come out of his lectures as from out of Notre Dame, with an entirely new idea about perfection.

Freud was also shifting his "ego ideal" from the precise and rigorous neuro-anatomical investigator Bruecke to the more expansive, charismatic, and psychologically oriented Charcot. In a letter to his fiancée, he revealed a new pantheon of heroes and confirmed his unflagging ambitions by revealing that he now felt that he could one day "possibly reach the level of Charcot."

For a person of Freud's prodigious talents, energies, and ambition, the decade following his time in Paris might be considered a period of quiescence. The neurological publications appeared at a diminishing rate; it was becoming clear that Freud was not going to convulse the scientific world by his neuroanatomical research or by his investigations of the neurological diseases of childhood. Freud found it necessary to abandon most of his work in the laboratory in order to marry, open a private practice, and launch what would eventually become a large family.

Not drawn to the daily practice of clinical medicine, Freud had also not been sufficiently stimulated by the laboratory benchwork of neuroanatomy and was disappointed by the lack of professional acclaim that he had obtained therein. To borrow a phrase from the psychoanalyst Erik Erikson, Freud's twenties had represented a kind of extended "psychosocial moratorium," during which he had been trying out various life roles and styles, in an effort to determine where he was most likely to succeed. But Freud was now on the scent of the problems that were to occupy him for the rest of his scientific career. Charcot had introduced him to a world of nervous disorders, which was in itself fascinating and for which he felt that he had finely honed skills of observation, classification, and explanation. Perhaps he would make his mark in the elucidation of neurotic behaviors.

One of his associates in Bruecke's laboratory had been Breuer, a somewhat older physician. Also a Jew whose father had escaped from the ghetto into "freer" Vienna, Breuer was a polymath, knowledgeable in art, philosophy, and politics. The two men took a liking to one another, and Breuer sweetened the emerging father-and-son relationship with occasional monetary gifts to the struggling Freud. As early as 1880, Breuer had been seeing patients who had exhibited hysteric symptoms. He shared with Freud some of the clinical impressions of the patients, in particular one Bertha Pappenheim (later immortalized as Fräulein Anna O.). Freud came to see that the scientific interests that he had acquired in Paris could be pursued in private practice in Vienna, and that the well-respected and kindly Breuer might prove an excellent partner in these investigations.

The case of Anna O. has become one of the best known in the Freudian corpus, though Bertha Pappenheim was actually seen by Breuer when Freud was still a medical student. The daughter of a wealthy Jewish merchant in Vienna, Pappenheim had developed a number of hysterical symptoms—including partial paralysis, a squint, visual disturbances, hallucinations, and an inability to detect any sensations—in the course of nursing her father during a fatal illness. Aware of her own bizarre clinical picture (though at a loss to explain it), she had sometimes resisted her symptoms, sometimes reveled in them, thus giving rise to a confused, dualistic personality.

Pappenheim received help from a number of physicians, among them Breuer and the famed psychiatrist Richard von Krafft-Ebing. Experimenting with the use of deep hypnosis, Breuer noted a remarkable event. As part of her hysterical condition, Pappenheim had found it impossible to drink any liquid. Once, when under hypnosis, she had expressed at some length her anger with an English lady companion. Just as soon as this furor had been vented, Pappenheim discovered that she was able to drink. She woke up from her hypnosis with her glass at her lips, and the symptom never returned. Revisiting this treatment, Breuer found that many of Pappenheim's symptoms could be traced back to emotionally charged events that she had experienced while nursing her father. Rather than acknowledging these often upsetting thoughts at the time, she had apparently suppressed them, but at the cost of activating a physical symptom. Enunciation of the once-suppressed memory, along with release of the affect that normally accompanies it, seemed sufficient to eradicate the symptom.

Using a term actually coined by Pappenheim herself, Breuer and Freud wrote about the "talking cure." On their analysis, a symptom would disappear once the patient had been able to reproduce in hypnosis the event that had given rise to it. Sometimes the event had been one of great moment—for example, the thought of abandoning a parent-in-need. At other times, the event was less momentous in itself but had occurred at a time when the patient was under great emotional strain. The relationship might also be symbolic—thus, Pappenheim's pain in her right heel was traced back to her fear that, when introduced into polite society, she might not find herself on the "right footing."

When the case was first published, Pappenheim was reported to be largely cured. Breuer and Freud explained how the unwanted memory had been pushed into the background or repressed and then had returned to consciousness:

> (The cure) brings to an end the operative force of the idea which was not abreacted in the first instance, by allowing its strangulated affect to find a way out through speech; and it subjects it to associative correction by introducing it into normal consciousness (under light hypnosis) or by removing it through the physician's suggestions.

Subsequent investigation indicated, however, that the cure had not been sustained. One day the patient had suffered from tremendous cramps. When asked why, she had declared: "Now comes Dr. B's child." According to Freud, Breuer recoiled from considering the full implications of this "transference" of feeling from the patient to the physician, of possible "counter-transference" from physician to patient, and of the strong sexual undertones of a hysterical pregnancy.

Impressed by Breuer's initial observation and report, Freud began to use the hypnotic method in his own treatment of hysteric patients. He was able to confirm the general picture reported by Breuer in the case of Anna O. Prodded by Freud, and also by some related publications by Charcot's student Pierre Janet, Breuer agreed to publish the findings about hysteria in a short coauthored monograph, *Studies in Hysteria* (1895). That volume reported five cases in detail and also put forth what Freud called an "unpretentious theory."

The theory propounded in *Studies* was what one might have expected from two physicians reared in the Bruecke-Helmholtz tradition. It assumed that a powerful affect had been dammed up, in hydraulic fashion; through a process of conversion, a symptom instead appeared, using the same quantity of energy that would have been expended had the affect not been suppressed. The therapy was cathartic: It released the dammed-up energy, thereby eliminating the symptom.

Breuer and Freud issued *Studies* together, but their friendship was already being sorely tested. As Freud was to comment: "The development of psychoanalysis afterwards cost me his friendship. It was not easy for me to pay such a price but I could not escape it." Breuer could accept a dynamic view of hysteria but was made acutely uncomfortable by the discussion of unconscious processes, the theme of transference between physician and patient, and especially by the apparent importance of sexual motifs and motives. Whenever possible, he preferred a purely physiological explanation; Freud was already searching for psychological motivation and was beginning to construct a comprehensive view of the mental apparatus that could give rise to symptoms, abreactions, and catharsis.

Freud received a preview of future responses to his work when he made a presentation in 1896 on "The Etiology of Hysteria" to the Vienna Society of Psychiatry and Neurology. In this presentation Freud confidently put forth the principal ideas he had been developing: that the hereditary view of neuroses, which he had imbibed from Charcot, was not adequate; that hysterical symptoms are overdetermined and only appear exaggerated because observers have such a limited knowledge of the motives from which they emerge; that men as well as women are susceptible to hysteria; and that at the foundation of each case of hysteria there exists one or more precocious sexual experiences. As he boldly declared: "Whatever cause and whatever symptom we take as our point of departure, in the end we infallibly come to the field of sexual experience."

This presentation shows Freud at the height of his powers: eloquent and seductive, armed with evidence from eighteen patients, poised to deal with possible objections, placing himself by an act of imagination in the mind of his skeptical audience. Yet, from what evidence we can secure, Freud's presentation met with little enthusiasm. Krafft-Ebing, the sexologist who presided at the meeting, said, "It sounds like a scientific fairy tale." Freud reported that he had been given "an icy reception from the asses." So despondent was Freud at this response that he gave only one more public medical lecture in Vienna during the remaining forty-two years of his life.

Freud was at a turning point. In his own mind he had arrived at the brink of the most important discoveries: He had demonstrated to the colleagues at the Vienna society "a solution to a more than thousand-year-old problem—a 'source of the Nile.'" Yet neither in Vienna nor elsewhere were his words attended to;

either they were ignored (in most places) or they were severely denounced as "absurd, wildly conjectural, irrational, unproved and unprovable." Once an individual who had been virtually canonized by his own family, appreciated and admired by peers and mentors, and able to master vast bodies of information in a veritable library of subjects, Freud had evolved to a most unhappy situation: one where his closest colleagues, like Breuer, were no longer willing to stand with him, while his closest family members, like his wife, could not possibly understand what he was claiming. If Freud—once so richly connected to the world— was to continue on his fateful path, he had to do so virtually alone.

LONELINESS AND CONFIDANTS

At earlier periods of his life, Freud always had one or two individuals to whom he felt very close and with whom he could share his innermost thoughts, fears, and aspirations. From Freud's own recollections and the comments of others, we can assume that his mother, his father, and his nurse fulfilled these roles during his childhood. Once he had begun his higher education, his friend Eduard Silberstein seems to have played the role of confidant. It is from hundreds of letters to Silberstein that we have secured our picture of the energetic, lively, and reflective young man Freud. Freud and Silberstein even created a secret two-person association, complete with its own hermetic communication code and insignia.

During the time of his unnaturally lengthy engagement, Freud poured out his soul on almost a daily basis to his fiancée, Martha Bernays. Even though Martha is unlikely to have understood Freud's scientific work, he spared her little detail about his efforts. And when, for some reason, Freud did not unload all of his feelings on Martha, he deflected them to Martha's sister, Minna, a somewhat more intellectual woman who never married and who became a member of the Freud household for over forty years.

Freud was able to split his ties to some extent, directing professional affinities to individuals like his mentors Charcot and Bruecke, and personal loyalties to family members, but he clearly appreciated the opportunity to form both scholarly and affective links to the same individual. Breuer was able to assume both niches for the young adult Freud: For this reason the tie was particularly precious to Freud, and its rupture especially painful. When Freud's ideas began to draw him away from Breuer (or vice versa), a sizeable void remained.

Fortunately, Wilhelm Fliess stepped into that void during the pivotal period of the 1890s. Fliess was a Berlin-based physician who, at the time Breuer introduced the two men to one another, was somewhat more established than Freud. Fliess had developed a highly complex and—to current sensibilities—entirely bizarre biological theory, in which the nose was the dominant organ, influencing all human health. A committed numerologist as well, Fliess believed that human life was governed by biorhythmic cycles of twenty-three days for men and twenty-eight days for women. He attempted, with at least superficial success, to explain every medical contingency through some kind of arithmetical operation involving combinations of these numbers.

For over ten years the two men were in almost constant contact, mostly by mail, though they attempted to hold self-styled "congresses" at least once a year, where they would meet at a resort and discuss their emerging ideas in more intimate surroundings. Both of the men were developing their notions in relative isolation, and both craved at least one companion who was worldly and sympathetic, though perhaps not wholly uncritical. Freud did not hesitate to flatter his Berlin friend, praising both his personal characteristics and the contents of his theories. Only rarely did he voice the slightest hesitation about his friend's often extreme views and claims.

In retrospect, it seems that Fliess played two vital roles for Freud throughout the 1890s. On the one hand, he was serving as a sounding board—perhaps the only one—for the full panoply of ideas that Freud was sprouting, on almost a daily basis, about human psychology. Without regular contact with Fliess, Freud would have been completely isolated intellectually. Second, and equally important, Fliess was serving as a nurturant, supportive figure, to whom Freud could feel personally close, and with whom he could share the most intimate thoughts as well, whether or not these related directly to the topic of his scientific speculation. From all indications, Freud played at least some of the same roles for Fliess.

Since Freud craved companionship and approval, this period of early middle-age proved very difficult. Freud sensed that this was the last time in his life when he had the opportunity to stand out and to realize his enormous potential—if he did not make his mark on the world now, at age forty, he would never do so. Signs of human limitations were impressing themselves on him. Financial survival was still a struggle, and sometimes the patients did not come. The vaunted professorship seemed as far away as ever. His father was ailing and died in 1896, when Freud was forty. Freud fathered the last of his six children in 1895 and, as far as can be determined, soon thereafter ceased to have sexual relations. He suffered a great deal of mental and physical turmoil, fears about death, depression, and addiction to nicotine, as well as a painful chronic stomach difficulty.

Freud was keenly aware of his loneliness and lack of acceptance. In 1887 he had written to his family: "One finds scientific support nowhere; rather there is an effort 'not to give you a chance' which you feel is very disagreeable." Seven years later, he wrote to Fliess: "I am pretty much alone here in the elucidation of the neuroses. They look upon me as pretty much of a monomaniac, while I have the distinct feeling that I have touched upon one of the great secrets of nature." In his letters to Fliess, Freud swings between elation and despair as he strives to evaluate the importance of his findings:

[April 2, 1896]: "If both of us are still granted a few more years for quiet work, we shall certainly leave behind something that can justify our existence."

[May 16, 1897]: "Oh, how glad I am that no one, no one knows. . . . No one even suspects that the dream is not nonsense but wish fulfillment."

[August 18, 1897]: "I have finished nothing; am very satisfied with the psychology, tormented by grave doubts about my theory of the neuroses, too lazy to think, and have not succeeded in diminishing the agitation in my head and feelings."

[November 25, 1900]: "I have resigned myself to living like someone who speaks a foreign language or like Humboldt's parrot."

To have lived through such a period is obviously draining—one senses that Freud is sometimes on the verge of a breakdown. In 1913 he reflected:

At that time, I had reached the peak of loneliness, had lost all my old friends, and hadn't acquired any new ones; no one paid any attention to me, and the only thing that kept me going was a bit of defiance and the beginning of *The Interpretation of Dreams*. On the other hand, to have lived through such a period and survived created a feeling of pride and perhaps even euphoria.

In his history of the psychoanalytic movement, Freud recalled: "When I look back to those lonely years, away from the pressure and preoccupations of today, it seems to me like a glorious 'heroic era'; my 'splendid isolation' was not lacking in advantages and in charms." It is striking that one encounters virtually the same words and affects—the heights and the depths—in the recollections of other innovators as they reflect on their subjective state on the eve of their greatest breakthroughs.

THE DOMAIN AND FIELD ON
THE EVE OF FREUD'S BREAKTHROUGH

To delineate the nature of a creative achievement in terms of the framework introduced in chapter 2, it is important to understand how the dominant field perceived critical issues in the relevant domain at the time. With respect to the clinical syndromes that fascinated Freud, we have already seen that interest in hysteria and in the neuroses was relatively new in the medical profession. Charcot had brought these conditions to public attention and had suggested their hereditary bases. Most medical practitioners in Western Europe had little interest in probing further; they saw these conditions either as hereditary degeneration, as malingering, or as a mark of some deep moral failing in the "victims."

Academic psychology was a new discipline. The first psychological laboratories had been opened in Germany and in the United States in the late 1870s. Inspired by the Helmholtzian tradition, early psychologists hoped that human nature could be modeled as precisely as the physical world and measured with the accuracy of mathematics. This effort to place psychology on a firm scientific basis had led to a focus on those behaviors that could be most reliably measured—time of reaction to stimuli, the capacity to make sensory distinctions of different magnitudes. Issues of intention and will, which Freud had encountered when attending the lectures of the philosopher-psychologist Franz Brentano, could not be tackled by these new scientists, for they did not lend themselves to ready measurement by newly fashioned brass instruments in the psychological laboratory.

Medicine and psychology were publicly delineated disciplines, with their own journals, professional organizations, and scientific procedures. The area that was capturing Freud's interest was not. No established procedure or institution existed for examining dreams, unconscious processes, or one's own psyche. On the one hand, these issues had long been of interest to reflective individuals, particularly artists. Dream analysis dated back to the Egyptians; self-analysis was part of many of the world's great religions; and reference to the unconscious could be found in many writers' works, dating back to classical times, even as it had reemerged as a prominent theme in the works of such nineteenth-century figures as Goethe, Schelling, Kierkegaard, Schopenauer, Herder, and Dostoyevsky. Yet, as common as these themes might have been in certain quarters, they were either ignored or considered taboo by most workers in medical, psychological, and natural scientific disciplines. When Freud declared an interest in these elusive phenomena, he was laying claim to territory whose existence his colleagues did not acknowledge.

THE KEY IDEAS OF THE FREUDIAN REVOLUTION

In any intellectual revolution, it is risky to pick out a single idea or theme that is most crucial. This risk is particularly acute in the case of a revolution like Freud's, where, as I shall argue, it is the particular *combination* of generative ideas that lends power and allure to the system.

Nonetheless, taking a leaf from the historian of science Gerald Holton, I contend that one can identify a central figure or theme in Freud's conception—one around which his other key ideas are readily organized. That focal idea is *repression*, the process (more technically, the defense mechanism) whereby certain potentially upsetting notions are withheld from consciousness. Freud himself confirmed the centrality of this idea in saying, "The doctrine of repression is the foundation-stone on which the whole structure of psychoanalysis rests."

A consideration of repression leads into the heart of the Freudian worldview. It becomes necessary to posit a collection of ideas striving toward consciousness;

a censoring mechanism that labels some as too disturbing to be allowed to consciousness and therefore consigns them to a purgatorial existence in the unconscious realm; and a conversion process, whereby the affect surrounding the disturbing idea can be converted into some kind of a symptom—a harmless one, like a verbal slip, or a more cataclysmic one, like a hysterical seizure. Only if the disturbing idea can be modified in some way does it have the potential to come to preconscious awareness and, ultimately, to enter the conscious level.

If repression qualifies as the central idea in the Freudian conspectus, the dream presents itself as the privileged route to an understanding of processes of repression and to the rest of the psychic life. Freud believed that his discovery of the power of the dream was the most significant of his life. He joked to Fliess that there ought to be a marble tablet on his house reading: "Here, on July 24, 1895, the secret of the dream revealed itself to Dr. Sigm. Freud"; and he called the dream "the royal road to the unconscious . . . insight such as this falls to one's lot but once in a lifetime."

During the 1890s, as his conceptual breakthroughs were taking form, Freud was working in four loosely coupled areas. These areas can be ordered, roughly, from Freud's initial to his final concern over the course of the decade; but of course, there is overlap. Each of the concerns can be discerned throughout this period and even extending beyond to his work in the early twentieth century. Each area was written about in different articles and pieces, but they come together with brilliance and decisiveness in *The Interpretation of Dreams* (1899/1900), which I discuss later.

The Neuroses

During the years after his return from Charcot's clinic, Freud directed his efforts toward the study of the various neuroses—hysteria, obsession, paranoia—and an attempt to unravel their mechanisms. A determined taxonomist, Freud devised a succession of organizational schemes. At one time he distinguished mechanisms of transformation of affect, as in conversion hysteria; displacement of affect, as in obsessions; and exchange of affect, as in melancholia. At subsequent times, he divided the neuroses into two primary families (repression and anxiety), into five different categories, and even into "actual" and "psychological" neuroses.

The neuroses called on different mechanisms of defense—the efforts of the psychic mechanism to deal with forbidding and charged material. Repression was a principal defense mechanism, but there were many others, ranging from sublimation to reaction-formation to projection, displacement, and inhibition. The student of clinical phenomena was to observe these various defense mechanisms at work and then help the patient undo the defense, so that the original triggering event could be recognized and the defensive mechanism then be dissolved.

For Freud these tasks of completing clinical descriptions and case studies, identifying etiology, classifying taxonomically, and searching for a cure actualized the model set forth by Bruecke in his neuroanatomical laboratory and by Charcot in the Salpêtrière clinic. Such tasks were relatively mainstream work in the domain, the kind most likely to be reported at meetings and published in medical journals. However, as Freud became increasingly convinced of the sexual etiology of the various neuroses, and as he began to describe their mechanisms in terms of psychic repression and unconscious mechanisms, even this relatively traditional work fell outside the purview of his scientific colleagues.

Psychology

In 1895 Freud embarked on what may have been the most peculiar scientific venture of his life: a lengthy, ultimately uncompleted and unpublished monograph, which he entitled "Psychology for Neurologists" but which was renamed "Project for a Scientific Psychology" or, for short, "The Project." In this monograph, penned in a feverish state over a few months, Freud attempted to lay out an entire neurological basis for the psychological mechanisms he was uncovering.

"The Project" became Freud's obsession. In place of the various piecemeal schemes he had concocted, he wanted to explain *all* the neuroses under one comprehensive framework, and, to do so, he concluded that he would have to construct an entire psychology that could encompass normal and abnormal, conscious and unconscious phenomena. Its purpose, in his own mechanistic words, "is to furnish a psychology that shall be a natural science; that is to represent psychical processes as quantitatively determinate states of specifiable material particles, thus making those processes perspicuous and free from contradiction." As he confessed at the time:

> A man like me cannot live without a hobbyhorse, without a consuming passion. . . . I have found one. . . . It is psychology, which has always been my distant beckoning goal and which now, since I have come upon the problem of neuroses, has drawn so much nearer. I am tormented by two aims: to examine what shape the theory of mental functioning takes if one introduces quantitative considerations, a sort of economics of nerve process; and to peel from psychopathology a gain for normal psychology.

"The Project" is demanding to read. One sees Freud groping with the issue of how to capture normal phenomena, like attention and perception, as well as the unconscious mechanisms that he had uncovered in his clinical work. Sometimes, in excitement, Freud simply strings together a set of predicates that are scarcely comprehensible. At other times Freud reveals his own despair at the magnitude of the task, the meager tools at his disposal, and the seemingly contradictory mission of laying bare what the psychic censors have withheld from introspection or consciousness. There is little point in summarizing "The Project," since ultimately much, if not most, of the more grandiose aspects were intentionally abandoned. But it is important to convey—by quoting a paragraph—the flavor of this complex, dense, and closely argued document:

> In the phi neurones, furthermore, the psi neurones terminate. To the latter, a part of the Qn is transferred but it is only a part—a quotient, perhaps, corresponding to the magnitude of an intercellular stimulus. At this point the question arises whether the Qn transferred to psi may not increase in proportion to the Q flowing in phi, so that a greater stimulus produced a stronger psychical effect. Here a special contrivance seems to be present, which once again keeps off Q from psi. For the sensory part of conduction in phi is constructed in a peculiar fashion. It ramifies continually and exhibits thicker and thinner paths, which end in numerous terminal points—probably with the following significance: a stronger stimulus follows different pathways from a weaker one.

As this passage indicates, Freud had developed an entire vocabulary—bound and unbound quantities, three kinds of neurons (for receiving stimuli, transmitting them, and carrying the contents of consciousness), conscious and unconscious processes, an economical view of nerve forces—for describing the mental apparatus. The view was Helmholtzian—a closed, fixed-energy physiological system. All psychic mechanisms had to be described in terms of neurological connections and conversions of energy states. In Freud's effort to state basic laws of the mind that are open to empirical verification—for example, neurons tend to divest themselves of quantity—the monograph has been called Newtonian in scope. Yet there was little to anticipate the actual procedures of psychoanalysis and little about the sexual bases of the neuroses; and only in the scattered passages on dreams and on consciousness are Freud's most dramatic clinical breakthroughs discernible.

At first Freud was pleased by his efforts, declaring that "the barriers suddenly lifted, the veils dropped and . . . everything seemed to fall into place." But Freud soon became disenchanted with "The Project." No doubt contributing to his disillusionment was the difficulty he encountered in coming up with a neurological explanation of repression. As he wrote to Fliess, he could not see how he had come to "concoct the scheme. . . . It seems to me to have been a kind of aberration." Not only did Freud not publish this manuscript, but he seems to have been ashamed of it, never referring to it again publicly and apparently hoping that any surviving copies would be destroyed by Fliess's heirs and by his own.

Yet I believe that "The Project" played a crucial role in Freud's development, if in a negative way. As a trained neurologist and neuroanatomist who continued to believe until his death that all psychic mechanisms must have a material basis, Freud felt virtually compelled to undertake such a monographic summation. In doing so, Freud was paying his dues to his chosen profession, to the domains of his training.

The creation of the new vocabulary, with its symbol systems, and the sketching of schematic diagrams that traced various neural connections and energy fields, was also an important endeavor for Freud; he was playing with ideas that could not be readily explained in the technical vocabulary of his time. If he was to avoid misunderstanding—translation of his points into inappropriate or outmoded concepts—he needed to create his own linguistic and graphic vocabulary, through which he could convey his exact meaning.

"The Project" was created largely for one audience—Freud himself. And yet, his sharing of it with Fliess introduces a theme that I encountered repeatedly in my study of creativity. At times when creators are on the verge of a radical breakthrough, they feel the need to try out their new language on a trusted other individual—perhaps to confirm that they themselves are not totally mad and may even be on to something new and important. This desire to communicate has both cognitive and affective aspects, as the creators seek both disciplinary understanding and unquestioned emotional support. Later I suggest that this desperate effort to communicate may hearken back in some ways to the initial communication link between mother and child and to youthful links between peers.

If psychology had been physics, if Freud had been Newton (or Einstein!), perhaps "The Project" would have evolved into Freud's most important treatise, thereby giving rise to the new field of psychological neurology or neurological psychology. Instead, however, this monograph proved to be a tour de force in a negative sense. Freud demonstrated to his satisfaction that the problems on which he was working could *not* in his time be solved in the language and with the methods of scientific neurology. If he wanted to continue mining this field, he would have to proceed in the manner of a psychologist, but one who invents his own forms, his own language—the latter, as it turned out, less technical, more akin to common sense, and almost completely cut off from neurological moorings.

Dreams and Self-analysis

Even before he had penned "The Project," Freud had become at least dimly aware that the road to understanding the mind—his mind, all minds—lay in the analysis of dreams. The actual insight had come in the summer of 1895—a time of creative efflorescence when, as Breuer remarked to Fliess: "Freud's intellect is at his highest. I gaze after him as a hen at a hawk." In the coming months Freud came to appreciate that dreams in normal persons reflected many of the various processes and mechanisms that he (and Breuer) had been observing, if through a glass darkly, in neurotic and hysteric patients. Here, too, one could discern a censoring mechanism, various kinds of disguises, forbidding ideas seeking to express themselves, and much latent sexual content.

By 1897 Freud had embarked in earnest on what may have been his most lonely and most important effort of explorations—his self-analysis. Using his nightly dreams as a point of departure, Freud allowed his conscious associations free rein and, in the process, examined in as dispassionate a way as possible the various ideas that came forth. This exercise in self-reflection not only yielded deep insights into Freud's own psyche; he came to believe that it could reveal as well certain forces and contents that occupied the consciousness of all human beings. The dream process also aided Freud in his lonely days: "Whenever I began to have doubts of the correctness of my wavering conclusions, the successful transformation of a senseless and muddled dream into a logical and intelligible mental process in the dream would renew my confidence of being on the right track." An inscrutable language was being deciphered exactingly.

Freud came to believe that all dreams contained some kind of a wish or fantasy. The dream was the disguised fulfillment of a repressed wish: a psychic means of carrying on some kind of a prior determination, concern, or desire. At times, as in the case of young children, the wish was apparent and undisguised— for a sweet or for a triumph over someone who had behaved in a mean way; with adults, the wishes were usually more complex and more likely to be disguised.

The dream was generally sparked by a thought from the previous day, often one the dreamer had not had the opportunity to explore. To gain access to this dream thought, it was necessary to peer through the "manifest (or surface) content" of the dream and to unravel its "latent (or underlying) content." An entire lexicon of symbols—though not one that could be applied in the absence of contextual information—was needed to decipher the text of the manifest content. The defenses that shaped the dream would include condensation, displacement, and various kinds of screens, each of which had to be patiently dissolved if the meanings of the dream were to be properly elucidated.

If the cases shown by Charcot intrigued Freud, his own dreams—and those of his patients—fascinated him even more. Each was a separate puzzle—and here, remember, was a man who *loved* puzzles. Each one needed to be unraveled, and as a dividend of its solution, revealed something not only about the dreamer but about human nature in its universal, unconscious manifestations. Freud eagerly allowed various associations to rise to consciousness and tried mightily to discern—if not to create—some coherence in the seemingly meaningless jumble.

The various dreams Freud analyzed have become so well known that a brief mention should suffice to trigger recall: the Irma dream, in which a patient receives an injection that makes her suffer and in which Freud wishes that he not be blamed for her suffering; the Count Thun dream, in which Freud confronts an arrogant political leader, thereby fulfilling the wish that he (Freud) might amount to something; the dream of the botanical monograph, where Freud secures the credit owed him for an earlier professional slight.

The process of analyzing and laying bare the "real" meanings of such dreams as these obviously engaged Freud's formidable intellectual powers and provided intrinsic pleasure as well. Now he had professional license to engage in the kinds of analyses that he had enjoyed since his youth. Yet it would be misleading to imply that this lonely physician found dream analysis a sheer pleasure. On the contrary: Analysis of dreams forced Freud to confront many unpleasant traits in himself (his vanity, his occasional cruelty, his jealousy); his ambivalent feelings, particularly with respect to his recently deceased father; and his sexual feelings, only a few of which he wrote about directly in *The Interpretation of Dreams*. It might even be suggested that Freud was only willing to confront the pains of self-analysis because he was deeply troubled; he himself craved the kind of "chimney sweep" or "talking cure" that others had gotten in earlier times from Catholic confession and that he was offering to his own psychoanalytic patients. As Freud had declared in a letter to Fliess: "The most important patient for me was my own person." Perhaps by this time, Freud was no longer dependent on any other individual—he had transferred the role of "sympathetic listener" to the psychoanalyst that he had created within himself.

Contents

As a scientist, Freud may initially have been drawn to the analysis of dreams because this activity would reveal to him the mechanisms of defense and of consciousness. But he soon made discoveries about the specific contents of dreams that proved vital to his further endeavors.

Certainly the most written about, and perhaps the most important, finding was the centrality of the Oedipal complex. Freud had discovered in himself deep and deeply ambivalent feelings toward his parents—ones dating back to earliest childhood. On his analysis, the young boy felt strong attraction, love, and lust for the mother, contrasting with jealousy, fear, and even hatred of the father. This amalgam of feelings would be translated into unconscious wishes to marry the mother and to kill the father. While first sensing these feelings in his own psyche, Freud soon concluded, based on his extensive literary background and on close analysis of his other patients, that these feelings were entrenched in the human sensibility. He proposed this complex as the basis both of the Oedipus myth of Greek times and the story of Hamlet from the medieval era. Unresolved Oedipal feelings lay at the root of much adult neurosis; Oedipal themes—or, in women, an analogous "Electra" complex—were encountered in the unconscious mental life of all individuals.

Freud had for many years suspected the importance of sexual factors in the etiology of the neuroses. He frequently invoked statements by his teacher Charcot, his close colleague Breuer, and Rudolf Chrobak, a prominent Viennese gynecologist—all of which pointed in the same direction. Now, in his analysis of dreams, Freud was confirming that sexual themes underlay the unconscious of all individuals, and that defense mechanisms were elaborated chiefly to deal with the unsettling and difficult-to-confront sexual themes.

Shortly after writing "The Project," Freud had experienced a traumatic shift in his own thinking. From the mid-1890s on he had attributed disorders in his adult patients to episodes of sexual abuse or exploitation during their early childhood years. Indeed, in his defiant address to the Vienna medical society of 1896, he made this claim explicit. However, in an apologetic letter to Fliess in 1897, Freud admitted that he had been mistaken. In many cases, it appeared, there had been no early sexual molestation by a parent or other elder; the seductions had been invented by the credulous mind of the young child.

Many commentators have seen Freud's about-face on this issue as a climactic turning point, and quite possibly a giant blunder. Some feel that Freud was a dupe for having ever believed that Viennese parents were engaging in sexual activities with their offspring; others, notably the controversial psychoanalytic scholar Jeffrey Masson, believe that Freud dishonestly eliminated the hypothesis of early seduction because it was proving too difficult for his contemporaries to swallow. In my view, while the change of mind obviously made a deep impression on Freud, it was not particularly crucial for the development of his theoretical or clinical ideas. Whether these early seductions had "actually" happened or had only "appeared" to happen did not materially affect Freud's thinking on pivotal questions; in either case, they had to be unpacked and dealt with. However, Freud's choice not to admit publicly so dramatic a reversal in his thinking does give one pause. Freud was prepared to change his mind; yet, proud and strongwilled, he was less prepared to acknowledge these changes.

Also evolving at this time was Freud's theory of infantile sexuality. The dream analyses and his self-analysis had convinced him that, from infancy, youngsters are subjected to strong sexual strivings—searches for pleasure of a psychic, as well as a somatic, nature. Every child passes through a series of libidinal stages in which the sexual energy is concentrated on specific bodily zones: initially the mouth, then the anal area, then the urethral area, and ultimately the genital area. This belief in infantile sexuality, perhaps as much as any of his other views, caused Freud to be ostracized. How could innocent children, living in the prim-and-proper Victorian-Hapsburg era, possibly harbor strong sexual feelings, even if these were only operating at an unconscious level? If Freud simply wanted to gain converts, he might well have tempered his claims on this topic.

A variety of other Freudian themes emerged at this time: the interest in memory and forgetfulness; the focus on jokes, slips of the tongue, and other revealing errors; and the recognition of primary and secondary processes, different forms of regression, and the psychic means of dealing with pleasure and pain. So, too, Freud's therapeutic mechanisms shifted in this decade: from electrical treatments, to hypnosis and suggestion, to the much less obtrusive methods of free association on a couch, with the analyst out of view and largely silent. But most of these ideas were to be played out chiefly in the twentieth century.

THE INTERPRETATION OF DREAMS: FREUD IN 1900

The ideas of the 1880s and 1890s all came together in Freud's magnum opus: *The Interpretation of Dreams*, a large volume published in 1899, though dated 1900. According to Freud, he had worked out all of the volume's major ideas by late 1896, but he had to take additional time to read background literature and then pen the lengthy manuscript. Freud knew it was his most important and original work: "No other work of mine has been so completely my own, my own dung heap, my own seedling and a *nova species mihi* on top of it." As he was to comment subsequently, the work contained "the most valuable of all the discoveries it has been my good fortune to make." In this book, Freud provided a detailed argument about why dreams represent the road to the unconscious, explained the mechanisms of dreams, and elaborated his views on the nature of the psychic apparatus.

Despite the apparent differences between *Dreams* and "The Project," the former can well be regarded as the latter's logical successor. *Dreams* is "The Project" shorn, for the most part, of its neurological underpinnings and terminology. And *Dreams* is "The Project" with a fascinating subject matter—human dreams—and a compelling plot—an explanation of the nature, sources, content, and mechanisms of dreams.

In the final chapter of *Dreams*, its most forbidding section, Freud explains "The Psychology of the Dream Processes." Freud describes different psychic systems: perceptual and motor centers; a function devoted chiefly to memory (which must retain traces); and a function devoted to perception (which must remain fresh and therefore possesses no memorial capacity). Once Freud explores the memory system, he breaks new ground. Memories are unconscious in themselves; however, dreams can provide the crucial clue about how the unconscious works.

Freud speaks of the need to assume two "psychic instances": (1) a mechanism (or censor) that criticizes; and (2) the material that is criticized. He also describes a new ensemble of systems: a continuum extending from perception to memory to the unconscious and preconscious systems. The impetus of the dream arises in the unconscious system, where the dream wish is harbored; it struggles to work its way into preconsciousness; it is thwarted during daytime by the censorship; but at night, through the weakening of resistance and the adoption of various disguises and compromise formations, it erupts into the dream life.

In the concluding pages, Freud draws on his model to account for diverse phenomena: the experience of fear, the excitation of pain, the achievement of complex thought, and the operation of repression, the dominant psychic mechanism. He declares that all psychoneurotic symptoms must be conceived of as wish fulfillments of the unconscious. An entire worldview has begun to be built on the insights and models constructed over the past decade.

An Indicator of Freud's Abilities

The Interpretation of Dreams reveals the strengths, as well as the limitations, of Freud's intellectual gifts. It is powerfully written, showcasing Freud's considerable literary gifts. It includes a panorama of sources, testifying to Freud's command of the scientific literature, the classical literature, and the political and cultural events of his own and other eras. With great dramatic power, Freud brings to life both the nature of psychic mechanisms and the features of individual dreams and dream personages, and he provides logical arguments and much supporting clinical data. On the other hand, there is nothing quantitative in the work, reflecting perhaps Freud's view of his own intellect: "I have very restricted capacities or talents. None at all for the natural sciences; nothing for mathematics; nothing for anything quantitative." Nor is there much consideration of evidence that might call Freud's principal assertions into question.

Perhaps surprisingly, Freud's vivid descriptions contain few spatial, visual-spatial, or bodily-kinesthetic images—unusual for a scientific work. Both those in the biological sciences, like Darwin, and those in the physical sciences, like Einstein, favored images in their own thinking; and in Einstein's case, the images often embodied the phenomena whose lawlike nature the scientist was striving to elucidate (see chapter 4). Freud's work, however, is almost entirely in the verbal domain, and his few simple diagrams add hardly anything to the persuasive narrative exposition. Linguistic in nature, Freud's scientific thinking involved some logical, but few spatial, components. Perhaps in explanation of this pattern of argumentation, Freud said, "I have an infamously low capability for visualizing spatial relationship which made the study of geometry and all subjects derived from it impossible to me."

Above all, *Dreams* reveals Freud's command of the realm of the personal. He is sensitive to the desires, needs, wishes, and fears of those whose dreams he analyzes; to similar elements in his own dreams; and to factors that influence all human beings, such as the strivings that define the Oedipus and Electra complexes. The interest Freud showed from childhood in the foibles of his family and in literary worlds and the dramatizing ability that came through in his youthful letters erupt in full force in his writings about dreams, as well as in subsequent psychoanalytic writings. These features contribute to the seductiveness and memorability of Freud's writings. Among scientific investigators, Freud stands out in his capacity to meld the realm of the personal to the realms of language and logical exposition—the mark of a prototypically effective social or behavioral scientist.

Freud's Scientific Approach

Freud's conceptualization and reconceptualization of his primary problem area resemble the pattern found among other scientific workers. His first detection of an anomaly occurred in his exposure to hysteric patients, whose behaviors could not be explained by standard organic accounts. Efforts to explain hysteria soon mushroomed into an attempt to explain the whole range of neurotic behaviors and, ultimately, in "The Project," into an effort to account for normal psychological behavior as well. The symbol system worked out in "The Project" may have been helpful to Freud (if not to Fliess), but he concluded that so technical and neologistic a system was neither necessary nor useful in communicating with a wider public.

Rather than elaborating on his own neuropsychological scheme, Freud instead made a lateral move: He drew on the phenomena and mechanisms of dreams in order to put forth his own views of behaviors and the unconscious. He created a new language with its own explanatory framework, composed primarily of words already in the German language and of a few simple schemes that could be described verbally, rather than having to be diagrammed. This framework proved generative enough for Freud to build on it for the remainder of his lengthy professional career; and, revealing the mark of a creative thinker, the scheme has provided ample fuel for the works of many other researchers and clinicians.

Freud's achievement differed from other scientific theorists' creations, like Einstein's, in its relatively loose formulation and presentation. Freud dealt with a whole ensemble of issues rather than with a focused problem or set of problems; and he gave no indication of crucial tests for his principal assertions. Indeed, in a manner more characteristic of humanistic studies, Freud's framework held the potential for use in diverse ways by a wide array of scholars.

Unlike many other investigators, Freud seems to have possessed intellectual strengths that would have allowed him to make a mark in a number of different spheres. While it is difficult to imagine Picasso as other than a painter or Einstein as other than a theoretical physicist, Freud could conceivably have been a significant biologist (in the Darwinian tradition), lawyer or jurist, or religious leader, and certainly a contributor to many areas of scholarship. Perhaps he did hit on the best domain for his own gifts, but it was certainly not the only conceivable one.

Initial Responses

With the publication of what Freud knew to be his master work, it was now time to see whether the world would indeed recognize the power of his discoveries. A work of such scope seemingly would have had an immediate impact on the field to which it was addressed. But as is well known, the first edition of *Dreams* sold only 351 copies in the first two years and was allowed to go out of print. It did receive its share of reviews, some sympathetic; but unlike, say, Darwin's *Origin of the Species,* neither the scholarly nor the public worlds paid significant heed to this book. Freud faced the possibility, in the aftermath of its publication, that he would remain obscure for the rest of his life. As he wrote shortly thereafter to Fliess: "Not a leaf has stirred to show that *The Interpretation of Dreams* meant anything to anyone. . . . The book's reception, and the silence since, have once more destroyed any budding relationship with my environment." Later he joked: "It seems to be my fate to discover only the obvious: that children have sexual feelings, which every nursemaid knows; and that night dreams are just as much a wish fulfillment as day dreams."

THE VIENNESE SETTING

The Vienna of more than a century ago, though not as populous as London or Paris, or as historically indispensable as Rome or Athens, may have surpassed its competitors in the vitality of its intellectual life. In fin de siècle Vienna, such redoubtable figures as members of the waltz-composing Strauss family, the philosopher Ludwig Wittgenstein, the novelist and essayist Robert Musil, the architect Adolf Loos, the city planner Otto Wagner, the essayist Karl Krauss, the librettist Hugo von Hofmannsthal, and the physicist Ludwig Boltzmann lived close to one another, hobnobbed at cafés, and exchanged views at meetings. As a well-read and cosmopolitan figure, Freud doubtlessly knew of these individuals and, at least in some cases, had personal contacts with them. The composer and conductor Gustav Mahler consulted Freud during a period of sexual impotence; Breuer was the physician to Freud's one-time psychology teacher Franz Brentano; and Freud greatly admired Vienna's leading playwright, Arthur Schnitzler, who treated in dramatic form many of the themes that Freud uncovered in his analysis of individual Viennese psyches. A recent account of Vienna captures the air: "Klimt and Wagner and Loos thus become tablemates of Freud and Mahler and Wittgenstein at an imaginary coffeehouse for a shining moment in the city that was the 'cradle of modernity.'"

Living in Vienna for nearly all of his life, Freud had decidedly ambivalent feelings about his city. He recalled with nostalgia the more rustic atmosphere of his native Moravia and certainly revealed a taste for the pastoral in his choice of vacation spots. In addition to whatever distaste he felt for the urban environment,

Freud also claimed to take offense at the scientific and intellectual life of Vienna, which he found to be authoritarian, narrow, and anti-Semitic. And he did not hesitate to remark: "Vienna has done everything possible, however, to deny her share in the origins of psychoanalysis. In no other place is the hostile indifference of the learned and cultivated section of the population so evident to the analyst as is Vienna."

It has been said that Freud revealed his Viennese nature, not least, in his stated dislike for his city. When, as an old man, he finally had to leave Vienna because of the Nazi *Anschluss*, he did so with regret. Whether Freud *actually* liked Vienna, however, is not germane for our purposes: It is difficult to envision his own scholarly career and output having taken place in a radically different environment.

Freud's ideas seemed to reflect, and perhaps to grow organically out of, the milieu in which he lived. The overt sanctimoniousness about matters of sexuality flew directly in the face of the sexual intrigue so common among middle-class Viennese, and probably among those in other social ranks as well. While the political and social tone veered toward the conservative, there was surprising tolerance for the articulation of avant-garde notions in the arts, so long as they did not directly threaten the political fabric. Rhetorical anti-Semitism abounded, but Jews were allowed to advance in the professions, particularly if they had been baptized. While disdained, criticized, and ignored, Freud did not have to fear overt censorship—at least, no more so than did the contemporary Viennese proponents of Zionism, urban planning, medical innovation, and artistic revolution. Freud benefited from the considerable distance between the official rhetoric and the actual conditions for creation that were permitted in Vienna; quite possibly, had he lived in an ostensibly more tolerant England, he would not have been permitted to publish his texts dealing with sexuality.

One could not expect an outpost of Hapsburg conventionality like Vienna to greet Freud's revolutionary statements with open arms; at most, the Viennese were likely to accept these themes when they had been "disguised" in Schnitzler's dramatic forms. Yet, at least to some individuals, the conditions Freud described were confirmed in their own daily observations, in their readings of the news, in their treatment of patients, and in their introspections about their own lives. It was these individuals who became Freud's first disciples, the ones who made their way to his Wednesday evening salon.

That Freud would have disciples, and that he would want to have them, was not a foregone conclusion. We have already seen that he was at times resigned to a continued life of oblivion. Like his friend Fliess, Freud might have continued to believe in his own ideas without proselytizing vigorously on their behalf. He could have remained an intellectual hermit. He might have withdrawn into his family life, retiring to the role of paterfamilias, or reverted to a comfortable Jewish milieu, playing cards and socializing with his friends in the B'nai Brith, which he had joined in 1897, the very period of his greatest loneliness. He might have sought converts but, like so many other aspiring revolutionaries, have failed to find or keep them. He might also, like the philosopher and educator John Dewey, have allowed individuals to gather in his name without shaping their activities; or, like the political philosopher Karl Marx, Freud could have tried to lead a movement based on his ideas and, in the process, have virtually destroyed it. Instead, Freud forged a different and far more successful course with his followers.

FREUD AS A LEADER: BROADENING OF THE NETWORK

In his launching and sustaining of the psychoanalytic movement during the early 1900s, Freud revealed facets of his personality that had long been dormant: his fascination with the military and his desire to lead an engaged and embattled unit. In terms of the triangular analysis introduced in chapter 2, we can say that Freud as an individual *talent* had outlined a revolutionary set of ideas; these ideas were significantly at variance with prevailing teachings in the *domains* of psychology, psychiatry, and allied fields; if these ideas were to acquire influence, Freud had to create a supporting *field* that could evaluate and disseminate the ideas.

The meetings at Freud's home of the Wednesday Psychological Society served many purposes. Certainly one of the most important was the opportunity it gave Freud to assume a leadership position. Earlier Freud had not been shy about speaking in public, giving a lecture, or defending his position, but he had done so in the same spirit as characterized dozens of other aspiring academics. Now, however, he had constructed a corpus of concepts and had even given a name to his school of thought—psychoanalysis. He was the unquestioned leader of this group, not only an authority in his own home but also as the father of a school.

Freud represented a set of ideas as well as a set of practices. All serious scholars are expected to develop original ideas; that is the basis of their profession. But only rarely are these integrally related to practice, and even more rarely do they lead to an entirely new mode of treatment. Like Gandhi halfway around the world, Freud could speak to a far larger potential community because he had developed techniques—free association, dream analysis, and therapeutic interventions—that could actually be used to help people; Freud addressed the treatment of diseased, unhappy people who craved a cure.

Subtle changes were taking place as Freud passed into midlife. For years, he had dreamed about going to Rome; but he had deliberately withheld this pleasure from himself, for a number of largely irrational, self-denying reasons. Beginning in 1901, Freud made several trips to Rome, partially as a reward for having stuck to his proverbial last and for having produced his major work. Freud had also dreamed of becoming a professor but had been reluctant to become enmeshed in politics in order to secure this promotion. Now he found himself able to bracket his excessive moralism, enter into political bartering, and succeed in obtaining a professorial appointment after seventeen years of waiting.

What did not change was Freud's tremendous productivity. Even as he was writing *The Interpretation of Dreams*, he was also working on two of his most important and well-received volumes: *The Psychopathology of Everyday Life* (1901), in which he analyzed the hitherto underappreciated nature of various kinds of verbal and practical "slips"; and his *Jokes and Their Relation to the Unconscious* (1905), a study of the various functions assumed by jokes and other vehicles of humor. He published his revolutionary *Three Contributions to the Theory of Sex* (1905), in which he elaborated on his ideas about sexual aberrations, infantile sexuality, and the transformations associated with puberty. A steady stream of case studies, with patients including artists, physicians, schoolboys, aristocrats, and paranoid personalities; papers on therapeutic technique, and eventually, more reflective papers on psychoanalysis ensured that there was much to talk about at the Wednesday Psychological Society and that psychoanalysis did not remain a frozen set of texts.

Freud proved to be a magnificent personal advocate for psychoanalysis. Overcoming whatever initial shyness or arrogance might have detracted from his presentations a decade or two before, and exploiting the verbal fluency apparent even in his early letters, he had become a riveting lecturer. With seemingly little preparation, he was able to address a spectrum of audiences in a personal, thoughtful, and wide-ranging way, drawing examples from history, art, his cosmopolitan reading, the current scene, and the concerns of the particular audience he happened to be addressing. He could anticipate their objections, even voice them articulately, and thereby disarm some of their reservations and criticisms. Clearly, many reflective young individuals were attracted to psychoanalysis by Freud's increasingly persuasive written briefs or by his charismatic person.

Father of the Movement

I referred earlier to Freud as the father of the psychoanalytic movement. Freud had called attention to the unsurpassed role of the father in every young male's life. The role of the father proved equally important within the psychoanalytic movement and created as much chaos among Freud's followers as it was alleged to engender in the classical nuclear Oedipal situation.

Of all the early converts, Carl Jung, a Swiss psychiatrist nineteen years Freud's junior, was without question the most important: He was clearly an intellect of the first rank; he had no ulterior motive in being attracted to the ideas of an obscure Jewish Viennese doctor; he was an attractive personality in his own right; and he represented the endorsement of another land, culture, and social and religious background. Freud felt validated in a way that he had never been before. After confessing about his many years of "honorable but painful isolation," he writes to Jung of "the calm assurance that finally took possession of me and bade me wait till a voice from the unknown answered mine. That voice was yours." Freud felt little hesitation in making Jung the most prominent member of his small band and offering him in 1910 the presidency of the newly formed International Psychoanalytic Association, a gesture he was soon to regret when tensions flared between these two formidable personalities.

Even though an enormous amount has been written about the strife within the ranks of the early psychoanalysts, it remains difficult for the outsider to determine how much of the conflict was normal, inevitable, and perhaps even healthy and how much of it reflected the idiosyncratic pathologies of the first leader and his early followers.

That there was a pattern seems clear. Initially Freud was seductively open and welcoming to individuals, and especially so to those from far away, from another country, from outside the faith. Freud wanted desperately for psychoanalysis to be not simply another movement for Jewish intellectuals, and he was willing to reward those gentiles who would stray under his tent. (His fawning letters to public figures who had a good word to say about psychoanalysis indicate that he was ever on the lookout for influential converts.) The welcome would remain in place for a while: Freud could be a patient teacher and could tolerate deviations from the canon on the part of otherwise attractive converts. Indeed, an influential and sympathetic critic sometimes was included in the inner circle and drawn into discussions about more troublesome or deviant members. Freud marked membership in this elite by the bestowal of a special golden ring bearing an ancient Greek figure.

Ultimately, however, Freud demanded loyalty—loyalty of person and loyalty of ideas. Those who were not satisfied followers eventually came to grief within the psychoanalytic movement. For some like Adler or Jung, the break came relatively soon and was filled with ill feelings; others, like Rank and Ferenczi, remained within the circle for many years before a painful parting of the ways. Only a very few individuals managed to remain close to the psychoanalytic movement and to its charismatic but demanding founder for the entire period of their involvement.

It does not take training in psychoanalytic theory to detect the effects of this involvement on Freud himself. He had already anticipated the difficulties between the father and the offspring in his writings about the Oedipal complex, dating back to the 1890s. The events in Freud's personal life and the gathering political storms in Europe may well have led Freud to a more overt concern with social issues. In *Totem and Taboo* (1912–1913), written around the time of the first defections, he spoke about the special powers that surround the taboo figure, the primal horde's growing compulsion to kill the father, and the ensuing struggle among the surviving brothers for leadership and power. To the outside world, it may have been a parable, but within the Freudian circle, it was virtually autobiographical.

Social and cultural issues came to play an increasingly important role in such critical texts as *Civilization and Its Discontents* (1930) and *The Future of an Illusion* (1927). Then in his final work, *Moses and Monotheism* (1939), Freud identified quite directly with the leader who had founded a new religion, only to be rejected by those to whom he had revealed the "true" path. This essay represented a remarkable (if perhaps not entirely conscious) turnabout, since Freud had made opposition to organized religion a cornerstone of his personal and scientific philosophies.

Despite, or perhaps even because of, these struggles and tensions, the first decades of the twentieth century represented a time of remarkable growth and spread of psychoanalytic ideas throughout the Western world. In 1908 the Wednesday Psychological Society became the Vienna Psychoanalytic Society, a model for many other such groups around the world. In 1909, accompanied by Jung and Jones, Freud made an important trip to the United States, where, at Clark University in Worcester, Massachusetts, he enjoyed the first recognition of his work beyond Europe. By 1910 there was an International Psychoanalytic Association, which included a number of national organizations under a common president; by 1920, when the first congress after the First World War was held in the Hague, there were active movements in the major Western European countries, Russia, and India, and two schools in the United States. Without doubt, Freud's personal leadership capacities contributed to the success of the movement. Indeed, it is at least conceivable that the name of Freud is much better known than his one-time colleague Jung and his long-time rival Janet, less because of the intrinsic superiority of his ideas than because of the brilliance and relentlessness of his campaign on behalf of their acceptance.

Freud's Growing Renown

Gradually Freud became a world-renowned figure, rivaling Einstein from the world of science, Gandhi from the world of leadership, and even the "genuine" stars of the movies and sports. He had "official friendships" with leading figures of letters, like Romain Rolland, Thomas Mann, and Arnold and Stefan Zweig, and "official correspondence" on war and peace with Einstein, as well as more personally tinged friendships with disciples like Lou Andreas-Salome, alleged at one point to have been Nietzsche's mistress. He had official enmities as well, of which his long-standing feud with Janet, his closest peer in European psychiatry, is the best known. He vigilantly monitored those who wrote about him and the psychoanalytic movement, cajoling when he thought that would help, denouncing when that seemed more strategic.

Less happily, their involvements with Freud proved costly for some individuals, particularly those who had broken with him. Freud's young protégé Viktor Tausk, despondent over his recent rupture with the unforgiving Freud, committed suicide; of the earlier followers, at least six others ultimately did the same. These facts represent our first evidence of the casualties that tend to befall those within the orbit of highly creative individuals.

Neither gaining converts nor seeing defections and suicides seemed to affect Freud as deeply as had his experience of earlier liaisons and disruptions with Breuer and Fliess. Part of this is simply age: As we grow older, it proves more difficult to develop strong emotional ties to new individuals, and we must harden ourselves to the increasingly frequent loss of our contemporaries. But I think that in Freud's case, three additional factors are at work. First, he had spent many years alone with his ideas; this hardening experience taught Freud not to become excessively dependent on other individuals. Second, Freud believed that he was in a sense maintaining relationships to others—they simply had become less personal, more institutional and intellectual. Finally, and here I become even more speculative, Freud saw himself as a military leader mounting a difficult campaign. As part of such a campaign, it becomes necessary to undertake risky operations; to make investments that may not pan out; and when necessary, to cut losses, regroup, adopt a new strategy, and return to the fray. Sometimes, great leaders sponsor others who can undertake these steps: Moses certainly benefited from the activities of his brother, Aaron, and Darwin willingly left campaigning on behalf of evolutionary theory to the biologist Thomas Huxley; for the most part, Freud had to serve as both general and first lieutenant.

But if Freud was a busy officer, he managed to save time for other pursuits. His large family continued to benefit from his interest and counsel. He saw patients for eight to nine hours a day, took a daily constitutional, maintained ties with friends and his B'nai Brith colleagues, read literature, collected antiquities, and wrote almost every night from eleven o'clock to one or two in the morning. Even to list his publications from 1910 to 1930 would take several pages. Certainly, Freud was true to the nineteenth-century bourgeois ideal of the tireless worker, who occupied himself in some productive way for nearly every hour of the day and berated himself mercilessly whenever he felt that he was sloughing off. While the calling of the scientist might have come into its own in Puritan England, it certainly was flourishing centuries later in Jewish Vienna.

Freud ventured well beyond the boundaries of case studies and clinical treatises. At the height of the First World War he penned six major theoretical "metapsychological" papers, all within a two-month period. And, picking up on themes first introduced in *Totem and Taboo*, he thrust psychoanalysis increasingly toward wider political and cultural concerns. Freud issued important and controversial works about group psychology, religion, politics, war, aggression, and "civilization and its discontents," the phrase he used for the book title. In branching outward, Freud was giving vent to that strong philosophical and cultural streak he had "ruthlessly suppressed" six decades earlier, when he had elected a career in medicine and the natural sciences. And he was addressing an audience that, in the aftermath of the traumatic Great War, hungered for an explanation of human destruction.

As mentioned earlier, in my study of creative individuals, I frequently found a "ten-year-rule" at work: A creative individual makes a breakthrough after ten years of work in a domain and then, depending on various factors, may or may not realize additional breakthroughs in subsequent decades. Freud fits very well as an instance of the first part of this rule: His work on dreams occurred almost exactly a decade after his initial apprenticeship in Charcot's laboratory. Without question, Freud's creativity was maintained for several decades thereafter; in this instance, he resembles artists like Picasso, Stravinsky, or Graham much more than he resembles the scientist Einstein. Whether the discoveries occurred precisely at decade-long intervals is somewhat more problematic, but one can at least point to Freud's initial move to social issues, around 1910, and his full-fledged involvement in political and cultural issues in the 1920s and 1930s.

While Freud's later, more speculative works were widely criticized, they were crucial in making Freud into a world figure, one of abiding interest to citizens of many nations and to scholars in numerous disciplines. So long as his writings were restricted to medical conditions and practices, he would belong to the world of the Havelock Ellises and Richard von Krafft-Ebings—iconoclastic physicians who had dared to write about sexual matters. But now, he was becoming part of the international belletristic tradition, to which earlier philosophers like Jean-Jacques Rousseau and John Smart Mill had contributed, and of which writers like Walter Lippmann, Bertrand Russell, and Henri Bergson were the contemporary representatives. Freud began to receive the recognitions accorded to world-class minds; among the many plaudits and degrees, the receipt in 1930 of the prestigious Goethe Prize (awarded to an outstanding writer in the German language) meant the most to him. Perhaps with tongue in cheek, Freud disparaged this phase as a "regressive development." He said: "My interest, after making a lifetime detour through the natural sciences, medicine, and psychotherapy, returned to the cultural problems which had fascinated me long before, when I was a youth scarcely old enough for thinking." But he clearly had entered this phase with gusto.

Freud remained active until almost the end of his life, seeing patients and writing works even after his forced migration to London at age eighty-two. His stoicism in the face of a debilitating cancer, the loss of his homeland, and the knowledge of his imminent death has been widely—and appropriately—admired. The continued survival, vitality, and output of Freud remained important to his associates in a personal way and to the rest of the world in a symbolic way, yet it seems safe to say that the history of his movement would have remained much the same had Freud died at any time after his sixty-fifth or seventieth year. For Freud had accomplished a feat virtually without precedent for a scholar-researcher in our time (Marx would be the closest rival): He left not only an impressive legacy of works to be read, studied, and argued with or followed but also an organized institution, the psychoanalytic movement, which could continue to build on his legacy after his death.

A Worldwide Following

Today, throughout the industrialized world, in countries from Argentina to Japan, as well as in developing countries like India, one finds trained individuals called psychoanalysts, who can trace their pedigree to the initial Freudian circle; associations, journals, and training institutes that call themselves psychoanalytic, Freudian (or, less frequently, Jungian, Adlerian, or Lacanian); and individuals in other disciplines (history to philosophy) and careers (artists, painters, critics) who consider themselves members of the psychoanalytic community. Freud did not anticipate all of them and might not have approved of some of them, but their existence is unthinkable without his work and his example.

Nor have the ideas themselves become stagnant. To be sure, there are the true believers, for whom every word of the master is sacred, and the vociferous critics, who disdain the entire corpus. But there are also many serious scholars and practitioners who attempt to glean what remains worthwhile and timely from the psychoanalytic movement, while nurturing and guiding it in productive directions, some of which could not have been anticipated by the members of the original Wednesday evening gathering.

A century later, controversy has not abated. In England and France, the discussions surrounding the works of Melanie Klein or Jacques Lacan raise all of the classical issues about psychoanalysis. In America, the tremendous interest shown in books by Janet Malcolm, about psychoanalysis as a profession and about the management of the Freudian archives, demonstrates a continuing fascination in matters Freudian. Freud's own integrity has been questioned regarding the accuracy of his case reports and his motives in treating certain wealthy patients. And when a new text or letter or journal or imbroglio surfaces, or an embargo is lifted, the familiar authorities emerge to pass judgment on it—fueling the pages of the *New York Times* and, perhaps, the imagination of Tom Stoppard or of Woody Allen.

Freud saw himself as a scientist and psychoanalysis as a science. He had the faith that, ultimately, his discoveries would be seen as having a neurological and chemical basis. (This did not prevent him—or others—from recognizing the artistic and philosophical nature of his work, and the conquistadorial facets of his personality.) While some aspects of psychoanalysis have received modest scientific support over the years, it seems fair to say the bulk of interest in psychoanalysis has come from outside of the scientific community and that most hard-nosed scientists do not take Freud seriously as a member of their fraternity. This situation would have disappointed Freud, but probably not surprised him; and he would have contended that, in the long run, the scientific basis of his principal discoveries would be confirmed. For my purposes, however, it makes more sense to regard Freud as a contributor to our knowledge about human beings, along with his heroes Shakespeare and Sophocles, and to our accumulated reflections on the nature of human society, along with Friedrich Nietzsche and Arthur Schopenhauer, whom he characteristically refused to read because he sensed that they had anticipated too many of his conclusions.

Freud now belongs to the world: I find it difficult to imagine that he will become a lesser figure in the foreseeable future. Such a reception is a remarkable feat for an individual who was unknown a century ago, who lived until the outbreak of the Second World War, and whose most powerful weapon was not the sword that he dreamed about as a child, but rather a mode of inquiry into the nature of dreaming and of the dream itself.

For my discussion he is emblematic—a stunning demonstration that one may attain the heights of creativity through the use of a particular intelligence: through the intrapersonal examination of one's own thoughts and feelings, and in his case, persistence even when no one else displays sympathy for or understanding of what one is doing. Freud then successfully redirected his energies and convinced an often hostile world of the plausibility of his discoveries. Proceeding from an initial fascination with the world, to the most isolated and hermetic of pursuits, and then back again to a conversation with multiple constituencies, Freud serves as a haunting reminder of the dual nature of creativity: a breakthrough within a particular domain that ultimately may speak as well to the interests and values of diverse human communities.

4

ALBERT EINSTEIN:
THE PERENNIAL CHILD

Einstein, 1898

IN HIS SCIENTIFIC AUTOBIOGRAPHY, which he insisted on calling his "obituary," Albert Einstein recalled a number of childhood events that had made a strong impression on him. When he was only four or five, his father showed him a compass. Young Albert was struck by the tenacity of the needle, which did not move even when its case was rotated. He contrasted the enigmatic compass with events from childhood that cause no surprise, such as the falling of tangible objects or the fact that the moon does not fall down. Einstein recalled a feeling of profound wonder, when, around the age of twelve, he received a little book dealing with Euclidean geometry. Certain surprising assertions—for example, that the three altitudes of a triangle will intersect at one point—could be proved so decisively from specifications that any doubt appeared to be out of the question.

Young Einstein exhibited another revealing tendency: He posed gritty questions and then pondered them at length. Perhaps most presciently, he asked himself around the age of sixteen what it would be like for an observer to move alongside a light wave—would the observer ever surpass the light wave? Sometime later, he wondered what would happen to the possessions of an occupant of a box that was falling freely down a long shaft—if removed from the occupant's pockets, would these possessions fall to the floor or just remain suspended in the air? This proclivity for imagining and pondering puzzles persisted. As the older Einstein was consulted regularly on cosmological issues, he kept raising the unsettling question of whether God would dare to play dice with the universe.

PUZZLES OF CHILDHOOD

Such questions are reminiscent of the type that young children have always posed—at least those youngsters who are not habitually "shut up" by elders. Children in the first five to ten years of life have ample opportunity to let their imaginations roam, to raise questions about phenomena that inspire doubt or awe, and then, at least sometimes, to pursue these questions for a while as they walk in the fields or fall asleep at night. In some cases, as we saw with Freud, these questions pertain to the nature of human relations—to situations involving powerful parents, unwanted siblings, paragons of good or evil; in other cases, as we shall observe with artists, the issues may be resolved in some kind of a nonlinguistic symbol system—What is the most haunting melody for this aria? What are the possibilities of a certain color? Can I dance the way that I feel? With Einstein, as with so many other children, the questions are of the sort that the noted psychologist Jean Piaget posed to youngsters: What causes the objects to behave in the way in which they do? Could rules of nature be altered, and with which consequences?

Einstein was aware of the parallels between his thought patterns and those commonly associated with children. He once asked, with perhaps undue self-deprecation:

> How did it come to pass that I was the one to develop the theory of relativity? The reason, I think, is that a normal adult never stops to think about problems of space and time. These are things which he has thought of as a child. But my intellectual development was retarded, as a result of which I began to wonder about space and time only when I had already grown up. Naturally I could go deeper into the problem than a child with normal abilities.

And it was Einstein who suggested to Piaget that he investigate children's intuitive notions of speed and time, thereby inspiring one of the psychologist's most illuminating lines of research.

To posit deep similarities between the mind of the child and the mind of the creative adult is a relatively recent, if not distinctly modern, phenomenon. Charles-Pierre Baudelaire, writing in the middle of the nineteenth century, explicitly tied the drawings of children to those of adult artists, dubbing the child "the painter of modern life." Only in the past century or so have artists, writers, and other creative persons shown a pervading interest in the symbolic products of young children. And Einstein stood out among natural scientists in his abiding curiosity about children's minds. He had once declared that we know all the physics that we will ever need to know by the age of three.

Yet, when he posed the questions that he did, Einstein was also speaking to—and for—his colleagues. Many professionals in the disciplines raise questions that make sense only to those who have had years of training in that domain (and Einstein constantly generated such puzzles). But it has been a particular burden of the physicist to pose the most fundamental questions about existence: When did the universe begin? What is the smallest unit out of which everything is made? How do we determine time? Can we transcend space? And it has been the special privilege of twentieth-century physicists—preeminently, Einstein—to begin to secure answers that have the feeling of permanence. Another redoubtable physicist of the era, I. I. Rabi, once declared: "I think that physicists are the Peter Pans of the human race. They never grow up and they keep their curiosity. Once you are sophisticated, you know too much—far too much."

A Not-So-Unusual Childhood

Many legends have arisen about the childhood of the man who became emblematic of brilliance in the twentieth century. Einstein is variously described as a late speaker, a dyslexic, a loner, a prodigy, a poor student, and a diamond in the rough. While shards of evidence can be accrued in favor of most of these statements, the truth appears less dramatic.

In some ways, Einstein's childhood resembled that of Sigmund Freud. Both came from upwardly mobile Jewish homes—families a few generations out of the ghetto, not yet established financially or professionally, but not subjected to egregious anti-Semitism and hopeful for a brighter future among the bourgeoisie. Residing in Munich, Germany, the Einstein family seems to have been unpretentious, fun loving, and relatively nonauthoritarian. Not unlike Jakob Freud, Albert's father, Hermann, was a kindly, relatively unambitious, and not very successful businessman who experienced his share of ups and downs. Einstein's mother, like Freud's, seems to have been the more cultivated and ambitious of the parents. Though she does not seem to have doted on young Albert to quite the extent that Freud's mother catered to young Sigismund, she did have, according to one biographer, "a touch of the ruthlessness" that Albert later exhibited on behalf of his interests.

Family members confirm that Einstein did speak relatively late and quite slowly, and was not a particularly verbal child. Unlike Freud, who was highly articulate and immensely interested in the world of other human beings, the young Albert showed—or as he once scrupulously modified it, "appeared to" show—a correlative interest in the world of objects. Father Hermann, in conjunction with his brother Jakob, manufactured a variety of electric appliances, and these piqued the curiosity of the child. Young Einstein loved to make constructions of all sorts. He built giant houses out of cards, ones that occasionally reached fourteen stories in height; he pored over jigsaw puzzles; and he was fascinated by wheels and all other objects with moving parts.

While not avowedly antisocial, Albert seems from an early age to have marched to his own drummer. He is supposed to have walked through the streets of Munich by himself as early as the age of three; he often played alone even when other children were around. For the most part he was quiet and thoughtful, but sometimes he exhibited powerful tantrums, including one episode when he hurled a chair at a tutor. Much of his time was spent pondering scientific questions and puzzles, in the company of a small circle of friends.

One feature of the young Einstein has not been much remarked upon, but I think it quite important. Like many emancipated Jewish families of the era, the Einsteins showed little interest in organized religion; they viewed themselves as "freethinkers." Young Albert, however, embraced religion and was quite scrupulous in his belief in God and in his observation of ritual. In adopting this religious stance, Albert was defying both his own family and the bulk of students in the Catholic school that he attended; his classmates presumably found his devout practices odd. I interpret the strong religious streak in the youthful Einstein as a sign of his potent spiritual needs, his interest in ultimate questions, and his capacity—if not his compulsion—to place himself in opposition to the conventional wisdom.

While Freud was a spectacular success in his school, Einstein sustained a much more vexed relation to formal education. He exhibited a strong dislike of the regimentation that characterized most German schools at the time. He particularly disdained the subjects requiring rote learning, and he revealed his contempt by performing poorly and acting defiantly in class. He could be arrogant about topics where he was knowledgeable, at one time occasioning a teacher to remark that young Albert had undermined the respect of the entire class by his rude manner.

But while Einstein was unhappy for much of the time in school, he certainly was an enthusiastic learner in some subjects. Once his uncle Jakob had introduced him to algebra and geometry, he devoured these topics on his own. He loved the beauty and order of geometry, the systematic proofs, the intimate connection between the diagrams and the reasoning; at the same time, he did not hesitate to argue with the geometry book, when proofs made no sense to him.

Adolescent Explorations

When Einstein was a young adolescent, his family offered regular hospitality to Max Talmey, an indigent Russian-Jewish medical student. Talmey took a liking to the young Einstein and gave him many books to read, including such classics as the works of Kant and Darwin. Noting the youthful Einstein's interest in physics, Talmey also furnished popular books on force and matter. Particularly influential was a series of volumes issued by an Isaac Asimov–like polymath named Aaron Bernstein. Apparently, Einstein absorbed from these books not only much factual information but also a basic scientific worldview: That view (echoing the one also adopted by the youthful Freud) was purely mechanistic and atomistic in its explanatory framework, and unreservedly optimistic in its assessment of the potentials of scientific research. Einstein devoured such works readily, enthusiastically, thoroughly, and also critically. Talmey recalls that "the flight of [Einstein's] mathematical genius was so high that I could no longer follow him." His early intoxication with formal religion was undone by this philosophical and scientific course of study.

A sign that Einstein was not set against all formal schooling occurred when, following a very difficult personal period in his early adolescence, he was given the opportunity to attend a "progressive canton school" in Aarau, thirty miles north of Zurich. This school was heavily influenced by the pedagogical philosophy of Johann Pestalozzi, who had encouraged a humanistic approach to subjects and had noted the centrality of visual understanding (*Anschauung*) in the mastery of concepts. Einstein liked the school very much, reveled in its emphasis on hands-on as well as theoretical science, made good friends there, and, as we might put it today, "turned around." Just a month before he died, Einstein re-called: "[The school] made an unforgettable impression on me, thanks to its lib-eral spirit and the simple earnestness of the teachers who based themselves on no external authority." The experience at Aarau (a school that preserved his student papers) showed Einstein that his idiosyncratic curiosity could be pursued within a supportive setting.

Based on his success at this school, Einstein was able to gain admission to the prestigious Zurich Polytechnic Institute, whose entrance examinations he had failed a year before. In the essay that accompanied his application, he wrote with deprecating self-knowledge: "Here are the causes which have led to this plan. . . . It is above all my personal disposition toward abstract thought and mathematics, lack of imagination, and of practical talent."

There is no better prognosticator of Einstein's ultimate career path than a pa-per he sent at age sixteen to his uncle Caesar in Stuttgart. In the five-page essay, "Concerning the Investigation of Ether in Magnetic Fields," he outlined what was known about electromagnetism with respect to ether, the hypothetical medium in which all waves were supposed to be transmitted. In this suggestive essay Einstein proposed to study the state of the ether in magnetic fields of all kinds, carrying out experiments that measure "the elastic deformation and the acting deforming forces." It was just such studies that experimental scientists were actually attempting at the time. Einstein concluded the paper by comment-ing: "I believe that the quantitative researches on the absolute magnitude of the density and the elastic force of the ether can only begin if qualitative results exist that are connected with established ideas."

Inasmuch as these ideas were being expressed by a young man who had imag-ined himself to be traveling alongside a light beam with a velocity equal to that of light, it is not fanciful to discern in the writing of this sixteen-year-old student the germ of the special theory of relativity, which was to come to fruition ten years later. Moreover, in contrast to most students (including brilliant ones like Freud), whose education begins in earnest at the university level, it is possible to think of young Einstein as already formed scientifically in the most important respects: He had arrived at issues in which he had an enduring interest, a scientific credo ab-sorbed from discussions and from popular readings, and a model of the pleasures of scientific work, gleaned from his family's business and from the congenial atmosphere at Aarau. Already Einstein combined the curiosity and sensibility of the young child with the methods and the program of the mature adult.

MASTERING THE DOMAIN

Einstein's somewhat unusual educational history continued during his years at the Zurich Polytechnic Institute. As a student, he was expected to master the physics and mathematics of the day by reading texts, attending lectures, going to the laboratory, and completing the standard problem sets and examinations. Einstein actually attended a wide variety of classes, including surveys of geography, the financial markets, Swiss politics, anthropology, geology, and the works of Goethe. However, he was not happy with the mainstream science courses.

In particular, Einstein was frustrated by the formal physics classes of Heinrich Weber, which were taught largely for engineering students. While Weber covered classical physics up through the work of his own teacher Hermann von Helmholtz, he ignored the seminal work of James Clark Maxwell and the issues of electromagnetism, which had already begun to fascinate Einstein. Accordingly, Einstein began to cut classes (fortunately, his classmate Marcel Grossmann loaned him excellent notes). The inveterate loner educated himself by reading Maxwell's work along with its refinements and reformulations by Heinrich Hertz and Hendrik A. Lorentz, as well as writings by other theoretical physicists like Ludwig Boltzmann and Gustav Kirchhoff.

In reconstructing the education of pathbreaking figures, one naturally focuses on those predecessors who were closest in spirit and achievement to the work of the discoverer; and so figures like Maxwell, Lorentz, and the great French mathematician Jules-Henri Poincaré loom large in the "prehistory" of Einstein's achievement. Yet, just as second-rate novelists sometimes render the most accurate picture of the era in which they live, more representative, if less illustrious figures may play a pivotal role in defining the problems and puzzles that capture the attention of young creative individuals.

The historian of science Gerald Holton has called attention to the writings of a hitherto obscure teacher of physics, August Föppl, with whose writings Einstein came into contact in the late 1890s. From all indications, Einstein thoroughly studied Föppl's *Introduction to Maxwell's Theory of Electricity*. From this survey, written deliberately so that students without formal training could follow it, Einstein may well have identified some of his most enduring concerns. Föppl helped Einstein to see that mechanics is a part of physics, and that an exploration of these topics extends to philosophical and epistemological questions that cannot be ignored. In a chapter on "The Electrodynamics of Moving Conductors," Föppl declared:

There can be no recourse to an absolute motion in space since there is absent any means to find such a motion if there is no reference object at hand from which the motion can be observed and measured. . . . The notion of completely empty space would be not at all subject to possible experience; or, in other words, we would first have to make a deep-going revision of that conception of space which has been impressed upon human thinking in its previous period of development. The decision on this question forms perhaps the most important problem of science of our time.

Einstein's classic paper of 1905, in which he laid out his principle of relativity, echoes much of the conceptual framework and even one of the thought experiments introduced by Föppl in this section of his popular treatise.

Thanks to his own thought experiments, his formal education, and his study of authors like Föppl, Einstein already had identified the set of issues that would occupy him for years to come: the relation between electricity and magnetism, the putative role of the ether, and conceptions of space and time, as formulated by a philosopher like Kant or a scientific thinker like Maxwell. Einstein later recalled:

> What made the greatest impression upon the student, however, was less the technical construction of mechanics or the solution of complicated problems than the achievements of mechanics in areas which apparently had nothing to do with mechanics: the mechanical theory of light, which conceived of light as the wave-motion of a quasi-rigid elastic ether and above all the kinetic theory of gases.

During this period, Einstein thought about these issues chiefly on the plane of empirical science; for instance, he pondered what experiments one might carry out to elucidate the nature and effects of the ether. In 1897, at age eighteen, he wanted to build an apparatus that could accurately measure the earth's motion against the ether of space. In 1901, after he had graduated from the Zurich Polytechnic Institute, he wrote to his friend Grossmann that he had envisioned a new and simpler method for investigating the motion of matter related to light ether. While his ultimate assaults on this issue would assume a more theoretical vein, young Einstein—largely self-taught—was already set on his life course. The philosopher Morris Raphael Cohen later noted: "Like so many of the very young men who have revolutionized physics in our day, [Einstein] has not been embarrassed by too much learning about the past or by what the Germans call the literature of the subject."

THE SCIENTIFIC BACKGROUND:
FROM GALILEO TO LORENTZ

Every creative breakthrough occurs within some specific domain or discipline. In the case of a Picasso or Stravinsky, identification of the domain is straightforward, and it is relatively easy as well to identify the contours of the innovations. With an individual like Freud, the identification of the domain proves much more problematic. Freud can be seen as not only contributing to neurology, psychology, clinical psychiatry, dream study, self-analysis, or even broader realms, like the delineation of human nature, but also ultimately charting a new domain, with its accompanying field.

In the term made famous by the historian of science Thomas Kuhn, domains like psychology and other social sciences are "preparadigmatic"; studies are carried out by researchers in the absence of agreed-upon bodies of knowledge, methods of investigation, or signs of progress. In contrast stand the paradigmatic scientific disciplines, which contain a relatively agreed upon corpus of knowledge, a consensual set of problems, recognized procedures for approaching them, and clear standards for judging new work.

Despite its limits, the notion of paradigm can be helpful in characterizing the overall history of a domain like physics. Indeed, the paradigmatic case of a paradigmatic discipline has been physics. One might speak of paradigms in the tradition of Aristotle and the medievalists; paradigms linked to Galileo and Newton; and twentieth-century paradigms, centering around Einstein's work in relativity theory and the work of numerous researchers in quantum mechanics. In each case, the earlier paradigm has been widely accepted by scientists for a period of time, with research and experiments being comfortably carried out under its aegis. Problems that begin to arise are at first ignored; then they become increasingly noticed, obtrusive, and troublesome. At such times, as Einstein himself put it much later, "it was as if the ground had been pulled out from under one, with no firm foundation to be seen anywhere, upon which one could have built."

At such times of crisis, one or more scientists eventually put forth a scheme, a set of principles, a theory that holds the promise of reconciling discrepant findings and somehow incorporating them within a broader framework, which subsumes much, if not most, of the earlier synthesis. In the particular instance I am examining, Newtonian physics becomes a special case within a more comprehensive framework called Einsteinian or relativity physics.

Nineteenth-Century Strides in Physics

It would be foolhardy to attempt to provide a full-scale survey of the background that led to Einstein's work at the beginning of the century, and I am not qualified to present such a historical scientific account. Still, if one is to obtain any grasp of the nature of the Einsteinian breakthrough, it is important at least to sketch some of the notions and dilemmas that confronted physicists during the latter days of the nineteenth century.

The revolution in mechanics brought about by Galileo's and Newton's work centered on a faith that all phenomena could be explained on the model of simple machines like levers or wheels. These scientists refused to accept that objects had places in the universe because they simply "belonged" there; rather, they searched for laws that could govern the behavior of all bodies—from falling apples to celestial bodies—and of all motion—constant, uniform velocity as well as the constant acceleration downward that characterized gravity. Galileo claimed that the laws of motion applied to any system of matter, provided that system was merely moving uniformly with respect to its neighboring system. He then questioned what would happen if one dropped objects from the masts of moving ships and noted the different ways in which objects appeared to fall when viewed from deck and from dock. We see Galileo wrestling here with the earliest intimations of relativity, as he attempts to align descriptions made with respect to one system to descriptions made with respect to other systems.

Newton sought to apply laws of mechanical motion not only to celestial and earthly bodies but also to optical phenomena, electromagnetism, and heat. He found that, for certain purposes, he had to speak in absolute terms. Accordingly, he posited "absolute motion," "absolute time," and "absolute space." To be sure, Newton sometimes had his doubts, for example, about whether any objects could be considered to be absolutely at rest. But ultimately the existence of these absolutes had to be taken on faith: In Newton's view they had to exist in Nature, even if only God could appreciate them.

Following the popularization of Newton's powerful ideas, scientists attempted to apply Newton's ideas in every physical realm (and ultimately in the psychological realm as well). In line with a strict causality, they believed that if they could fully understand a mechanical system (that is, positions and velocities of all entities contained therein), they could predict the future of that system. As it happens, this possibility of prediction of future motions is sometimes called the *relativity* principle of mechanistic physics, because it deals only with relative, and not with absolute, motion. As Einstein's biographer Philipp Frank expressed it, Einstein's achievement was to discover that this relativistic principle holds even when Newtonian "absolute" mechanics are no longer valid, for example, when great speeds are attained.

Newton's findings did allow very accurate extrapolations about the movement of celestial bodies. But one area of application gave rise to much controversy. Counter to what Newton had believed, light needed to be thought of not as composed of corpuscles or particles that behaved according to the laws of motion, but rather as wavelike, with light vibrating just as sound vibrates in the air. Scientists found it necessary to posit an ether, a kind of medium through which the oscillations of the waves of light had to be transmitted. They then began to ask questions like this: Can one detect the motion of objects through the ether, for example, that of the earth as it revolves around the sun? Does the ether impede the progress of objects that move through it, and is there any dragging effect?

Another enduring concern of physicists in the post-Newtonian era was the relationship between electricity and magnetism. In the 1830s the brilliant self-taught English physicist Michael Faraday had discovered principles of electromagnetic induction; he examined magnets and current in relative motion and argued for the existence of electromagnetic lines of forces. Faraday developed the concept of a field as the medium transmitting energy; the field was a region of space in which certain physical conditions were created and through which forces were transmitted. Interest in the field led to a concern with the medium through which waves could travel, leading again to the questions about ether.

James Clark Maxwell sought to supply a mathematical basis to Faraday's discoveries, and linked the theories of electricity and magnetism with the wave theory of light. The thrust of their joint contribution was to overturn the Newtonian notion of instantaneous actions operating at a distance in favor of the field as a fundamental variable in its own right. Energy could be located in time, and its effective forces could be described as vectors of stress at any point. Maxwell explicitly rejected the notion of absolute time and space, declaring that

> all our knowledge, both of time and space, is essentially relative. Position we must evidently acknowledge to be relative, for we cannot describe the position of a body in any terms which do not express relation. . . . There are no landmarks in space; one portion of space is exactly like every other portion. . . . We are, as it were, on an unruffled sea.

As we have noted, Einstein was fascinated with Maxwell's discoveries, and he referred to their widespread implications as "a revelation." Maxwell's achievements were closer to Einstein's interest than anything else of which he had read, perhaps because they shed light on issues that had preoccupied Einstein since his playful childhood activities and daydreams. Soon he proceeded beyond the secondhand account provided by Föppl to read Maxwell, Hertz, Kirchhoff, and other authorities firsthand. Einstein was struck by the fact that an ether theory implied the existence of well-defined rest frames, at rest relative to others. Yet the search for a state of absolute rest was destined to be unsuccessful. Maxwell-Faraday electrodynamics showed the existence of electromagnetic phenomena detached from every ponderable matter—they are waves in empty space that consist of electromagnetic fields. If mechanics was to be maintained as the foundation of physics, Maxwell's equations had to be interpreted mechanically. Yet Einstein noted that "mechanics as the basis of physics was being abandoned, almost unnoticeably, because its adaptability to the facts presented itself finally as hopeless."

Heinrich Hertz had taken the Maxwell-Faraday perspective one step further. In 1888, having confirmed the existence of electromagnetic waves, he set out to explain these in terms of a physical theory. He soon realized that this would be difficult to accomplish in terms of mechanistic physics, yet he could do so simply by using Maxwell's equations governing electric and magnetic fields and charges. Here, following Maxwell, Hertz made a point of crucial importance: that one could arrive at equations not from experience but rather as hypothetical assumptions, whose plausibility would be determined by the number of natural laws they encompass.

The writings of the physicist-philosopher Ernst Mach shook young Einstein's wavering faith in mechanics as the final basis for all physical thinking. In Mach's view simplicity and economy of thought were the key to a physical theory; one should only employ propositions from which statements regarding observable phenomena can be deduced. Mach criticized Newtonian mechanics, claiming that it contained no principle that was self-evident to the human mind. Newton had merely organized observations, and so his principles and predictions would be correct only if the experiences described happened to be true. Particular targets of Mach were Newton's expressions of "absolute space" and "absolute time," which could not be defined in terms of observable quantities. As Mach put it: "All masses and all velocities, and consequently all forces, are relative. . . . Every single body of the universe stands in some definite relation with every other body in the universe." Rejecting the notion of absolute space, Mach rewrote a Newtonian account in terms of observable, definable phenomena: "Every body maintains its velocity both in magnitude and direction, relative to the fixed stars, so long as no forces act upon it"—substituting for the undefined "absolute space" the definable "fixed stars."

In an effort to solve vexing issues about the ether and the field, scientists began to think in terms of relevant empirical investigations. According to one line of reasoning, to any observer who is moving with respect to the ether, the velocity of light will be greater or less, depending on whether the direction of the propagation of light through the ether, and the motion of the observer, proceed in opposite directions or in the same direction. And so, for instance, if the earth moves through the ether without dragging that along in its revolutions around the sun, its velocity relative to the ether should be observable by measuring the velocity of light relative to the earth in different directions. This chain of reasoning gave rise to a number of crucial experiments, among which by far the best known is the Michelson-Morley experiment of 1887.

In 1887 Albert Michelson and Edward Morley conceptualized the ether as a current streaming past the earth at approximately twenty miles a second. They asked whether the velocity of light varied in different directions owing to this posited motion of earth through the ether. On earth the velocity of light in the direction of the motion of the earth should vary slightly from the velocity in the opposite or in a perpendicular direction.

The Michelson-Morley experiments were undertaken in a variety of ways. The basic idea was to split a beam of light into two perpendicular "pencils of light," and these pencils, or half beams, traveled back and forth diagonally an equal distance by multiple reflections across the presumed ether, before being recombined into a small telescope eyepiece. If the ether flow had an effect like normal mechanical effects, the two returning rays would be out of phase. However, the experiments revealed conclusively that light had the same velocity, whether moving in the direction of the earth's rotation or perpendicular to it; it does not matter whether the source or the observer moves. This result was difficult at best to square with the existence of the ether; all of the efforts to reconcile Michelson-Morley (and similar experiments) with the existence of an ether had a distinctly ad hoc quality to them.

The various theoretical and conceptual modifications offered by individuals like Mach and Maxwell, along with the empirical demonstrations by investigators like Michelson and Morley, combined to strain the Newtonian synthesis to its limits. Even as the notion of the unconscious can be said to have been "in the air" by the latter part of the nineteenth century, so, too, the principal strands of relativity theory should have been detectable to able young students. And, indeed, some raised questions about the limits of our knowledge of nature and argued that understanding probably could not occur entirely in a mechanical vein; others felt that perhaps a more acute logical analysis would lay the groundwork for a new science.

Two men stand out in the effort to push beyond mere intuitions of problems and possibilities toward the full-blown embracing of a relativistic perspective. Hendrik A. Lorentz, the preeminent Dutch physicist of the day, showed that Maxwell's equations remain invariant if they are conceptualized mathematically by what have come to be called Lorentz equations: that is, the same equations applied when making a transition from a vehicle at rest in the ether to one moving uniformly relative to it. The transformations allow one to find the space and time coordinates of events in one system if they are known in the other, and if the relative speed of these two systems is known. However, in this system, two simultaneous events in one system are not simultaneous in another. Thus, the transformation connects in a new way the space and time coordinates of an event in one system with the time and space coordinates of the same event in another system.

One problem arose: Variables of the Lorentz transformation could not coincide with the actual coordinates of space-time in the new frame of reference. So Lorentz had to treat space and time as kinds of fictional variables, for example, defining a "local time" with the aid of a specific variable. Increasingly uncomfortable with the notion of the ether as an explanatory variable, Lorentz came to regard it as absolutely fixed and not affected by matter; but, committed epistemologically to the existence of the ether, he found himself unable to challenge it in a definitive way.

The other great anticipator of relativity theory was the French mathematician and savant Jules-Henri Poincaré. Poincaré had coined the phrase "principle of relativity" to stand for science's failure to determine the earth's absolute motion. In a remarkable paper of 1898 he asserted that "we have no direct intuition about the equality of two time intervals. . . . The simultaneity of two events or the order of their succession, as well as the equality of two time intervals, must be defined in such a way that the statements of the natural laws be as simple as possible."

Two years later, he asked challengingly whether the ether really exists, and he ridiculed efforts, in the wake of Michelson and Morley's negative results, to rescue the notion of the ether by ad hoc methods. Exploring this line of thought, Poincaré considered two observers in uniform relative motion who wish to synchronize their clocks by means of light signals. He pointed out that they can only mark "local time," and "as demanded by the relativity principle [the observer] cannot know whether he is at rest or in absolute motion." Poincaré speculated that "perhaps we must construct a new mechanics, of which we can only catch a glimpse . . . in which the velocity of light would become an unpassable limit . . . but we are not there yet."

Clearly, the conceptualizations in the two centuries since Newton and, in particular, in the two final decades of the nineteenth century had brought about clear advances in understanding, even as they had raised issues of staggering complexity. To some, in 1905, the electromagnetic theory was proceeding in the correct direction, and Lorentz's theory seemed to serve as a plausible basis for a unified-field theoretical view of nature. But the kinds of questions raised by Poincaré, and the doubts and ad hoc assertions embedded in Lorentz's work suggested, at least to one young thinker, that a far broader change in the way of thinking was needed. It was most likely to come from someone with the ability to raise fundamental questions, someone steeped in the findings of recent physics, but not yet too entrenched in its current points of view—a mind at once young and mature.

EINSTEIN'S "OBJECT-CENTERED" MIND

In the previous chapter I described the mind, personality, and mode of operation of Sigmund Freud, roughly a contemporary and, ultimately, an acquaintance of Einstein's. Whereas, from an early age, Freud's interests were directed toward other human beings, Einstein's interests centered around the world of objects and the physical forces around them. As a youngster, he would typically prefer playing with toys or reading a book to mixing with other children. To adopt a commonplace distinction, Freud was far more person centered, Einstein correlatively object centered; perhaps more precisely, Freud was fascinated by the relations among individuals, while Einstein saved his ardor for the relations among objects. Einstein once declared that he had sold himself body and soul to success, being in flight from the "I" and the "we" of the "it." Yet, perhaps paradoxically, Einstein had good friends whom he retained for many years and may well have been more genuinely likeable than was Freud in his later years. Also, in apparent distinction to the sexually inhibited Freud, Einstein displayed an overt interest in young women, who often showed interest in him.

While at the Zurich Polytechnic, Einstein formed an enduring friendship with the young mathematician Marcel Grossmann, a tie that evolved as well into professional collaboration some years later. An even stronger tie was formed with a classmate, Mileva Marie, whom he eventually married and from whom he was divorced some fifteen years later. Once, at the patent office, he befriended Michelangelo Besso,* a young engineering colleague with whom he had constant discussions about his developing ideas and with whom he remained in regular contact until their almost coincident deaths fifty years later. Perhaps none of these friendships had quite the passion and intensity of some of Freud's relations with his male colleagues, but they proved more enduring and largely free of the tension and paranoia that eventually colored nearly all of Freud's highly charged professional relations.

Worthy of special note were the friends who composed a small group nick-named the Olympiad. Once he had settled down in Bern, after finishing his for-mal studies, Einstein began to meet regularly with a talented polymath, Maurice Solovine, and a younger friend from Zurich, Conrad Habicht. The Olympiad members decided to pursue a systematic program of reading that included works in philosophy (Mill, Hume, Spinoza), mathematics (Riemann, Poincaré), and other scientific texts (Karl Pearson, Ernst Mach). Of all the writings, said Solovine, Poincaré's *La science et l'hypothèse*, in which the notion of absolute time was directly challenged, "profoundly impressed us and kept us breathless for weeks on end." The Olympians would hike, camp, swim, and then converse feverishly as they made their way home. Discussion of personal aspirations and fears, as well as the expected joking around, were also features of the exchanges. A few years older than the other men, Einstein was the natural leader of the group. Though Einstein married in 1903, the group continued to meet until 1905, when both Solovine and Habicht moved away from Bern.

We might hypothesize that Freud needed the companionship, personal ties, and affective affirmation that comes from a close associate like Fliess, while Ein-stein appreciated the opportunity to try out his ideas on others—such as mem-bers of the Olympiad, Besso, and his physicist-wife, Mileva—and to benefit from their feedback. One might even go so far as to say that, without the kind of stimulation and critique offered by a trusted friend or lover, these men might never have completed their innovative work. (Einstein thanked Besso explicitly for a conversation that led to the special theory of relativity; and scholars have re-cently speculated that Mileva may also have aided in the development of his most original ideas.)

Yet the need for support is not equivalent to a dependence on others for the core of one's ideas. In neither case does one get the feeling that the final work dif-fered in crucial respects because of the collegial feedback. Both Freud and Einstein had a strong sense of where they were headed, and it would probably have been difficult for another individual to get them to alter their course in a major way.

---

*Often referred to as Michele.

Einstein did not seek loneliness, but unlike Freud, he did not find it a threat. He was quite happy to be on his own from earliest life and did not crave companionship. This lack of craving for another person may well explain why neither of his marriages was a success and why his relations to his two sons were also unsatisfactory. In working out problems, Einstein once recalled, "I lived in solitude in the country and noticed how the monotony of quiet life stimulates the creative mind." He went on to comment, with some nostalgia: "There are certain callings in our modern organization which entail such an isolated life without making a great claim on bodily and intellectual effort. I think of such occupations as the service [*sic*] in lighthouses and lightships."

Einstein was remarkable for his powers of concentration; he could work uninterruptedly for hours and even days on the same problem. Some of the topics that interested him remained on his mind for decades. For relaxation he turned to music and to sailing, but often his work would continue during these moments as well; he usually had a notebook in his pocket so that he could jot down any idea that came to him. Once, after the theory of relativity had been put forth, he confessed to his colleague Wolfgang Pauli, "For the rest of my life I want to reflect on what light is." It is perhaps not entirely an accident that a focus on light is also the first visual act of the newborn child.

Einstein felt that he did not have great mathematical gifts and deliberately chose not to take courses and to continue in that area.

> The fact that I neglected mathematics to a certain extent had its causes not merely in my stronger interest in science than in mathematics but also in the following strange experience. . . . I saw that mathematics was split up into numerous specialities, each of which could easily absorb the short lifetime granted to us. . . . In physics, however, I soon learned to scent out that which was able to lead to fundamentals and to turn aside from everything else, from the multitude of things that clutter up the mind and divert it from the essential.

This capacity to pick out important issues dovetailed with Einstein's search for the most general possible conception. "In a man of my type," he declared, "the turning point of the development lies in the fact that gradually the major interest disengages itself to a far reaching degree from the momentary and the merely personal and turns toward the striving for a mental grasp of things."

Like any other individual, Einstein had to pass through phases of development until he was a mature thinker. Not even prodigies are born whole! His first few papers were not extraordinary, though they were readily accepted for publication in the major physics journal of the day. Yet from very early on Einstein evinced a definite scientific style. Dating back to high school, his notebooks were terse; he stated the results of problems, or the points of articles, precisely and succinctly. He was attracted especially to theoretical physics because he craved the opportunity to reduce the incredibly complicated occurrences of nature to the most general physical principles.

By his own testimony, then, Einstein was a person interested in the phenomena of the physical world that he could express in mathematical terms. He was a perfectly adequate writer, but language as such held little interest for him; he often lamented his meager talents in mastering foreign languages and in retaining verbal materials accurately. In contrast, his gifts of logical-mathematical and spatial intelligence were notable. He could readily assimilate the "mental pictures" invented as models by other scientists. The contrast with Freud's strong cognitive suits is striking indeed.

Einstein seems to have had a special gift not always available to scientists—the gift of envisioning problems and situations of relevance, and of carrying out those vivid and revealing mental puzzlements, called *Gedanken* (or thought) experiments. In the opening of this chapter I described some of the phenomena that entranced the youthful Einstein and remained on his mind for many years. Einstein was able to probe these examples much further and to think readily *and generatively* of numerous variations of his imagined spaceship or train or free-falling box. The ability to keep these invented spatial configurations in mind and to operate on them in diversely instructive ways played an indispensable role in Einstein's original scientific thinking. These were his symbol systems of choice. In reflecting on these activities of "puzzling out," he discerned the central features of his thinking:

> The words of the language, as they are written or spoken, do not seem to play any role in my mechanism of thought. The psychical entities which seem to serve as elements in thought are certain signs and more or less clear images which can be "voluntarily" reproduced and combined. . . . From a psychological viewpoint this combinatory play seems to be the essential feature in productive thought. . . . The . . . elements are, in my case, of visual and some of muscular type. Conventional words or other signs have to be sought for laboriously only in a secondary stage, when the mentioned associative play is sufficiently established and can be reproduced at will.

In addition to noting these components based on different sensory systems, Einstein also emphasized the important role played by imagining and fantasizing: "When I examine myself and my methods of thought I come to the conclusion that the gift of fantasy has meant more to me than my talent for absorbing positive knowledge." Einstein obviously enjoyed creating and exploring worlds within his own mind; but unlike the worlds contrived by pure mathematicians, these worlds always bore a resemblance to, and were governed by, the principles of physical reality.

While spatial visualization and imagination were important to him, Einstein invented concepts like "the relativity of simultaneity" that certainly did not lend themselves to a simple mental picture. Indeed, his innovation sprung out of his capacity to integrate spatial imagery, mathematical formalisms, empirical phenomena, *and* basic philosophical issues. As Philipp Frank described it, Einstein reveled in a plurality of modes of representation:

> When Einstein had thought through a problem, he always found it necessary to formulate this subject in as many different ways as possible and to present it so that it would be comprehensible to people accustomed to different modes of thought and with different educational preparations.

The historian of science Arthur I. Miller indicates how Einstein was able to combine different modes:

> It cannot be overstated that the emphasis on visual thinking among German-speaking scientists and engineers circa 1900 was widespread. Yet in 1905 it was Einstein who combined visual thinking with Gedanken experiments and quasiaesthetic notions with dazzling results.

As for other aspects of his working style, Einstein could be assertive and tough skinned. Referring to himself, he quipped to his friend Grossmann, "When God created the ass, he gave him a thick skin." While not gratuitously nasty, he did not hesitate to make his criticisms known—so much so that Besso claims to have rescued him from insulting Max Planck, Germany's leading physicist, in an early paper and thereby to have preserved the possibility that the two scientific titans might one day become friends. Similarly, Einstein's treatment of Boltzmann's theory of gases in the draft of his thesis was so critical that he eventually withdrew it. Einstein found himself congenitally so much in opposition to authority that he may have taken extra steps to challenge senior figures, particularly during his younger days. He took a certain pride in not knowing the literature of the field in any great detail. And he could not hide his disdain for scientists who shrunk from the most ambitious challenges: "I have little patience with scientists who take a board of wood, look for its thinnest part, and drill a great number of holes where drilling is easy."

Einstein conceded that much had to be given up to become the kind of scientist he wished to be. Only monomaniacs, he confessed to his friend Besso, made scientific discoveries. And such an individual also had to be daring—a "rank opportunist," to use his own words; "reckless," to use the word that the physicist Robert Millikan applied to one of Einstein's breakthrough 1905 papers.

Certainly, in some aspects of scientific personality, Einstein resembled Freud. Both men were ambitious, dogged, daring, and willing to stand alone and even welcome combat. Yet they had chosen—or had found themselves—in domains as different from one another as two scientific endeavors could be. Working in physics, Einstein was involved with problems that had been examined for decades and whose delineation was relatively clear, though the nature of their solution remained obscure. Working on the very issues that obsessed him were the finest minds of his time, including Lorentz and Poincaré. If neither of these older savants had "cracked" the issue of relativity, it is virtually certain that someone of Einstein's generation—if not precisely his blend of genius—would have soon done so. Indeed, Einstein, not known for giving undue credit to others, himself expressed the view that his French colleague Paul Langevin would have put forth the special theory, if he himself had not. As is characteristic of a domain in which problems can be well delineated, and solutions readily recognized, Einstein had done much of his most important work by the age of thirty and was world famous before he was forty.

Here the contrast with Freud is stark. At age twenty-six, Freud had just finished medical school; at age thirty he had just completed his study sojourn with Charcot; at age forty, he was still an obscure practitioner in Vienna, decidedly uncertain whether anyone would ever know his name, let alone esteem his work as pathbreaking. It was clear neither to him nor to others in which domain he was working, nor who the proper judges of his work should be. And, as I have suggested before, Freud perhaps had to create both a domain and a field before the merits of his work could be properly appreciated. Einstein certainly changed a domain, but the field remained as it had before, though it was the younger members of the field who were most likely to recognize the extent of their contemporary's remarkable achievement and eventually to lament his failure to transcend that achievement.

THE SPECIAL YEAR AND THE SPECIAL THEORY

In 1665 and 1666, while in his early twenties, Isaac Newton left the University at Cambridge for a sojourn in the quiet village of Woolsthorpe. There, working essentially alone, he developed the calculus, arrived at major insights about light and color, and started on the path to the discovery of the law of gravity. On the basis of that work he constructed the first modern synthesis about the physical world, a mechanically based view in which the smallest particles as well as the largest celestial bodies all move in accordance with the same mathematical principles. Newton was to recall later: "In those days I was in the prime of my age for invention and minded Mathematics and Philosophy more than at any other time." It is not a coincidence that Einstein revered Newton and kept his picture on the wall above his bed.

The only period in the history of physics at all comparable is the year 1905; the figure of Newtonian proportions is Albert Einstein. In less than a year

Einstein wrote and published four major papers, each of which constituted a significant contribution to our knowledge of the physical world, one of which was to win him the Nobel Prize seventeen years later, and another of which introduced the basic notions of his most important discovery—the theory of relativity. In what follows I give some of the background to the composition of that paper, introduce its major revolutionary ideas, and then examine the process whereby the theory—and its creator—became known and accepted throughout the world.

Einstein's Entry into the Work World

Upon his graduation from the Zurich Polytechnic, Einstein had hoped to secure a teaching job at that institute or elsewhere. However, neither his own efforts nor those of his father, who intervened poignantly on his behalf, sufficed to get Einstein an academic position. It is indeed incredible that a young man of such talent was academically unemployable just a few short years before he made his epochal discoveries. (One wonders whether the same blind spots exist in the field today.) As is well known, Einstein did land a job as a patent officer in Bern, Switzerland; when not evaluating inventions, he worked on physics.

Though not Einstein's first vocational choice, the patent office work proved reasonably satisfying. He met regularly with his comrades of the Olympiad, got married, and began to raise a family. He often recalled this period, when he could concentrate totally on his work, as the happiest in his life. Indeed, perhaps making a virtue of necessity, Einstein later maintained: "A practical profession is a salvation for a man of my type; an academic career compels a young man to scientific production and only strong characters can resist the temptation of superficial analysis."

Still, without the benefit of hindsight, an observer of the young Einstein might well have considered him somewhat of a failure: After all, he had not graduated successfully from the gymnasium; he had not gained entry to the polytechnic on his first try; he lacked influential mentors or sponsors; he had failed to secure an academic job; he had not completed his dissertation. More likely than not, it seemed, he would remain in obscurity in the patent office.

The Early Revolutionary Papers

Einstein's papers in the first years of the century were not all gems; in fact, he declared some of them worthless. However, even the less meritorious of these papers revealed a telling concern with the most central issues of physics. In the early 1900s Einstein worked on statistical mechanics, the law of thermal equilibrium, and the second law of thermodynamics in the tradition of Boltzmann. More concretely, he sought to prove that atoms have a definite size and to discern the nature of the links between the molecules of a liquid. He wrote to Grossmann in 1901: "It is a wonderful feeling to recognize the unifying features of a complex of phenomena which present themselves as quite unconnected to the direct experience of the senses."

What is true of physics at its best may pertain as well to papers about physics. It is only on the surface, suggests Gerald Holton, that Einstein's papers of 1905 appear disparate. Three epochal papers, written but eight weeks apart, seem to occupy entirely different fields of physics: an interpretation of light as composed of quanta of energy; an explanation of Brownian motion that supports the notion of the atomic nature of matter; and the introduction of the "principle of relativity," which reconfigured our understanding of physical space and time. However, Holton indicates that all three papers arise from the same general problem—fluctuations in the pressure of radiation. Holton also notes a striking parallel in the style of the papers. Einstein begins each with a statement of a formal asymmetry or some other incongruity of a predominantly aesthetic nature; he then proposes an elegant principle that at once dissolves the observed asymmetry, eliminates redundancy, and leads to one or more empirical predictions.

It would be ludicrous to suggest that these papers, written in so brief a compass, reflected simply the thought of the preceding days. In a letter to his friend Carl Seelig, Einstein commented: "Between the conception of the idea of this special relativity theory and the completion of the corresponding publication, there elapsed five or six weeks. But it would be hardly correct to consider this as a birth date, because earlier the arguments and building blocks were being prepared over a period of years, although without bringing about the fundamental decision." Einstein does not here explain the reference to a "fundamental decision." One can infer from other writings that the decision was stimulated by a conversation with Besso and hit upon when he awoke the next morning. Einstein decided to abandon the notion of the absolute meaning of the simultaneity of the two events; an absolute definition proves impossible because "there is an inseparable connection between time and the signal velocity." We know that the special theory of relativity emanates from a *Gedanken* experiment Einstein had been puzzling over for a decade, that much of his reading as a student had been dedicated toward unraveling the puzzles of space and time, and that many of his letters document his continuing quest to explain the electrodynamics of moving bodies. With little question, similar spade work had been done for the other important papers of this year, though, as with Newton, one should not underestimate the rapidity of thinking that characterizes a mathematical-scientific brain at its prime.

Einstein's special relativity paper is written with disarming simplicity and directness. There is little scholarly paraphernalia—no citation of the literature, only one acknowledgement (to his friend Besso), no debate with other experts who had been toying with ideas of relativity. No drafts of the paper have been found. The physicist Hermann Bondi declared that to author such a paper "is to leave as disembodied, as impersonalized a piece of writing as anybody might be willing to read, knowing that others have to read it if they wish to know what has been achieved. The paper is very likely to tell the reader almost nothing about how the result was found." One has the feeling that Einstein reached his conclusions after much thought and simply wrote them down as they appeared most logical and straightforward to him. He subsequently indicated that, even after many years of discussion, he would change little in its argument or presentation.

The Essence of the Special Theory of Relativity

The mathematics of the paper require some technical background, but the essential assertions can be set forth in a reasonably direct fashion. Many scholars, including Einstein himself, have provided primers to the special theory of relativity, and I have relied on them for the examples and analogies used here.

At its very simplest, one can read the 1905 paper "On the Electrodynamics of Moving Bodies" as Einstein's answer to the *Gedanken* experiment of the middle 1890s. As Einstein recalled, "From the beginning it appeared to me intuitively clear that, judged from the standpoint of an observer (attempting to pursue a beam of light with the velocity of light), everything would have to happen according to the same laws as for an observer who, relative to the earth, was at rest. For how, otherwise, should the first observer know that he is in a state of fast uniform motion?"

Einstein pondered a paradoxical incompatibility posed by two assumptions: (1) the speed of light is a constant; (2) laws are independent of the choice of the inertial system. Daring to confront these assumptions head-on, he came to the radical and unexpected conclusion that it would be impossible to travel as quickly as the speed of light because only light can go at that speed. Acceptance of the velocity of light as a limit (186,000 miles per second) challenges one of the tenets of physics, and indeed of common sense. Ordinarily, if one is traveling at a certain speed, and one is placed on an object that is itself traveling at a certain speed, then one's motion amounts to the sum of those two speeds. But if a light source moves with respect to an observer with near the speed of light, and then also emits a light pulse, the speed of the light pulse with respect to the observer is still the same. It is more accurate to say that no material body can move at the speed of light. The fallacy in the old principle of adding velocities lies in the assumption that the duration of an event is independent of the state of motion of the system of reference in which the duration is measured. Hence, Newtonian mechanics does not hold at these speeds.

Viewed somewhat more formally, the theory hinges on two assumptions: (1) laws do not depend on which of two coordinate systems in uniform relative motion they refer to; and (2) every light wave is found to move in such a coordinate system with the same speed, whether emitted from a stationary or a moving source. Thus, all phenomena are governed by relativity principles, and all systems in relative uniform motion are equally privileged. Lorentz's notions of general time and space and absolute motion are meaningless. But the principle of relativity can be made compatible with the postulate of the constancy of the velocity of light. The Lorentz equations thus can be reduced to the assertion that electric and magnetic forces do not exist independent of the state of motion of the system of coordinates; the equations should apply to a moving frame of reference as well as to the invariance of the velocity of light.

From this novel procedure and set of assumptions, startling implications follow about our perceptions of time. Two events simultaneous for observers in one system are not simultaneous for observers in a second system moving uniformly relative to the first. As Einstein observed: "We cannot attach any absolute significance to the concept of simultaneity but . . . two events which, viewed from a system of coordinates, are simultaneous, can no longer be looked upon as simultaneous events when envisaged from a system which is in motion relatively to that system." Put vividly, while elements A and B may still appear to be simultaneous in one reference frame, there exist others where A occurs before B, and still others where B occurs before A.

Some of the stranger implications have caught the popular imagination. For example, systems have their own clocks, and moving clocks will change their rhythms while in motion. A clock attached to a system that is in relative motion will be observed to run more slowly than one that is stationary with respect to us. Rods appear to contract in the direction of their motion when they are observed to move from rest into uniform motion—or, expressed more generally, until the observer has been specified, the length of the rod is not determinate. With time and distance now conceptualized as relative, the central notions of physics—velocity, acceleration, force, energy—must be rethought. For instance, the mass of a moving body increases with the body's velocity relative to its observer. Finally, and of epochal consequence, if an agglomeration of masses is formed or falls apart with the emission of radiation, the sum of the masses is altered: hence, the famous equation $E = mc^2$—a simplification of a result reported here by Einstein.

The Reorienting of Physicists' Views

Einstein saw his 1905 paper not as revolutionary but rather "as an amazingly simple summary and generalization of hypotheses which previously have been independent of one another." Contrary to what others, with some abetting from Einstein, were to maintain, the experimental demonstration by Michelson and Morley of the null effects of the so-called ether did not play a significant role in Einstein's discovery. However, Einstein's general reading, which no doubt drew on results of ether-drift experiments, had made it clear to him that nothing was to be gained by positing an ether. As he put it, "The introduction of a 'luminiferous ether' will prove to be superfluous inasmuch as the views here to be developed will not require an 'absolute stationary space' provided with special properties." If there had been an ether, then one would have been able to add the velocity of light to the velocity of the earth, or subtract the motion of the earth, if it had been moving in the opposite direction. And so, dismissing this possibility, Einstein was able to use the constancy of the velocity of light as a point of departure, rather than as a deduction from experimental results.

It is important to underscore what is going on here. The standard procedure in physical science is to make observations or to collect systematic data and to derive principles and theories therefrom. Einstein was essentially reversing this procedure. Operating at a higher level of abstraction, he was asserting basic laws of physics—for example, the constancy of the velocity of light independent of relative motion—and drawing empirical inferences and links to other laws from such basic principles.

To be sure, Einstein did build directly on the theoretical work of several predecessors. He expanded what Galileo had applied to mechanics to include light and other electromagnetic phenomena; he extended Newton's claims that one could predict a future motion from an initial position by taking into account the motion of the inertial system itself. And he removed the notion of simultaneous effects at a distance with his critique of the notion of absolute simultaneity.

The electrodynamics of Faraday and Maxwell, with its theory of fields of forces, seemed more comprehensive than Galileo's or Newton's mechanical theories. These field theories explained actions at a distance more convincingly than the "matter-and-motion" notions of the earlier era. But a new mechanics that would apply to particles moving with great speeds had to be built. Using the Lorentz transformations allowed one to determine the space and time coordinates of events in one system if they are known in the other and if the relative speed of these two systems is known. Mechanical and field theory could now be joined, using the new relativity conceptualization.

Yet, if there was continuity from earlier work in physics and mechanics, and if portions of the theory had been lurking in different quarters, it seems undeniable that Einstein accomplished an amazing feat. By sheer power of logical-mathematical reasoning and visual-spatial imagination, aided by a willingness to take a bold stance and reexamine first principles, the twenty-six-year-old patent official completely reoriented the way in which physicists—and, eventually, nonphysicists—thought about reality. What others might have been willing to accept only if demonstrated through experiments, Einstein instead took as his starting premise and simply reasoned therefrom. As he recalled: "Shortly after 1900 . . . I despaired of the possibility of discovering the true laws by means of constructive efforts based on known factors. The longer and the more despairingly I tried, the more I came to the conviction that only the discovery of a universal formal principle could lead us to assured results."

Contrasts in Scientific Revolutionaries

Here one confronts directly the mental powers and the state of mind needed to effect a revolution in science. On the one hand, one must be sufficiently steeped in the findings and principles of the domain, or the chances that one will reinvent what has already been discovered are too pronounced—indeed, such duplication of already known results happened to Einstein on more than one occasion early in his career. With respect to issues of relativity, however, he had been addressing the topics for a decade and knew them intimately. As a complement, one must be willing to advance beyond the facts and even the conspectus of the domain, and dare to lay out a wholly new approach to the issues. Such audacity becomes less likely with increasing age. Einstein retained the willingness to take a stab into the unknown, one of the main ties to the spirit of the young child that remained within him for many years.

It is worth contrasting Einstein's achievement with those of other scientific revolutionaries, like Darwin and Freud. While from the broadest perspective, all three were involved in profound innovation, the differences among them merit comment. Darwin introduced a model and a few basic assumptions designed to encompass data about the distribution and variation of species now and in the past. Freud called attention to a previously neglected realm of thought and behavior, and he suggested some of the processes by which that unconscious realm operated. Rather than replacing existing paradigms or joining portions of earlier syntheses, both Freud and Darwin cast off in new directions, and the success of their efforts would be evaluated primarily in terms of the fruitfulness of their models for explaining traditionally recognized, as well as subsequently discovered, phenomena. Moreover, because the domains in which they were working were virgin territory, and because they themselves had articulated many of the questions on which they and others would subsequently work, both men were able to contribute for the remainder of their relatively lengthy lives.

Einstein, in contrast, was involved quite directly in response to previous schemes: the long-standing work of Newton, the more recent efforts by Maxwell and Faraday, the contemporary concerns of Lorentz and Poincaré. Turning their variables into a fundamental assumption, he recast basic intuitions about space, time, and other concepts of the physical world. As historian Arthur I. Miller has shown, Einstein addressed particular claims of Boltzmann, Hertz, Poincaré, Mach, and other scientists by "resolving problems in a Gordian manner—by inventing a view of physics in which certain problems do not occur . . . by realizing the necessity for a demarcation between data and the mental constructs that are concepts or axioms, in order to pluck out of the air the version of space-time that was out there beyond our perceptions."

Einstein's breakthrough was classic in that it sought to unify the elements of a physical analysis, and it placed the older examples and principles within a broader framework. But it was revolutionary in that, ever afterward, we have thought differently about space and time, matter and energy. Space and time— no more absolute—have become forms of intuition that cannot be divorced from perspective or consciousness, anymore than can the colors of the world or the length of a shadow. As the philosopher Ernst Cassirer commented, in relativity, the conception of constancy and absoluteness of the elements is abandoned to give permanence and necessity to the laws instead. A single line now ran from philosophy through physics to human psychology. Paradoxically, however, because the domain was so much more heavily traveled and because other scientists could (if so inclined) assimilate the new paradigm, it was not evident that Einstein could continue to make fundamental contributions for the remainder of his life.

RELATIVITY: IMMEDIATE FATE

Einstein did not consider his paper on the electrodynamics of moving bodies revolutionary; he reserved that word for his "heuristic" paper on quantum theory. But according to his wife, he expected that the relativity paper would attract attention—and probably spark controversy—and was disappointed when there was no mention of it in the first few subsequent issues of *Annalen der Physik*. Finally, Max Planck, the editor of *Annalen*, wrote for clarification of some obscure points, but he clearly had understood the papers.

Discussion of the reaction to the relativity paper constitutes a whimsical application of the theory itself. From one perspective, the reaction was slow: only one early rejoinder in a major journal; an academic "feeler" some three years later; the first invitation to speak at a major conference four years later. It took six years before a serious textbook was devoted to the special theory of relativity; many individuals confused it with Lorentz's electrodynamics and called it the Lorentz-Einstein principle. Except in Germany, it was hardly discussed at all until 1912.

Yet, when one considers that the author was an unknown patent officer operating outside the academic centers of Europe, and that he wrote at a time when communication was far less extensive and rapid than today, the reception seems quite swift. In addition to Planck, several of the most important young physicists heard about the paper, read it promptly, and reported its profound effect on their thinking. As early as 1908 the renowned mathematician Hermann Minkowski declared in a popular lecture:

Gentlemen! The ideas on space and time which I wish to develop before you grew from the soil of experimental physics. Therein lies their strength. Their tendency is radical. From now on, space by itself and time by itself must sink into the shadows, while only a union of the two preserves independence.

These words resounded in scientific circles everywhere. Other signs of recognition emerged. Almost immediately Einstein began to get letters addressed (inappropriately) to Professor Einstein at the University of Bern; in 1909 he was called to a professorship at Zurich, where he had been unable to secure a job less than a decade before; he received an honorary degree from the University of Geneva in 1909 and was nominated for the Nobel Prize as early as 1912, though he did not receive the prize until a decade later.

We may say, then, that all things considered, Einstein received rapid and largely favorable press. Yet it has long fascinated Einstein scholars that the individuals who were themselves closest to the discovery of relativity, and who had exerted the greatest influence on Einstein's own thinking, themselves never embraced the theory in its totality. In my terms, the most influential members of the field resisted the new discovery. In a posthumously published document, Mach revealed that the theory had left him cold and that he had little sympathy with its dogmatic adherents. Planck, while enthusiastic in part, preferred to talk about the principle, rather than the theory, of relativity; and he saw the theory chiefly as a generalization of Lorentz's work.

More bewildering was the almost complete silence with which Poincaré greeted the theory in the seven years before his death. I have already noted that Poincaré anticipated many of the points and even some of the phraseology of relativity theory. It is known that he was familiar with Einstein's work, and certainly he should not have had any difficulty in understanding its major points. Many have speculated about Poincaré's lack of reaction. The most plausible hypothesis—that Poincaré resented being scooped and was jealous—cannot be discarded but is unlikely insofar as he seems otherwise to have been a generous person and to have commended Einstein's talents. What seems more likely is that Poincaré was by temperament a gradualist, and he may simply have been reluctant to accept a theory put forth in so revolutionary a manner, with its implications spelled out so starkly. Poincaré did not want to abandon Newton entirely, particularly for an unproved replacement, though he was himself happy to raise provocative questions. Poincaré may have seen the relativity principle as an empirical assertion, to be revised in light of experimental data, rather than as the foundation for a new conception of physics. As a mathematician, Poincaré was little impressed by the *Gedanken* experiments of the physicists; he was more interested in the rules and conventions governing the equations than in how they might actually represent (or alter one's conceptions of) reality.

Hendrik Lorentz's responses are equally puzzling. He and Einstein eventually became friends, and Einstein came to venerate Lorentz as he did few other individuals. Lorentz, too, had glided close to some of the ideas of relativity, but unlike Poincaré, had never been comfortable using its language and its concepts; basically, in contrast with Einstein and Poincaré, he did not enjoy philosophical discussions. While he had invoked relativity terminology in earlier papers, he had done so because experiment demanded it, not because he felt intuitively that this language was correct. And where Lorentz had reached conclusions with a relativity flavor through his use of equations and his analyses of data, Einstein instead proceeded from postulates that yielded the equations and the data. Much of Lorentz's effort can be seen as an attempt to save classical physics from within, even if this effort required a great deal of ad hoc thinking. Gerald Holton provides a vivid image:

> Lorentz' work can be seen somewhat as that of a valiant and extraordinary captain rescuing a patched ship that is being battered around the rocks of experimental fact, whereas Einstein's work, far from being a direct theoretical response to unexpected experimental results, is a creative act of disenchantment with the mode of transportation itself—an escape to a rather different vehicle altogether.

Like Planck, Lorentz ultimately came to speak favorably of relativity, in a way in which neither Poincaré nor Mach ever did. Yet he had much invested in the concept of the ether, and even more in the classical physics of the nineteenth century. It was simply too much for Lorentz, older and more set in his way of thinking, to embrace relativity fully.

Such examples support Thomas Kuhn's assertion that revolutionary new scientific ideas are rarely embraced by the older generation, the dominant field with its vested interests: Acceptance of the new paradigm awaits the maturing of a new, less entrenched generation of scientists. In a compelling illustration of this point, review of notes from a seminar held at the University of Göttingen in the summer of 1905 and attended by the leading physicists of the period reveals no anticipation of the themes of Einstein's epoch-making paper—even though it had already been submitted for publication.

In contrast to the muted response by major theorists, experimenters readily displayed an interest in Einstein's work. Perhaps the best known was Walter Kaufmann, who announced results that appeared to disprove Einstein's and Lorentz's theories. Quite revealingly, Einstein almost never responded to such empirically based attacks. Not only had he become less interested in a theory based simply on experimental findings, but he had a sublime confidence that his very generally applicable theories *had* to be correct. This view stood him in good stead for the first half of his scientific life.

Reactions to relativity theory differed across cultures, too. As the historian Stanley Goldberg points out, there was little discussion in Britain (home of the ether theory), France (dominated by Poincaré), or the United States (still a scientific backwater). The theory was quite vigorously discussed in Germany and German-speaking countries, not only because it had been written in German but also because the scientific establishment was much more democratic and fluid. Interestingly, once the theory became known, it was also widely debated in the United States.

In the course of the twenty years following its publication, relativity theory—in both its special and its general guises—came to be quite widely accepted within the physics community. There were exceptions, of course; but as the years passed, these had less to do with science and more to do with political considerations.

The persistence of a rough consensus about a theory like relativity shows that the domain of physics is different from others, such as psychology. Now, a century after Freud wrote his works and half a century after the heyday of behaviorism, anything but consensus exists about the merit of either of these psychological revolutions. Such an underdeveloped science features conflicting schools, rather than a normative paradigm that the field as a whole acknowledges and that all members situate themselves around. In contrast, when physicists as a group moved on to quantum-mechanical concerns, they did not desert individuals who would defend relativity; instead, attention turned to issues not adequately handled by either classical Newtonian or "classical" Einsteinian physics. Similarly, with the introduction of evolutionary theory, Darwin helped move biology toward the status of a developed science: In the 135 years since *The Origin of Species* (1859), we have passed through the Darwinian synthesis, the neo-Darwinian synthesis, and the current struggle between gradualists and those who believe in evolutionary discontinuities. In this respect, biology is closer to the example of physics than to that of psychology.

Einstein's Career Advances

After its first few uncertain years, Einstein's career steadily progressed. By the time of the first international physics colloquy, the Solvay Conference of 1911, Einstein was clearly considered a first-rank physicist—an estimable colleague of Planck, Poincaré, Lorentz, Ernest Rutherford, and Marie Curie. As early as 1912, he joined a number of leading scientists, including Sigmund Freud, in endorsing a statement that opposed metaphysics and declared that philosophy should evolve in a natural manner out of the sciences.

In short order Einstein's call to a professorship in Zurich was followed by a prestigious professorship in Prague, a return to Zurich, and then an unparalleled position in Berlin: Einstein was appointed to the directorship of the Kaiser Wilhelm Institute for Physics, with a high salary and the option to teach only as much as he chose; he could now devote himself fully to pure research, surrounded by the best physicists in the world, including the venerated Planck.

Einstein does not appear to have been a person consumed by ambition, in the sense of Freud; he was not caught up in petty feuds or vengeful rivalries; he did not systematically bother to collect (nor did he dramatically destroy!) his personal papers until 1920. But Einstein did value the appropriate working conditions, colleagues, and recognition; so he accepted this position in a land for which, since his unhappy school days, he had had little fondness. This decision would inject him into the limelight in ways he could have never anticipated.

BECOMING A WORLD FIGURE: TWO STRUGGLES

At a considerably younger age, but roughly contemporaneously with Freud, Einstein became a world figure. By the 1920s his face was known throughout the world. Of course, the actual content of his writings was understood by only a tiny fraction of those who knew his name, his face, or the tide of his theories; at one time, it was seriously maintained that only a dozen or so individuals in the world understood relativity theory! It was quipped that eight of them lived in Berlin, even as it was maintained by Einstein's British admirer, the astronomer Arthur Eddington, that he was not sure there were even three who fully understood.

Despite periodic complaints, Freud clearly valued his world fame and regarded it as a kind of revenge against the many who had ignored or derided him for so long. Possibly because fame and recognition had come to Einstein early in life, possibly because he did not crave public approval or the limelight, Einstein was not fully at ease with the glare of publicity. It meant that for the last half of his life—roughly from his early forties—Einstein had to struggle mightily to make time for his work midst the many competing demands on his time. As it happens, there was a struggle of equivalent ferocity within Einstein's own scientific work, as he sought to alter the direction in which his cherished domain of physics appeared to be proceeding.

The Phenomenon of Fame

Many have speculated about the reason for Einstein's enormous fame during the 1920s and thereafter. No single factor can explain it. Certainly, the fact that he was the most illustrious scientist in the world was a start, but how many can name the most illustrious scientist in the world today? The fact that he put forth radically new conceptions of space and time, and did so under the seductive label of "relativity," contributed as well: Never has the more influential "quantum theory" gathered a comparable popular following.

Einstein's notoriety seems to have been due at least as much to his qualities as a person. His physical appearance—arresting, casual, and perennially bohemian, with his informal dress, unruly hair, and bushy mustache—made him instantly recognizable, readily lovable, and faintly risible. There was something enduringly childlike in Einstein's appearance and manner and in his veiled disregard of "grown-up" standards. Even when quite old, he never lost the carefree manner of the child, who would not permit society's conventions or the elders' frowns to dictate his behavior. His willingness to make faces, mug for the camera, and stoop down to play with children or animals conflicted refreshingly with the public stereotype of the ivory tower professor. That irreverent image remains on the sweatshirts widely sported today, more than a half a century after his death.

But the single element that contributed most to Einstein's fame was the worldwide reaction to an important empirical investigation carried out in the aftermath of the First World War. As part of his general theory of relativity, published during the First World War, Einstein had predicted a gravitational red shift and the deflection of light of 1.7 seconds of arc for rays of starlight grazing the sun. British scientists decided to travel to the southern regions of Brazil and Africa to observe a solar eclipse. Their precise measurements confirmed Einstein's theory. Einstein thanked Lorentz, who had sent Einstein the confirming telegram; he then notified his own mother that the British expeditions had proved the light deflection near the sun.

At a joint meeting of the Royal Society and the Royal Astronomical Society, the epoch-making results were announced. The astronomer Frank Dyson reported: "The results of the expedition . . . leave little doubt that a deflection of light takes place in the neighborhood of the sun and it is of the amount demanded by Einstein's generalized [*sic*] theory of relativity as attributable to the sun's gravitational field." The president of the Royal Society announced that Einstein's work was "one of the greatest—perhaps the greatest—of achievements in the history of human thought." The philosopher Alfred North Whitehead, present at the august occasion, declared that "a great adventure in thought had at length come safe to shore."

Newspapers, most notably the prestigious *London Times*, declared that Einstein was a man for the ages. One hundred books on relativity appeared within the year. All over the world lectures on relativity were delivered. A prize of thousands of dollars was offered for a popular article that could explain the theory in three thousand words. As his biographer, Banesh Hoffmann, comments: "Einstein was world-famous. This essentially simple man, a cloistered seeker after cosmic beauty, was now a world symbol, a focus of widespread adoration—and of deep-rooted hatreds." Apparently, the fact that the theoretical musings of an exotic visionary could anticipate the precise measurements of an eclipse confirmed upon Einstein a unique mantle.

With his newfound world fame, Einstein had to deal with understandings and misunderstandings of his theory. Sometimes he was gentle and patient, as when he tried to explain the main ideas of the theory to reporters:

If you will not take the answer too seriously and consider it only as a kind of joke, then I can explain [my theory] as follows. It was formerly believed that if all material things disappeared out of the universe, time and space would be left. According to the relativity theory, however, time and space would disappear together with the things.

Sometimes he was firm, when, for instance, an art historian tried to connect relativity theory to cubism. And when scientists with a political agenda baited him with willful distortions of his major claims, he remained stoically silent.

A Role in World Political History

Science was Einstein's first love, but from the time that he had moved back to Germany at the end of 1913, he had found himself enmeshed in the international political situation. In less than a year Germany was at war. As a Swiss citizen, he did not have to take any part in the war effort, but Einstein went beyond neutrality, actually voicing his misgivings about the German role in the First World War. Indeed, at this time, Einstein, who had been opposed to militarism since his childhood, revealed his enduring commitments to internationalism, pacifism, and socialism.

Throughout the 1920s Einstein allied himself with pacifist efforts and also supported Zionism. He sustained a love-hate relationship with Germany, taking pride in the scientific and cultural greatness of its past, but looking with horror at the emerging signs of totalitarianism. Germany represented both the promise and the peril of humanity.

As the specter of fascism rose in Europe, it was apparent that Einstein, with his leftward leanings and his strong identification with Jewish causes, could not remain at his post in Berlin. With some relief he accepted a permanent position at the recently founded Institute for Advanced Study in Princeton, where he would once again be paid a generous salary and be allowed to work as he chose— a universal symbol of the unfettered pursuit of knowledge in our time.

As his one-time colleagues in Germany denounced "Jewish" physics and his writings were burned, Einstein more outspokenly criticized the Nazis. Once the Hitlerites started a war that threatened Western civilization itself, this lifelong pacifist appreciated that the Allies would have to wage total war. In one of the most well-publicized and pivotal political actions ever undertaken by a scientist, Einstein signed a letter to President Franklin Roosevelt in 1939. Drafted by the physicist Leo Szilard, this terse communication brought to the administration's attention the possibility that extremely powerful bombs might be constructed if one could successfully set off nuclear chain reactions in a large mass of uranium. Einstein noted that Germany, clearly aware of this option and then the leader in nuclear physics, had already stopped the sale of uranium. Within six years, the most powerful weapon known to this planet would be devised and detonated, based on Einstein's formulations about the relationship between mass and energy.

Already a part of scientific history, Einstein made a move to being part of world political history as well. Whatever fame Einstein had achieved as a scientist was now magnified by his catalytic role in creating the weapon that won the war and that was to structure the ensuing cold war. Now aging (he turned seventy in 1949), Einstein probably achieved more of a legendary status than any other person of his era. Even the greatest of scientists felt awe in his presence. Certainly, he was equal in stature to the most prominent political leaders, like Chaim Weizmann and David Ben-Gurion of Israel, and the most prominent thinkers, like Bertrand Russell and Sigmund Freud. Indeed, Freud and Einstein conducted a well-known, if inconclusive, exchange on the topic "Why War?" Referring to the nonmeeting of their minds, Freud quipped: "He understands as much about psychology as I do about physics, so we had a very pleasant talk."

THE GENERAL THEORY OF RELATIVITY

To complement the struggle between his status as a scientist and his role as a world political figure, Einstein experienced another equally profound struggle— over the nature and direction of physical science. The ten years following the publication of the special theory of relativity had been productive, though sometimes difficult, for Einstein. As early as 1907 he was already considering the relationship between the phenomena of gravity and acceleration. He worked with notable success but under enormous pressure, conditions that occasioned a nervous breakdown at one point. During this period Einstein developed his far more comprehensive and challenging theory of general relativity. This theory required more sophisticated mathematical tools than had the earlier work and constituted a more elaborate and technically dazzling achievement. Yet, as with the special theory, Einstein worked primarily by developing the implications of his intuitions. In the earlier instance intuitions had concerned systems in uniform rectilinear motion; now the intuitions extended to systems that had uniformly accelerated motions and more complex properties, including arbitrary motion.

As described by Einstein's associate Leopold Infeld, Einstein began by posing for himself a series of questions about a man enclosed in a falling elevator. A glass elevator in a skyscraper falls freely, with uniformly accelerated motion. In it, observers are making measurements and taking notes, in tandem with those observers outside the elevator. Note that the two systems now differ, since the falling elevator is accelerated in contrast to the world outside.

Within the elevator an operator drops a compact and a lipstick tube. Judged within the elevator, the lipstick and compact remain at rest, since all objects, including the elevator, fall with the same acceleration relative to the earth. If the operator pushes the compact, the compact then moves with uniform motion in the direction in which it was pushed until it collides with the wall of the elevator. Thus, within the elevator, the observer who does not detect any evidence of the gravitational field concludes that all bodies in the system remain at rest or move uniformly until disturbed by forces, by the walls of the system, or by a collision with the bottom of a shaft.

Einstein proposed that, judged from within, the falling elevator and its contents constitute an "almost inertial" system. The system is not fully inertial because sooner or later the lipstick will collide with the wall, and sooner or later the elevator will crash at the bottom of the shaft. But to the observer outside the elevator, all elements—elevator, operator, compact, and so on—are falling with the same acceleration caused by the earth's gravitational field. Thus, there are now two systems: the almost inertial system in the elevator, and the observer's system, which is subject to gravity. Accordingly, one needs to consider two systems moving with accelerated motion relative to each other. Any transition from one system to another necessarily involves the appearance of a gravitational field in one of them and its disappearance in the other. And so, if one wishes to include systems whose relative motion is not uniform, one must take into account the phenomenon of gravity, which is the link connecting systems in nonuniform motion. The gravitational field can be created or wiped out by choice of a proper reference system.

From such reasoning the general theory of relativity develops. The *Gedanken* experiments continue to become more complex, as one drills holes in the elevator and sends light through them. The trajectory of the light now has one appearance to the individuals within the elevator, and a different appearance to the observers stationed outside. Because gravity affects light, just as it affects material particles, the observer will see the gravitational field deforming the straight path of the light wave.

As with the special theory, exotic facts and predictions flow from the perspective of general relativity. Though the meaning is more technical than my phrasing suggests, space needs to be considered as curved, and Euclidean geometry does not apply in the gravitational field. Clocks are seen as proceeding more slowly if set up in the neighborhood of masses. The course of light rays must be bent. The orbital ellipse of a planet undergoes a slow rotation in the direction of motion. And so on . . . As the popular science writer Lincoln Barnett puts it: "The universe is not a rigid and immutable edifice where independent matter is housed in independent space and time; it is on the contrary an amorphous continuum, without any fixed architecture, plastic and variable, constantly subject to change and distortion."

Such deflection is important because, as noted before, it can be observed on a cosmic scale. During a solar eclipse one can measure stars in the neighborhood of the sun and observe light rays as they pass near the edge of the sun on their journey toward the earth. So that such measures could in fact be carried out, Einstein had to master various geometries and then modify them appropriately. These techniques allowed the astronomers observing the eclipse in 1919 to determine whether light rays bend in a gravitational field and whether this effect agrees quantitatively with the predictions of relativity theory.

The propounding of the theory of general relativity assured Einstein's place as one of the greatest scientists of all times. Not only had he imagined and solved problems of profound complexity, but he had made a daring prediction that was proved correct. Einstein knew this and felt secure in his achievement. Indeed, while the world toasted the confirmation of the theory in the eclipse studies of 1919, Einstein was calm, even blasé. As he commented to a student who saw him that day, he *knew* that the observations had to come out in the way that they did; and he was confident that other scientists would come to this recognition as well. Still, Einstein may not have been quite so casual as these remarks suggest; recently discovered correspondence documents that as early as 1916 he encouraged a student to provide experimental data in support of the general theory.

OUTSIDE THE MAINSTREAM

In producing the special and general theories of relativity at the time and in the manner in which he did, Einstein effectively represents a pattern that characterizes creative activity across domains. Following a decade of mastering his chosen domain, Einstein, still young, took a decisive step and thereby reoriented work in physics. Following another decade of work, in which many individuals pondered the implications of his first radical step, Einstein took a second decisive step. The general theory built on his original slicing of the Gordian knot: At the same time, it was far more comprehensive, relating the original insight to remaining areas of the discipline, particularly to gravity.

Yet, as early as 1913 it had become clear that physics was moving in a new direction—one not at direct odds with relativity theory but one that was to make Einstein increasingly uncomfortable. Work was proceeding with respect to quantum-mechanical accounts of matter. The line of investigation launched by Planck in 1900 and strongly advanced by Einstein in his Nobel Prize–cited paper of 1905 was rapidly becoming an entire worldview within physics.

Events came to a head in the mid-1920s with a series of breakthroughs by young physicists like Erwin Schrödinger, Louis de Broglie, Paul Dirac, and Werner Heisenberg. The slightly older, distinguished Danish physicist Niels Bohr supported their efforts with great articulateness. Bohr described the quantum-mechanical consensus that operational meanings alone should be trusted, that an acceptance of probabilities should replace the search for causal laws, and that all observers should recognize the effects of the act of observation on what is observed.

Einstein understood the quantum-mechanical claims, and though he admired many of the results obtained, he did not like the theory. He felt that the world could be explained in a more profound way that did not deny classical causality and challenge the *possibility* of a complete scientific account. In one of his most famous utterances, he insisted that "God does not play dice with the universe." For Einstein, science had to have order in its smallest detail. "I find the idea quite intolerable that an electron exposed to radiation should choose *of its own free will* not only its moment to jump off but its direction. In that case I would rather be a cobbler, or even an employee in a gaming house, than a physicist," he wrote to his colleague Max Born. Einstein and Bohr respected each other enormously and debated "complementarity" against "objective reality," publicly and privately, for over thirty years; but neither ever appreciably influenced the other's point of view. Still, in the cordiality with which they treated one another, they constitute a striking contrast to the relationships between Freud and his antagonists.

In the ensuing twenty years there occurred one of the most poignant events in recent scientific history. Never losing confidence in his beliefs, but increasingly alone, Einstein fought a rearguard action against quantum mechanics. Einstein's many friends continued to venerate him, and to listen respectfully to what he had to say publicly and in his writings; but they felt he was locked within his own world, out of touch with the new paradigm of physics. Einstein was aware of this estrangement and even joked about it. He struggled valiantly, but unsuccessfully, to construct a unified field theory that would synthesize quantum and relativity work. To most observers, this appeared a doomed undertaking. General relativity explained the large-scale working of gravity and cosmology, but it could not apply appropriately at the atomic and subatomic levels. Quantum mechanics was needed to explain the nature and composition of elementary particles from which the world of matter and radiation would be constructed.

Clearly, Einstein felt pain at the growing disconnection between his own work and that of many of his colleagues. Also, the disjunction between the adulation over his work in the relativity era and the pained silences that greeted his work in the quantum-mechanical era must have been at least a little difficult for him to accept. He was probably sustained by the conviction that he had been alone but correct in the past, and by his quasi-religious belief that some kind of grand design *must* exist, even if it proved temporarily—or even permanently—inaccessible to mortals.

INTUITIVE AND REFLECTIVE WISDOM

Can one do first-rate work in the natural sciences after the age of forty? Of course, very few scientists of any age conceive of powerful new theories, but the few that manage the feat seem to do so when they are very young. In a manner reminiscent of Einstein, each of the principal architects of quantum mechanics made his breakthrough while still in his twenties; and while many continued to lead active scientific lives for the ensuing half century, none was associated with subsequent discoveries of equivalent magnitude. In contrast, in nonscientific or in less structured scientific domains like psychology or pre-Darwinian biology, one can continue to make major contributions for several decades.

It could be that the sheer brain (or computational) power of scientists declines precipitously after their twenties, but the capacity of these individuals to be superb scientific critics tends to undermine that hypothesis. It could be that the competition for time becomes acute, and that these scientists, now recognized as world class, no longer have the years of splendid isolation it takes to produce truly revolutionary work.

But in my own view, it is a particular combination of youth *and* maturity that allows the most revolutionary work to take place in the sciences, and such an amalgam can only occur during a relatively small window of the life span. In the natural sciences, where a domain is well delineated, one can only hope to make a significant advance if earlier work in that domain has been quite thoroughly assimilated. Einstein could not have gone beyond Maxwell, Lorentz, and other physicists if he had not understood their contributions, including their strengths and limitations. At the same time, however, too much time and experience thinking in a certain way can prove uncongenial to any innovation; both Poincaré and Lorentz may have been too set in their mental habits to countenance an entirely novel approach to issues of space, time, and related matters.

The fresh consciousness that allowed Einstein to proceed beyond them in 1905, and then to advance yet further in the next decade, represents the gift of youth; once past, it cannot be recaptured. In a prescient remark, made when he was but thirty-seven, Einstein said of Mach: "It is not improbable that Mach would have discovered the theory of relativity, if, at the time when his mind was still young and susceptible, the problem of the constancy of the speed of light had been discussed among physicists." Einstein himself declared to a friend: "I have thought a hundred times as much about the quantum problems as I have about general relativity theory." He remained passionate on the subject, but this immersion did not in itself suffice to stimulate a breakthrough.

Perhaps the amalgam of youth and maturity is an identifiable feature of creative scientific genius. But it may well be only a necessary, and not a sufficient, feature. Einstein was fortunate, first, in that the questions he pondered during his youth turned out to be relevant to the physics of his day and, second, in that his gifts of spatial and visual imagination could advance his scientific work. Had the same person been born twenty years later, his own talents and worldview might well have proved ill-suited to the demands of a quantum-mechanical era in which spatial abilities proved less decisive than logical-mathematical powers.

Einstein's remarkable combination of youth and maturity struck many observers. On the one hand, Einstein retained throughout life his interest in intriguing phenomena of the world. Hoffmann comments:

No matter what he was doing, science was always present in his mind. When stirring tea, he noticed the tea leaves congregating at the center and not at the circumference of the bottom of the cup. He found the explanation and linked it to something unexpectedly remote: the meandering of rivers. When walking on sand he noted with wonder what most of us have known unthinkingly; that damp sand gives firm footing although dry sand and sand immersed in water do not. Here, too, he found a scientific explanation.

Yet in matters of art, he displayed the conservatism of age, never embracing any of the twentieth-century modernisms that were created in his shadow. He loved the music of the baroque and the classical periods, the art of the medieval and Renaissance periods. And as early as 1911, when scientists were beginning to talk about the need for countenancing probabilistic behavior at the atomic level, Einstein was already displaying a conservative streak.

There are two principal ways in which a brilliant young scientist can continue to do revolutionary work in middle and older age. One way is to work in a developing discipline, like a social science, where the domain is loosely constructed and there are accordingly numerous directions in which one can proceed. Interestingly, in an apparent effort to prolong their intellectual youth or to gain fresh leverage in middle age, a number of productive physical scientists have moved to the biological sciences or even into psychology or cognitive sciences. The other way is to make a single discovery so momentous that one can live off its intellectual capital for the remainder of one's active life. This is what Darwin did; and, in a sense, Freud's last years were spent exploring diverse implications of discoveries made at the relatively late age (for a practicing scientist) of forty. Of course, the greater the number of gifted young individuals currently at work in the same domain, the less likely it is that one can continue to harbor the domain for oneself. The difference between the lackluster caliber of Freud's first followers, as against the quality of Einstein's first readers, is telling in this regard.

Paradoxically, Einstein, in his later years, *became* Lorentz and Poincaré. He remained a brilliantly insightful thinker and critic and retained the respect of his colleagues. But like his redoubtable predecessors, he was not able fully to embrace the new ways of conceptualizing physical entities. So he embraced classical physics, though this time the classical variety was one of his devising, rather than the more ancient brand of Newton and Galileo. Perhaps this is why, in his last writings, he recognized Lorentz's and Poincaré's contributions to relativity in ways he had never been able to before.

But while Einstein's contributions to physical science were modest after 1920, he continued to grow in his understanding of the issues raised by his work and in his capacity to connect science to the rest of life. Einstein's enduring evolution, indeed, was as a human thinker in general, rather than as a scientist. I believe that Einstein continued to occupy a unique position in the life of his times because he became our emblem not merely of the brilliant scientist but also of the mature and reflective human being—a man wise about science and its place within life. The special genius of early life consists of brilliant intuitions shrewdly and quickly followed up. Another kind of understanding can continue to grow throughout an active life—the kind that merits the term *reflective wisdom*. Though we may tend to associate such wisdom with political and religious leaders like Lincoln or Gandhi, I believe it can be found occasionally in scientific figures like Einstein as well.

Indeed, in his last thirty years, Einstein turned his considerable energy and talents to issues that more properly concern the wiseman or philosopher: the practice and the allure of science; the nature of epistemology; the psychology of thought, including personal thought processes; the relationship between common sense and scientific thought; the role of religion; the existence of God and of the world; and the role of an aesthetic sensibility in the sciences. In each of these areas, Einstein put forth provocative ideas: for example, that science and epistemology were integrally linked, that scientific thinking was but an extension of commonsense thought, that the scientist and the artist both seek to escape from everyday life, and that it was his task as a scientist to discern the central elements of the scheme God had diabolically, but not inexplicably, put forth. He declared forthrightly: "I want to know how God created this world. I am not interested in this or that phenomenon, in the spectrum of this or that element. I want to know his thoughts, the rest are details." Perhaps none of Einstein's philosophically oriented statements were wholly original. But they were put forth with a confidence, a coherence, and an evocativeness that render them among the most powerful personal philosophies of our time.

In my view, Einstein's conceptions of the scientific, religious, and aesthetic aspects of experience were consistent: They emphasized both the decipherability of the world and its essential rationality, order, and harmony. Like Freud, this creator of the modern era himself remained a child of the Enlightenment. What began as powerful intuitions in his childhood ultimately became a respectable and comprehensive philosophy. Still, Einstein was not completely blind to the irrational and irregular in life. He commented in the preface to a biography by his son-in-law: "What has perhaps been overlooked is the irrational, the inconsistent, the droll, even the insane, which nature, inexhaustibly operative, implants in an individual, seemingly for her own amusement. But these things are singled out only in the crucible of one's mind." Einstein would not have looked to science to explain these wrinkles (perhaps this accounts in part for his skepticism about

Freud's psychology); determining the eternal principles governing the physical universe was a sufficiently formidable assignment.

Einstein's longtime secretary Helen Dukas once declared: "If Einstein had been born among the polar bears, he still would have been Einstein." I do not believe for a moment that Einstein's genius would have been equally realized across domains; theoretical physics as it had evolved at the beginning of the twentieth century was clearly the optimal area for a man of his gifts (and limitations) to tackle. Yet, it may well be that Einstein's dogged interest in the problems he envisioned, and his desire to discern a relation among the various spheres of life, might have emerged, bearlike, even if he had become a musician, a rabbi, or an engineer.

The Sharing of Personal Visions

In the last decades of his life, Einstein was constantly asked for his views about political and social events in the world. Perhaps surprisingly, he responded to such requests. He put forth his views about a wide range of issues: the military and peaceful uses of atomic energy, the fate of Jews in Israel and elsewhere around the world, the need to punish Germany and monitor it continuously, the relationship between Jews and Arabs, the desirability of eliminating all weapons, and the decline of civil liberties in America during the McCarthy era. Few would dispute that Einstein's views on these matters were less trenchant and original than his views on scientific and philosophical matters; and yet, they too contributed to the coherent worldview he was developing.

All these biographical and philosophical considerations add up to a complex, but not completely inexplicable, state of affairs. Einstein was a man of seeming contradictions: an individual in some ways young, in other ways mature beyond his years; a nonbeliever who spent much time thinking about God; a pacifist who stimulated the production of the most deadly weapon in history; a scientific radical who spent his last years seeking to refute the radical new scientific paradigm; a scientist whose own standards as a theoretician were quintessentially aesthetic; an individual obsessed by the physical world, who pondered timeless matters as well as the concept of time, yet also one who devoted many hours to addressing the mundane problems that beset the humans of his era.

Despite these apparently mixed messages, Einstein's scientific genius, his aesthetic and religious sense, and his involvement with the world's problems may add up to a coherent human being. Reflecting on Einstein's many strands, the historian of ideas Isaiah Berlin suggests that the kind of direct, intuitive leap needed to make sense of the physical world may be of a different order than the sensitivity to limits, nuance, and compromise that govern the world of human beings. Berlin suggests that "in the case of seminal discoveries—say of imaginary numbers, or non-Euclidean geometry, or the quantum theory—it is precisely dissociation of categories indispensable to normal human experience, that seems to be required, namely a gift of conceiving of what cannot in principle be imagined nor expressed in ordinary language." Like Newton and Copernicus, Einstein sustained a vision of a unified, harmonious, physically caused world. This dissociation led both to Einstein's genius in the world of physics and his inspirational, but ultimately less successful, forays into issues of world order.

Berlin notes, as well, the apparent disconnection between Einstein as a private person, with a distinctly limited capacity for close relations with other individuals, and Einstein as a world citizen surprisingly accessible to strangers and deeply involved with the problem of humankind. Here again, however, there may be an underlying link. Perhaps, in a manner reminiscent of Gandhi, Einstein renounced his family, distancing himself from his sons and even denying the paternity of his first-born daughter. His very lack of involvement with family and friends freed him for a broader connection with the entire world, whose physical nature he had done so much to decipher. Einstein himself seems to have shared an intuition about the Faustian bargain that he had forged:

> My passionate interest in social justice and social responsibility has always stood in curious contrast to a marked lack of desire for direct association with men and women. I am a horse for single harness, not cut out for tandem or team work. . . . Such isolation is sometimes bitter but I do not regret being cut off from the understanding and sympathy of other men. . . . I am compensated for it in being rendered independent of the customs, opinions and prejudices of others and am not tempted to rest my piece of mind upon such shifting foundations.

Einstein as an original scientific thinker was essentially finished by the time he was forty. But as one who reflected on science, philosophy, psychology, human nature, and world problems, he was able to continue productively for the rest of his days. Very few individuals have added to our sum of knowledge on these latter topics; Freud and Gandhi are among the aspirants in recent times. But especially on those topics where Einstein had some direct familiarity, he was able to provide needed and even treasured illumination. One would have to look hard for more reasoned statements about the nature of scientific work and its requisite thought processes. Also, Einstein's worldview and the principled stands that he took, while perhaps not for everyone, rightly inspire admiration. In domains less tightly structured than physics, the transition from childhood intuition to reflective wisdom occurred naturally and appropriately.

INTERLUDE ONE

For the first of three times in part II, I propose to step back and review the picture of creativity that is emerging. In each case, I examine a somewhat different form of creativity as linked, roughly, to science, the arts, and then the world of "live" performance.

In the case studies of Freud and Einstein, I have noted various similarities and differences in terms of biographical facts, personality, and views of their own work. Both men were preeminently scholars, thinkers, and academics—individuals who devoted their lives to the construction of knowledge. It is easier to apply the term scientist to Einstein than to Freud, particularly in light of the more speculative Freud of later life. (Einstein drew a much sharper line between his science and his philosophy.) Yet even those who might question the ultimate scientific status of Freud's contributions concede that he was trained as a workaday scientist and that he viewed his own work approach as scientific.

At the broadest level, scholars are to describe and, insofar as possible, explain aspects of the world—the physical world, the social world, the world of the mind. Humanistic scholars focus on specific concrete phenomena such as works of literature or art, lives of individuals, or historical events and use the methods that seem appropriate to these foci. Scientific scholars, in contrast, attempt to develop systems that explain kinds of entities and sets of events and, to the extent possible, allow prediction of what will happen to these objects or events under specifiable consequences.

Much of the work of scholars, including scientists, concerns the tackling of questions or problems that are furnished by the disciplines within which they work. Biologists taxonomize new species or investigate specific enzymes; historians research specific treaties or religious groups; anthropologists observe and create models of unfamiliar cultures. Part of the work undertaken by Freud and Einstein falls comfortably under the rubric of normal science: for example, Freud's taxonomy of kinds of hysteria and Einstein's attempts to measure the size of molecules.

But the more challenging aspect of the scientist's calling transcends the solution of problems already posed by others. What we see at work in Einstein's formulation of relativity theory and in Freud's explorations of the unconscious is better thought of as system building. Having considered the concepts and phenomena in a domain, and having found current views to be inadequate, scientists in a sense go back to the drawing board. Freud investigates a phenomenon ignored by his colleagues—the structure and processes of the unconscious—and develops a new model and vocabulary for elucidating it. Einstein demonstrates the inadequacy of earlier efforts to deal in a consistent manner with notions of simultaneity, time, space, and motion; cutting the Gordian knot, he introduces a set of bold propositions, argues for their logical cogency, and then considers the implications that follow from these propositions.

Although Einstein might have come up with Freud's discoveries, or Freud with Einstein's, it is not likely. As I have sought to show, these men had different kinds of minds and operated with different kinds of symbol systems. Freud was strong in the linguistic and personal intelligences: A shrewd observer of human nature from early life, he thought primarily in terms of language and built a system that was primarily linguistic-conceptual, with hardly any spatial or logical content. Most of his system could be explained in terms of common language and demonstrations that became intuitive. Efforts to convert the Freudian system into a set of logical propositions have been vexed; the Freudian enterprise proceeds by descriptions of new cases, and by the creation of new terminology and revised linguistically based models. In the last analysis, Freud was a portrait artist, and the field renders its judgment in terms of the power of the portraits he fashioned.

In sharp contrast, Einstein's linguistic skills were modest and his interest in the personal sphere strictly limited. Rather, as befits a physical scientist, his thought was rich with visual-spatial images and possible experiments. He could readily relate these experiments to mathematical formalisms and to concepts that existed within a tight logical-mathematical structure. Laypersons could admire Einstein's achievements but were destined to understand them at best in a superficial way.

Ultimately, Einstein put forth logically linked statements about the nature of the universe. Scientists reacted to his system chiefly in terms of their evaluations of the logical rigor and the explanatory power of his interlocking propositions. In the end, it proved possible to test directly some of the implications of the special and general theories of relativity, and these confirmations reinforced the tenability of Einstein's formulation. But Einstein indicated more than once that he was prepared to rest on his theories' logical and aesthetic foundations.

In a sense, both Freud and Einstein can be said to have invented systems or schools of thought; one can be a Freudian or an Einsteinian. These systems are very different from one another; but they are even more different from the visions created by the artists we will encounter in the next section of part II; and they differ just as much from the performances created by the individuals described in the final case studies in this part. Freud and Einstein also introduce a number of themes pervading my discussion. Though both were quite isolated during the years of their greatest breakthrough, they benefited from the cognitive and affective support of either a single individual, as with Freud, or of a small group of friends, as with Einstein. Both transcended initial disappointments and persevered, perhaps even gaining some pleasure from the controversies involving them. And they both sacrificed much in order to focus exclusively on their work: As an ascetic, Freud even renounced sexual relations, while Einstein was unwilling or unable to sustain a meaningful family life.

There are also instructive differences. As a scientist who had opened an entirely new way of thinking in a field that still lacked a dominant paradigm, Freud could continue to innovate as long as he lived. In contrast, Einstein, easily as revolutionary at the outset, made his contributions in an already well-articulated domain. Though he continued to "do physics" for the remainder of his life, he was overtaken relatively soon by individuals who were younger and more flexible than he—by the Einsteins of the next generation. The contributions of his later life lay more in the areas of philosophy and social wisdom than in his beloved physics. Both men saw themselves travel from obscurity to world fame: Einstein put up with this gamely, while Freud was much more of an agitator, for he saw himself as the leader of a military campaign that still had far to go.

It is fitting for this study of the modern era that both Freud and Einstein were very interested in childhood. Freud made the events of early childhood the principal propellant of the emotions and personality of later life. Einstein esteemed the mind of the young child, granting it powerful intuitions about physics, and, as mentioned, he encouraged his Swiss colleague Piaget to investigate children's thinking in the physical realm.

On the surface, neither Freud nor Einstein is easily described as childlike or childish, in the manner of Picasso or even, at times, of Gandhi. They were, after all, members of the German bourgeoisie and they dressed and acted accordingly. There are occasional pictures of Einstein clowning, to be sure, and even Freud was well-known as a teller of jokes; but it would be straining to think of either scholar as an individual who never grew up.

At a deeper level, however, we can discern the links to childhood that seem to run through the lives of highly creative individuals. As Einstein often pointed out, the problems he pondered were those that children spontaneously raise, but that most adults have long since stopped thinking about. And while the issues that preoccupied Freud are less likely to arise naturally in the minds of young children, they are just the ones that dominate the actual lives of children: not only phenomena like dreaming, joking, and sexual play but also psychological processes like displacement or condensation or substitution. Only individuals still in touch with the experiences of childhood could have unraveled these phenomena; and, I would propose, only those living at the beginning of the modern era would have explored them in the systematic yet generative way that they did.

5

PABLO PICASSO:
PRODIGIOUSNESS AND BEYOND

Picasso, 1904

ONE HAS TO GO back two centuries, to the case of Wolfgang Amadeus Mozart, to encounter an individual so prodigious in the first decades of life, and so masterful in his maturity, as Pablo Picasso. In both cases, the child exhibited unusual artistic gifts during the first years of life; was strongly encouraged by his father, who was himself a practicing artist and teacher of the art form; and had surpassed his father and other local masters by the time of early adolescence. Each one moved to the European centers of artistry where, within a few years, he was accomplishing his art as well as anyone else alive; and he then spent the remaining years following his own artistic inclinations, whether or not they happened to coincide with prevailing tastes. Of course, Picasso lived far longer than did Mozart, and he achieved in life a success and acclaim that eluded Mozart until a century after his death. But even those who had little affection for Mozart's music, and even less for his quirky personality, conceded his early prodigiousness and his later genius.

THE PHENOMENON OF THE PRODIGY

The term *prodigiousness* connotes a gift that borders on the miraculous. Even if one refuses to believe in miracles and looks only to probabilities, the kinds of performances exhibited by the young Wolfgang Amadeus Mozart or Felix Mendelssohn, by the youthful Picasso or the English painter John Everett Millais, are astounding. The customary assertion is that prodigiousness occurs only in certain domains, and prodigious performance—defined, roughly, as adult-level performance displayed by a child—is in fact far more likely to occur in music, mathematics, or chess than it is in literary or scientific studies. Also, prodigiousness has seemed to be more of a male than a female phenomenon, though the nigh universal focus until now on the education of apparently talented young males makes it impossible to determine the extent to which the predominance of male chess whizzes or mathematical masters is a cultural phenomenon.

Without addressing the issue of gender differences of prodigiousness, I believe that a significant genetic or neurobiological component exists in the prodigy: something in the structure or functioning of the nervous system of Mozart, of the chess player Bobby Fischer, or of the mathematician Carl Gauss that made it preternaturally easy to gain initial mastery of the patterns involved in musical tones, the configuration of chess pieces, or the possibilities of numerical combinations, respectively. But even those observers struck by the neurobiological "preparedness" of the prodigy should recognize the cultural aspects of the phenomenon of prodigiousness.

As David Feldman has shown, the prodigy must exhibit promise in an area that is valued by the culture and in which children's relevant behaviors are at least noticed. If graphic expression is not valued in a culture, if children's scribbles are routinely disregarded and discarded, there will be no drawing prodigies. By the same token, when a culture begins to attend to children's precocious performances in a domain—as has happened with visual artistry in contemporary China—one may discover unexpected gifts. Quite possibly the most gifted graphic prodigy in history turns out to be a Chinese girl named Wang Yani, whose extraordinary drawings from early youth confound the aforementioned common characterization of the prodigy: a young Western male working in chess, music, or mathematics.

Over and beyond cultural interest in and support of a domain, a prodigy always represents a "co-incidence" of factors. That is, one needs not merely a "prepared" child and a "welcoming culture" but also a tremendous amount of social support: good teachers, attentive parents, ample opportunities for performance and display, relief from competing responsibilities, access to avenues of publicity, and a sequence of hurdles that are acknowledged in the domain and over which the child has an opportunity to bound. The child "at promise" in a domain is simply one who can negotiate these steps more rapidly, and with less strain, than others in his or her cohort. We see all of these factors at work in the life of the young Picasso.

Despite his or her unusualness, the prodigy will encounter obstacles. Especially during early life, the prodigy needs one or more adults who will ease the way, provide opportunities, defend against criticism, offer satisfying explanations (or rationalizations) for real or imagined setbacks, and direct energies and talents in productive directions. No matter how snug the fit between the child and the surrounding domain, one cannot expect a naive youth to be able to deal with the vagaries of the surrounding field. The prodigy tends to work in ignorance of what is going on at the forefront of the domain and, while often extremely gifted in mimicry, cannot be expected to go beyond conventional practices. Indeed, the prodigy will focus on his or her own interests, on pleasing "significant others," or on mastering the common code of the domain, rather than engaging in a genuine dialogue with the leading innovators of the time, or with exemplary figures drawn from history.

Prodigies almost invariably encounter a rude shock sometime during the second decade of their lives. Until this point they have appeared as portents from the gods, to be marveled at by an audience predisposed to admire rather than to scrutinize. But as one-time prodigies begin to resemble adult practitioners of their domain, their preternatural giftedness no longer suffices; now they must be able to compete with the masters of their era, as well as other young persons who were not prodigies, but who have caught up with them. At the same time, especially in the West, they begin to confront the fact that they have until this time been the vessel of a pushy parent's, a forceful teacher's, or someone else's ambition, which does not necessarily coincide with their own long-term best interests. Now, however, they must—or at least *ought*—to take hold of their own lives, and this seizure of initiative may place them on a collision course with those who have, until now, "managed" their careers.

The transition from youthful prodigy to adult master is a vexed one; sometime during adolescence nearly all prodigies today suffer what the musician-psychologist Jeanne Bamberger has termed a "mid-life crisis"; and many, perhaps most, prodigies do not fulfill their youthful potential. Picasso and Mozart are obviously exceptions in this regard. It is perhaps more common to encounter the situation described by the composer Hector Berlioz: Of the once prodigious Camille Saint-Saëns, Berlioz quipped, "He knows everything but he lacks inexperience." Making original creative contributions to a domain emerges as an enterprise quite different from mastering the domain as it has been practiced in the recent past.

PICASSO AS PRODIGY

When talented children go on in later life to make enduring contributions to the domain, many stories will be told of their youth, including stories they themselves tell. We can hardly be certain that all the accounts of the young Picasso are true, but of his *relative* prodigiousness in the domain of drawing there can be little doubt.

Early Talent Revealed

Born in backward Malaga, Spain, in 1881, son of an academic painter of modest talent, Picasso is reputed to have begun to draw by the time of his first word—which, indeed, is said to have been *piz* for *lapiz*, or (drawing) "pencil." As a young child, he drew incessantly and with increasing skill. His longtime friend Gertrude Stein observed that "Picasso wrote painting as other children wrote their a b c . . . drawing always was his only way of talking." There is no reason to doubt that he passed through the early stages of drawing in the standard order—scribbles, pure geometric forms like circles, simple figures like suns and flowers, "flat" compositions, and the like—but his first preserved drawings, from the age of nine, reveals a boy who had already achieved considerable control of line and composition. Picasso's father specialized in drawing pigeons: Like many Spanish males, he especially loved the bullfights, and so early Picasso works often feature those two subjects. But Picasso was fascinated by the world of human beings, and from the beginning of the second decade of his life, his drawings show an unusual emphasis on the human form and on the wide range of human emotions.

Clearly, Picasso had sufficient manual dexterity to draw copiously and competently. He also had the knack of starting a work from some arbitrary point, proceeding in a seemingly desultory manner, and yet ending up with a coherent work. He evinced skill in noticing visual details and arrangements, thinking in spatial configurations, remembering virtually every live and painted scene that he had ever witnessed, and attending to the world of other human beings. In terms of multiple intelligences theory, Picasso's precocity was most striking in the visual-spatial, bodily-kinesthetic, and interpersonal areas. Such gifts are what one would expect in a visual artist, but artists differ from one another in whether their strengths are more pronounced motorically, visually, spatially, or with respect to the world of human beings. Part of Picasso's prodigiousness stems from the fact that he was gifted across the range of relevant skills and could draw on them synergistically. Here the comparison with Mozart is justified, for Mozart also displayed enormous talents across the musical board—performance, composition, and the dramatization of human action in theatrical works.

Difficulties with other areas of learning are not a necessary concomitant of prodigiousness, but they occurred with Picasso. He hated school, tried valiantly to avoid attending, and performed poorly when he did show up for class. He had difficulty learning to read and write and even greater difficulty in mastering numbers. He seems to have wanted to treat numbers as if they were visual patterns rather than symbols for quantities, for example, seeing a pigeon as an arrangement of eyes as 0s, wings as 2s, and the baseline as the sum of the figures; art historian Mary Gedo believes "he anthropomorphized the numerals and was distracted by his own fantastic perceptual abilities." Without backdoor connections, extensive tutoring, and frank cheating, he might well never have made it through grade school. Neurotically, he would not attend unless his father was with him, remained nearby, or promised to return at a certain hour.

It has been said that Picasso never mastered certain scholastic skills and had trouble with abstract thinking. While it is unnecessary to draw so severe an inference, the disparity (or, in my terms, the asynchrony) between his consummate artistic skills and his meager scholastic acumen clearly distressed the young Picasso. Forever after, this gap tainted his relationship to the world of scholarship and to those who were intellectuals (rather than artists).

The juvenilia of artists may harbor clues to their later accomplishments. From 1890, when he was nine, Picasso's family saved virtually every scrap of paper on which he drew, as well as numerous notebooks, school books, and other surfaces on which his graphic impressions were placed (figure 5.1 is the earliest preserved painting by Picasso). We can observe from this record not only Picasso's dogged efforts to master all of the formal aspects of the animate and inanimate models he encountered; far more important, we encounter his compulsion to experiment—to try out varied compositional arrangements, to draw

FIGURE 5.1.
The Picador (1889–90), oil on wood. Picasso's earliest known painting.
©1993 ARS, N.Y./SPADEM, Paris.

the same object from diverse angles, and to capture contrasting emotions, including ones that were highly evocative and dramatic.

These notebooks reveal a Picasso at early adolescence who can already draw most any object in his surroundings with skill, wit, and originality. There are all manner of flora, fauna, and manmade objects, as well as wonderfully alive portraits of the young and the old, the healthy and the diseased, the grotesque and the sensual. While the explicitly erotic and the gratuitously violent are not yet featured, hints of them can already be discerned.

But perhaps even more important as harbingers of Picasso's later artistry are the ways he fragments and distorts aspects of form and scale, the depictions garlanded by numerical and alphabetical symbols, the proliferation of *trompe l'oeil*, visual puns, outrageous caricatures, incongruous juxtapositions, and the like. We simply do not know whether Picasso consciously thought back to these graphic experiments when, two decades later, he and Georges Braque were inventing cubism; but at least at an unconscious level, Picasso was able to draw on this reservoir of youthful experimentation.

Such experimentation proves crucial in the early work of an artist. There are far more youngsters who master the established practices in a domain, with greater or lesser rapidity, than there are youngsters who are already fiddling with unfamiliar arrangements and guiding their domain in new directions. In the area of music, for example, for every young child who experiments with the musical canon—who "decomposes" the music that he or she is supposed to perform—there are dozens who simply play as they are taught to play, and who never even consider adopting a divergent thematic or rhythmic course.

Picasso's father appears to have been a distinctly unremarkable painter, to whom any thought of directly challenging the canon would have been extremely upsetting. (It was he who insisted that Picasso attend the traditional academies in Barcelona and Madrid so that the son could, presumably, also become a provincial painting teacher.) Thus, Picasso did not become a rebel by modeling his father's behavior (any more than Mozart acquired his iconoclasm from his pedestrian and law-abiding father). One might search for other role models for Picasso's irreverent independence, ranging from his strong-willed mother to his financially secure uncle Salvador.

But it seems equally likely to me that this willingness to experiment is more of an endogenous factor: one that arises from a temperament that seeks arousal, from sheer pleasure in working with the medium, from a confidence in one's own emerging powers, and, perhaps less happily, from the asynchrony between ease in one's artistic medium and difficulties with standard scholastic practices. If one cannot succeed in those pursuits where students are supposed to succeed, one may combat personal frustrations and confound family members by blazing a trail in one's areas of strength.

Formative Experiences and Haunting Memories

Another facet of Picasso's childhood may have constituted the most important contributor to his ultimate artistic accomplishment. I refer here to a series of traumas in Picasso's youth that seem to have had an enduring impact. Of course, every child undergoes traumatic episodes, and it is not possible to quantify how upset a child becomes and how lasting the wound. But from every indication, Picasso was a highly sensitive young individual who remembered with stunning vividness the appearances of events and persons from his first years.

In particular, Picasso was shaken by an earthquake that occurred when he was three and that resulted in his being taken away from home, in the middle of the night, to avoid a fire. Picasso remembered half a century later the sight of his alarmed mother, with her kerchief over her head, and his father, cape thrown over his shoulder, scurrying away to save the family. A second event was the death of a young sister, Conchita, when Picasso was fourteen. Conchita contracted diphtheria and, over a period of weeks, underwent a painful, highly visible (and audible) death by gradual asphyxiation. Young Pablo not only was deeply distraught by this experience but seems for some reason to have held himself accountable for this death and wanted somehow to atone for it. It has been speculated that, at the time, Picasso had promised God that he would stop painting in gratitude if Conchita's life were saved; and since this bargain had not been accepted, the deeply superstitious Picasso felt both free to do whatever he wanted in his professional and personal lives and concomitantly guilty at this hubristic seizure of power. Such a "bargain with the divine" recurs in the lives of several of our creators.

Picasso also harbored strong, ambivalent feelings about the surviving members of his family. He was jealous of his younger sister, Lola, who took away some of the attention he had been receiving; and he was angry at his mother, Maria, for beginning to dote on his sister. Picasso's mother admired her son, thought him beautiful, and harbored great ambitions for him, but she showed no real understanding of what he could accomplish; her domineering personality seems to have clashed with the equally headstrong personality of her son. He was impatient with his wealthy uncle Salvador, a powerful man who supported much of Picasso's education financially, thereby engendering the mix of gratitude and bitterness that so often attends acts of charity. (One thinks of Freud's resentment of Josef Breuer's largesse.) He felt frustrated by the five overweening women (four spinster aunts in addition to his mother) who lived in the house where he was growing up.

And most fundamentally, Picasso had decidedly ambivalent feelings about his father. On the one hand, he loved him desperately and depended on him on virtually an hourly basis up through early adolescence. On the other hand, he felt his father was a weak and compromising figure, who curried favor with people to hold a job, who tried to manipulate events to help his son's career, and who spurned artistic risk and, indeed, embraced the very academic and bourgeois values the adolescent Pablo was beginning to disdain. Picasso found it necessary to denigrate his father and even to render an unflattering portrait of him. While it is uncertain whether Picasso's father actually stopped painting when his son surpassed him, it is a matter of record that Pablo dropped his father's surname, Ruiz, and elected to become known to the world by his mother's name of Picasso.

Picasso's Childhood Works

The standard story told about Picasso is that he was a painting prodigy who effortlessly surpassed all the other artists in his milieu. This story has recently been challenged by Picasso's acute biographer John Richardson, who acknowledges Picasso's early talent but feels that it was not that extraordinary, and who stresses how hard Picasso had to work to achieve success in painting. In Richardson's view, "It would seem that Picasso—unlike certain composers (Mozart, for instance)—conforms to the rule that no great painter has ever produced work of any serious interest before puberty." Richardson also questions some of the legends about Picasso's scholastic failures and his early painting triumphs, claiming that Picasso (and his hagiographers) had a compulsion to portray their young genius in unjustifiedly adulatory and heroic terms.

It is worth considering Picasso's own testimony on the matter. In a conversation with his friend Brassai (the pseudonym of the painter Gyula Halasz), Picasso declared:

> In contrast to music there are no prodigies in drawing/painting. What one can consider an early genius is actually the genius of childhood. It disappears at a certain age without leaving around traces. It is possible that such a child

will one day become an artist but he will have to begin again from the beginning. I did not have this Genius, for example. My first drawings could not have been hung in a display of children's work. These pictures lacked the childlikeness or naivete. . . . At the youthful age I painted in a quite academic way, as literal and precise that I am shocked today.

And in a well-known—if enigmatic—aside, at an exhibition of children's work, Picasso once quipped, "When I was their age I could draw like Raphael, but it has taken me a whole lifetime to learn to draw like them."

In such remarks Picasso is at once placing children's work on a pedestal and, at the same time, distancing his youthful self from such a romantic view of artistry. It is true that the works preserved by Picasso's family support the notion of Picasso as an aspiring academic painter and not as a charming naif. However, we lack drawings of Picasso from the first eight years of his life, and so we cannot determine how his early works resembled those of other children; moreover, his experiments, caprices, and marginalia may have been as crucial in his artistic development as his more formally conceived canvases. My own conclusion is that Picasso's drawing during the first decade of life was unusually skilled, rather than frankly precocious, but that no word short of *prodigy* can describe his spectacular progress over the next several years. With these reservations, we may legitimately term him a prodigy.

The Honing of Talent

As almost always happens to a gifted youth who completes some initial training, Picasso outgrew (and outdrew) his first teachers. The story of his formal artistic education is reminiscent of Mozart's—a series of virtually unnecessary enrollments in formal classes, contempt for mediocre teachers, unpleasant personal clashes, and a reversion to self-education, at the feet of favorite masters from earlier epochs. Picasso matriculated first at the academy in Barcelona: He passed the entrance exams readily, rarely attended classes, proved completely unable to deal with the rules and regulations, and left soon afterward. (It must be difficult for students to remain in school when they think they are more competent than the teachers— and when that assumption proves correct!) Uncle Salvador then sent him on to the Academy of Fine Arts in Madrid, where the same pattern of alienation ensued. The restless Picasso learned far more in the galleries of the Prado and in the streets and bordellos of Madrid than in the staid classes of the academy.

Unhappy in Madrid, Picasso at age seventeen returned to Barcelona. Here for the first time, he stepped out into a wider world. He met older painters like Isidre Nonell and Roman Cases;* he joined a circle of young bohemian artists,

---

*Sometimes referred to as Isidro Nonell and Ramon Casas.

writers, and intellectuals that included the painter Carlos Casagemas and the poet Jaime Sabartés; and, thanks to his talent, tact, and charm, he soon became the leading figure of a *tertulia*, or circle of cronies, who hung around the cabaret Els Quatre Gats (The Four Cats). He continued to work tirelessly and found an outlet in a publication called *Arte Joven* (Young Art), for which he became the principal illustrator. He was exposed to—and readily absorbed—ideas in the political and the artistic realms. These avant-garde and anarchist sentiments matched his own—those of a talented but alienated youth.

During this stay in Barcelona, Picasso's talent was shaped and became more profound. On the one hand, he continued his self-education, absorbing the lessons of the old masters like Francisco José de Goya, Diego Velázquez, and Francisco de Zurbarán, as well as the new modes associated with impressionism and with cabaret art of the Toulouse-Lautrec mode. At the same time, he went beyond glib and facile renderings to probe the psychology of sailors and dockworkers, the emotional undercurrents of landscapes and street scenes, as well as the often charged human intercourse manifest in family portraits and scenes of night life. He sought to enter into the world of suffering and poverty. Obsessed with the beguiling appearance of a striptease artist, he would return to his room and spend all night trying to capture her moods and stances. He dealt more directly and powerfully than before with the themes of sex, death, and anarchy; and he drew himself faithfully, as well as in a dizzying array of disguises and costumes. His virtuosity in capturing the mood, the personality, and even the thoughts of his subjects, and in doing so with wit and penetration in a few strokes, made a deep impression on observers.

His associates soon realized that Picasso was an individual of unusual powers and sensitivities, an instinctive leader even when he was silent. And for perhaps the first time in his conscious life, Picasso began to feel—in a manner reminiscent of the young Freud—that there were no limits on what he could accomplish. Having achieved this realization and risen to a leading position among the young talents of his homeland, it was only a matter of time before, like so many other gifted youths, he migrated to the center of art in the Western world.

THE YOUNG ARTIST IN PARIS

While important artistic communities could be found in Vienna, Berlin, and other capitals, no other European city featured the appealing combination of artistic, intellectual, scenic, and romantic elements associated with Paris. The nineteen-year-old Picasso was ineluctably drawn there, just as Stravinsky, Freud, Eliot, and even Gandhi felt called to Paris for their respective reasons. And it is hardly surprising that, having made his mark in the French capital by the age of thirty, Picasso would center the rest of his long life in a city (and a land) that have so long been associated with visual artistry and the fabled *vie bohemienne.*

By the time that Picasso arrived in Paris in the artistic district of Montmartre, impressionism, the preeminent artistic breakthrough of the later nineteenth century, was already history. What had begun as an antiestablishment movement— one that spurned formidable classical subjects, the grand emotions, and the ideal of photographic realism—had now become the virtual norm: The use of ordinary subjects, the "scientific" experimentation with lights and textures, and the desire to capture transitory visual impressions were now commonplace. And like other once-controversial movements, impressionism had in turn spawned its own set of reactions, among them postimpressionism and expressionism. Picasso encountered not only the prototypical impressionist works of Claude Monet, Pierre-Auguste Renoir, and Camille Pissarro and the neoimpressionist works of Georges Seurat and Paul Signac, but also the expressionist work of Vincent van Gogh, the primitivistic canvases of Paul Gauguin, the socially and politically tinged posters of Henri de Toulouse-Lautrec, the massive, evocative sculptures of Auguste Rodin, the symbolist work of Eugène Carrière, and the spirited illustrative art that adorned journals and commercial advertisements. No single school was dominant: No less than in the fields of physics and psychology, restless change was in the air.

A quick study and an instinctive mimic, Picasso readily picked up the features of the various aesthetic models he encountered. He spent many hours in galleries and absorbed just about everything that he saw, being able subsequently to call it up seemingly without effort. And yet, by the age of twenty, Picasso no longer produced simply derivative works. Indeed, in the first years after Barcelona, he passed through a set of periods in which he produced works bearing their own distinctive mark. His works were marked by a directness, hardness, and pathos that rendered them readily recognizable. If Picasso had died at the age of twenty-five, he would not have been thought of as a revolutionary; but like the short-lived Toulouse-Lautrec and Seurat, he would have been recognized as a painter who had articulated in paint his unique visions of the world.

At first Picasso's life in Paris was anything but idyllic. (I think particularly of Picasso when I read Milosz's evocative poem that serves as a frontispiece to this book.) As a non–French-speaking foreigner with relatively little education, he felt lost in the intimidating cosmopolitanism of Paris. He was very upset by the poverty and disease that he saw, and while he eventually became hardened to it, the miserable existence experienced by so many took a toll on this sensitive and anarchistically oriented youth. Tending to be superstitious, he feared the infirmities, sexual diseases, and blindness that he encountered daily. At first unable to sell his works, he lived in poverty. At one time he even contemplated suicide. Though in some ways estranged from his family, he was still dependent on them emotionally and financially and felt the need to return home, especially when it grew cold in his barren flat. Between 1899 and 1904, he shuttled repeatedly among Paris, Malaga, Barcelona, and Madrid.

Given that many estimable artists never received acclaim during their lifetime, Picasso's rise to prominence occurred with surprising speed. His virtuosity and versatility were soon recognized. He mounted a first exhibit when scarcely twenty, and Félicien Fagus, a critic for *La gazette d'art*, commented on the purity of his painting, his love of color and subjects, his curiosity about everything, and the enormous range of influences he had managed to absorb. Having praised the gifted youth ("they say that he is not yet twenty and that he covers as many as three canvases a day"), Fagus put forth a shrewd note of warning:

> Danger lies for him in this very impetuosity which can easily lead him into a facile virtuosity. The prolific and the fecund are two different things, like violence and energy. This would be much to be regretted since we are in the presence of such brilliant virility.

As is well known, Picasso's first recognized phase came to be called the blue period, and it featured paintings that portrayed the dire circumstances of Parisian life: beggars, sad couples, poor families—clear "types" who are nonetheless depicted with individuating features. He seemed intrigued by the loners, the broken families, the juxtaposition of crowds and anonymous figures in a teeming metropolis; what one critic termed "the beauty of the horrible," another labeled "this sterile sadness," and a third called "a negative sense of life." He even went to the prison at Saint-Lazare so that he could exhaustively study the prostitutes.

As Picasso's circumstances improved somewhat and he began his first serious love affair, the pink or rose period followed. The works of this time were somewhat lighter, featuring the circus life and personages who, if not serene and at peace, were at least in less obviously dire straits. The dominant theme had shifted from unalloyed poverty to the bohemian artistic life. We might think of these two periods together as Picasso's first mature phase, the time when he had developed a distinctive, but not yet truly innovative, style.

By 1905, Picasso's personal situation had improved substantially. As had happened five years earlier in Barcelona, a group of artists, writers, and intellectuals—*la bande Picasso*—came to coalesce around the gifted young Spanish artist. The writer Guillaume Apollinaire, the critic André Salmon, and the poet Max Jacob were but three of the young Frenchmen who found Picasso to be an attractive figure. They were energetic, lively, sarcastic, ambitious, and ready to conquer. Picasso's old Spanish friend Sabartés also joined the group and began what was essentially a lifelong service to his admired countryman.

Picasso particularly appreciated the companionship of poets and writers, whose interests and skills complemented his own. They helped him articulate what he was trying to accomplish, gave him suggestions about where to direct his considerable energies, informed him about the world of ideas, and promoted his work to the rest of the world; at the same time, these people of the word marveled at Picasso's virtuosity, his ability to capture with a few perfectly placed strokes those aspects of character they were seeking to portray in natural language. In the case of Apollinaire, the relationship was especially symbiotic: The two men seemed to think in the same way, to have complementary interests and imaginations, and to develop along parallel artistic tracks: In John Richardson's view, they "served as each other's catalyst to an extent unparalleled in the history of art and literature."

The Aftermath of a Tragedy

Against the background of early difficulties and the first intimations of success, the most traumatic event of Picasso's early adult life occurred. Picasso's close friend from Barcelona, a gifted but tormented painter named Carlos Casagemas, had accompanied him to Paris. They shared lodgings, possessions, friends, and even a lover. Casagemas became depressed about his life circumstances, and especially his love life (he seems to have been impotent); and on an occasion when Picasso was in Spain, Casagemas attempted to kill his lover and then succeeded in killing himself. To compound the tragedy, Casagemas's mother died thereafter—according to some reports, she dropped dead immediately upon hearing of her son's suicide.

Throughout his long life, Picasso experienced enormous difficulty in dealing with death. For the most part, he sought to deny it altogether. He would not speak about those who died, he refused to attend their funerals, and he feared any person or agent who might bring about disease, advancing age, or death. None of these reactions are wholly abnormal, of course, and they may have even been characteristic of superstitious Spanish persons of his era; but, they were combined in Picasso's case with a fascination with violence, death, and tragedy, as conveyed in the bullfight, in international political conflict, and in the sordid settings of Montmartre, which he studied with Charcot-like clinical precision.

For a while, Picasso put Casagemas's tragic suicide at some distance from himself. (He did comment at the time, "Thinking of Casagemas being dead makes me paint in blue"; and he made some portraits of his friend, complete with bullet hole in the temple.) John Richardson maintains that "whether or not Picasso bears any blame for abandoning his friend, the guilt he felt toward Casagemas, like the guilt he felt toward his dead sister . . . and toward the father he wanted to kill, would provide his art with just the catharsis it required."

FIGURE 5.2.
La vie (1903), oil on canvas (77 ½ x 51).
Picasso's early masterpiece of the blue period.
© 1993 ARS, N.Y./SPADEM, Paris.

In any event, two or three years later, Picasso executed a painting that, in many ways, was the largest and most important of his young life. This painting, a blue period piece called *La vie,** is set in an artist's studio (see figure 5.2). It depicts two central figures—presumably Casagemas, with a wound in his temple, and his lover—across the canvas from a stern-faced, ominous Madonna-like figure nursing a child; on a small canvas in the background, two female nudes huddle together. The female lover rests dependently on the man, who appears as if he wishes to be removed from the scene. Perhaps Casagemas is pondering whether he should embrace or kill his lover.

From all indications, *La vie* represented Picasso's efforts to come to terms with Casagemas's life and death, and his own swirling reactions to it. More broadly, it touched on his own feelings about death—for example, the loss of his beloved sister; his attitudes toward artistic productivity and sterility; and his relations to family and to women, about which he would remain markedly ambivalent for the remainder of his life.

La vie *as a First Defining Work*

Picasso was a virtuoso artist, painting, drawing, sculpting, or engraving hundreds of works during nearly every year of his long life. Yet, beyond question, Picasso saw these works as differing in importance; and from time to time, he would proceed

---

*Here and elsewhere, I refer to works by the names that are most commonly used in nontechnical writings about works of art.

FIGURE 5.3.
Family of Saltimbanques (1905), oil on canvas (83 ¾ x 90 ½).
Picasso's climactic painting of the rose period.
© 1993 ARS, N.Y./SPADEM, Paris.

to work on a canvas that for him represented a summation of some sort—an oratorio among songs, a novel among short stories. The canvas would be bigger, the preparation would take longer, there would be many drafts and sketches, the themes (and even the tide) would be portentous, and, above all, the visual and thematic elements would draw together strands that had emerged more fragmentarily in previous works. We see this process at work in Picasso's most famous paintings: *Family of Saltimbanques* (1905—see figure 5.3), *Les demoiselles d'Avignon* (1907),

FIGURE 5.4.
The Three Dancers (1925), oil on canvas (85 5/8 x 55 7/8).
A defining work from the post-Cubist period.
© 1993 ARS, N.Y./SPADEM, Paris.

Three Dancers (1925—see figure 5.4), *Guernica* (1937—see figure 5.10), *The Charnel House* (1945), and several others, often completed at decade-long intervals. The prototype for these summary, synthetic, or defining works is *La vie*.

Examinations of the preliminary sketches and radiographic studies of the final canvas reveal Picasso's considerable experimentation in the preparation of *La vie*. At various times he placed within the work a self-portrait, a portrait of the young woman with different facial features, other paintings in the background, a birdman and a nude, and a portrait of his father. The work as a whole borrows much in spirit from an El Greco–like painting of 1901, called *Evocation*, that represents the burial of Casagemas and his ascent to heaven.

As Mary Gedo shows in a detailed study of *La vie*, every figure and stance can be discerned in Picasso's works of the previous few years. It is their particular combination—and the powerful statements they convey about love, life, chastity, sin, alienation, and death—that makes *La vie* a defining work. Gedo also points out the extreme fluidity in the early sketches of the major personae; Picasso switches graphic identities, stances, and moods constantly until he finally arrives at the combination he wants to preserve. For instance, the extent and even the possibility of pregnancy varies from one sketch to the next; an artist (again, initially resembling Picasso) is included and then removed from the canvas.

Perhaps most revealingly, the relation between the two central figures is varied, from affection to distance to a physical attack by the man on the woman. John Richardson relates the figures of *La vie* to those represented on tarot cards, stressing that each is to be treated just as ambiguously: "Like all the major arcana in the Tarot, it has a positive or negative meaning . . . according to which direction the card faces when dealt." In such revisions, analogous to Freud's drafts of "The Project" or to Eliot's early attempts with *The Waste Land*, one sees at work the thinking, the symbol-suffused problem finding and problem solving, of a graphic artist. A range of symbols and intelligences are mined, yielding a rich network of meanings and, ultimately, a work of singular condensation.

It is risky to single out works of art, in a life as long as Picasso's, as turning points. And, indeed, one could probably choose any number of Picasso's paintings as critical for his own development and for the way that the world has come to think of his work. Nearly every one of Picasso's more significant works represents a culmination and an anticipation: Most of them draw ingeniously on what he has seen and what he has done in the directly preceding years. At the very least, however, one can speak of *La vie* as emblematic, symptomatic, and prototypical of the changes taking place in the artistry of the gifted young Spaniard.

With Picasso and other great artists mining their respective media, their defining works succeed in bridging events and emotions of deep personal significance with themes and images of universal scope. This idea has been so often stated that it sounds banal, but it is no less true for its familiarity. Like other defining Picasso canvases, *La vie* probes the depth of Picasso's own feelings about those closest to him—family members, friends, and lovers—and does so in such a way that he can even touch the feelings and understandings of those who live in entirely different circumstances.

In painting a work like *La vie*, Picasso was increasingly going his own way. No other works of the period lend themselves readily to a comparison. Still in his early twenties, Picasso is already in the class of the greatest painters of the previous generation, like van Gogh or Paul Cézanne, and in the class of the one contemporary with whom he would continue to compare himself, Henri Matisse. Soon he would proceed even further in his own development, toward a form of painting that would establish new standards for the new century.

LES DEMOISELLES D'AVIGNON: TOWARD AN AVOWEDLY EXPERIMENTAL STYLE

By 1905, Picasso was already a relatively well established artist in Paris, beginning to command sums for his work, with dealers interested in representing him, as well as customers on whom he could rely. Especially for one who had been a prodigy, and who had made the difficult transition to a new country and to an intensely personal aesthetic style, it could have been extremely tempting to remain in the same style, to cultivate a following, to ride the wheels of success in the direction that they were already spinning. But there was something in Picasso—perhaps the same impulse that led him to disassemble forms in the notebooks of his youth—that prevented him from ever resting on his laurels; instead, he felt compelled perpetually to face new challenges and scale new heights—as well as risk unprecedented depths—professionally and personally. Such relentless drive, of course, has characterized our other creative titans and may indeed be their defining characteristic.

Apollinaire claimed that there were two kinds of artists: the "all-put-together" virtuoso who draws on nature, and the reflective cerebral structurer, who must draw on himself. Mozart could serve as prototype for the first, Beethoven for the second. As a prodigy, Picasso epitomized the first type, but, claimed Apollinaire, he was able to convert himself into the second kind of artist: "Never has there been so fantastic a spectacle as the metamorphosis he underwent in becoming an artist of the second type." Picasso himself sensed this antinomy when he complained to Gertrude Stein, "If I can draw as well as Raphael, I have at least the right to choose my way and they should recognize that right, but no, they say no."

Once again, a certain simplification is involved in identifying pivotal or defining works. One candidate, the *Portrait of Gertrude Stein* (1906), stands out because of the artist's process: Picasso asked his subject to remain for over eighty sittings; then he went away for the summer, annihilated the recognizable facial features, and finished the portrait away from Stein, substituting masklike features for realistic ones. (Chastised because the portrait did not look like

FIGURE 5.5.
Les demoiselles d'Avignon (1907), oil on canvas (96 x 92).
Picasso's early creative breakthrough.
© 1993 ARS, N.Y./SPADEM, Paris.

Stein, Picasso reputedly came back with one of the notable artistic one-liners of the century: "Don't worry, it will.") Another candidate from the same year, the *Two Nudes*, displays two muscular, corpulent, sculpturelike figures, whose identity is more of a type than of particular people; the viewer beholds a sprawling mass produced by raw swathes of paint. And there is Picasso's *Self-Portrait* of 1907, in which facial features resemble geometric forms rather than contoured living flesh—another instance of masklike portraiture.

But no one, simply viewing these paintings, could have been prepared for the shock of 1907's *Les demoiselles d'Avignon* (see figure 5.5), which many see as the most important painting of the century and one of the critical turning points in the history of any art form. Like other of Picasso's defining canvases, *Les demoiselles* summarized works of the previous years; but more so than any others, it looked forward to a whole new style of painting, in which the forms assumed by the subject matter would carry the aesthetic weight.

Much ink has been spilt over the possible graphic sources of *Les demoiselles*, the landmark work that in part inspired cubism. Clearly, the work did not arise outside the context of long-term and recent art history, and intimations of artists ranging from Ingres, Delacroix, and El Greco to Manet and Gauguin can be discerned therein. Picasso was influenced by Iberian art, which he had recently been studying, and by the powerful human figures portrayed by *les fauves* artists, especially Matisse. Picasso was also almost certainly affected by African masks he had seen at the Parisian Trocadero, though Picasso's own testimony on the supposed linkages between *Les demoiselles* and African tribal art shifted over the decades. The large-scale compositions of nude figures recently unveiled by his competitors Matisse and Derain also may have stimulated and motivated Picasso, who fiercely competed with the titans of his age. Recently, evidence has surfaced that Picasso had seen Degas' private drawings of a brothel and that these as well were incorporated into this defining work.

But probably the single most important artistic influence on Picasso (and, for that matter, on both Matisse and Braque, the two contemporaries who meant the most to him) was the work of Cézanne, their direct predecessor. Cézanne was neither a great draftsman nor a masterful colorist, but he had seen deeply into the nature of painting, at least as the twentieth century was to conceive it. Cézanne came to see painting as a formal understanding: formal in that issues of form were determinant and in that the geometric forms implicit in all perceptible matter served as the bedrock for graphic activity. "You must see in nature the cylinder, the sphere, the cone, he had declared."

Particularly in his later years, Cézanne abandoned perspectival rendering and instead fused the planes of the foreground and background into an ensemble of overlapping forms and colors. And in his search for a permanent pictorial reality, he turned his back equally on the complex of color, emotion, and sensation that had been prized, in their respective ways, by the impressionists and the expressionists. A major retrospective on Cézanne held in 1907, the year after his death, profoundly affected his younger contemporaries: "Do I know Cézanne? He's my one and only master," Picasso told Brassai several decades later.

In recent years there has been a terrific boon to Picasso studies—the discovery of approximately 175 notebooks kept by Picasso from 1894 to 1967. Included in this amazing find are no less than eight notebooks devoted to preliminary sketches for *Les demoiselles*. Moreover, some of the forms that eventuate in *Les demoiselles* have been investigated in even earlier notebooks, as Picasso engaged in countless deformations and caricatures in an effort to depict the human form. And in late 1991, the only existing preparatory oil sketch was discovered.

Picasso's mode of preparation is fascinating. Themes can be followed for years and even decades, as Picasso attempts variations on his familiar subject matter: bulls, horses, the two kinds of women (whom he dubbed "goddesses" and "doormats"), households and household objects (when Picasso moved to new lodgings, he would spend up to a month sketching his new surroundings), and renderings and transformations of his own past works as well as those of other artists he admired. Picasso supposedly never forgot any work he had seen or made and could play with forms several decades old.

With the possible exception of a movie of the artist painting, made in the mid-1950s, Picasso's notebooks constitute the best single route into the workings of his own versatile and richly stocked mind. They present a running notational record of the issues and problems that engaged him; in that sense they serve as a convenient iconic representation of the images that were presumably swirling around in his head. One is reminded of his friend Sabartés's testimony that this record extends to his personal life: "We would discover in his works his spiritual vicissitudes, the bows of fate, the satisfactions and annoyances, his joys and delights, the pain suffered on a certain day or a certain time of a given year." Picasso said succinctly: "My work is like a diary." It was appropriate that the first public exhibition of some of these notebooks was called "Je suis un cahier . . . " ("I am a notebook").

Like any epochal work, *Les demoiselles* can be described in numerous ways and from varied perspectives. Perhaps most simply it is a large (approximately eight feet by eight feet) and devastating portrait of five prostitutes, arranged in an arresting composition, each assuming a somewhat different position across the canvas, each appearing totally isolated from the others. Their faces are blank and vacant; their eyes, wide and black, stare directly out, independent of the orientation of the face; their bodies, virtually flat and without form, are only partially sketched in. There seems to be a struggle between the soft human flesh of the women and the harsh forms and angles with which they are depicted. The precise thematic message of the painting remains obscure, but clearly the portrait, allegorical in tone, is meant to be a savage characterization of the way of life of these women. (An earlier tide was *The Wages of Sin*.) The aesthetic message is equally clear: Picasso has declared war on the notion of painting as realistic or uplifting and has insisted on the priority of the formal features urged by Cézanne. But whereas Cézanne's themes were either neutral (still lifes) or benign (bathers, card players), Picasso's motifs were as searing as those treated earlier, in *La vie*, or later, in *Guernica*.

Few of Picasso's paintings give the feeling of having been completely finished; in most cases, one can find earlier versions that might have sufficed, as well as subsequent paintings that could be regarded as later versions of the canvas in question. But with *Les demoiselles*, this feeling of incompleteness is particularly telling. On the one hand, there is copious planning for the final work in each of its particulars—much "thinking in the medium." Yet earlier versions obviously had proceeded in a more narrative and less stark direction. As often happened in the evolution of Picasso's work, the trend was toward more paring away and more direct statement; indeed, such trends seem manifest in the notebooks as a whole over this period.

Initially, Picasso had included a sailor at the center (possibly one being initiated into sex) and a medical student at the left of the painting; at certain times the student was holding a skull, at other times a book. These figures were presumably designed to illustrate the range of male clients and the variety of attitudes toward a life of prostitution—or perhaps to capture different aspects of Picasso's own physical and psychological being. The features of the brothel were more clearly delineated as well (including a curtain drawn to allow an inward look), and for a while, there was a homunculus or embryonic creature in the sketches. The ultimate deletion of various moralistic "commenting figures" allowed Picasso to portray the life of prostitution more starkly. The depiction and even the number of female figures also changed over time; it appears that the two figures on the right were added later to the canvas and that their faces, originally Iberian in appearance, had been changed so that they look like African masks. One enduring feature is the raised "announcing" arm on the left figure: Intriguingly, a similar outstretched arm recurs in *Guernica* thirty years later.

Whatever the reasons for these changes in content and tone, the final work achieves tremendous power. Seeking to account for its enduring effectiveness, art historian and critic Timothy Hilton maintains that "there was never before—and never has been since, one might add—a painting with such a system of internal torques, volume made into a twisted or scything line." The work serves as a culmination thus far of the various experiments Picasso had been engaged in privately in his notebooks—multiple views, paired opposites, circular and jagged forms that crashed, and soft and harsh colors that clashed: Picasso seemed to be testing how much he and others could take in one canvas. The biographer Roland Penrose suggests that the work holds such power because Picasso was himself involved in a battle, a conflict, between the subtle charms of his rose period and the formal pulls of a more geometric form of art, and that the battle is actually enacted on the canvas, thus epitomizing a pivotal moment in the history of Western art.

Picasso's ambivalence toward the work was heightened by the reaction of his acquaintances and friends. Even those who had admired his earlier work did not know what to make of *Les demoiselles*. Reactions ranged from confusion or mystification (Gertrude and Leo Stein, Braque) to downright rage (Matisse). Only two dealers, Daniel-Henry Kahnweiler and Wilhelm Uhde, showed interest. Kahnweiler later recalled: "What I'd like to make you realize at once is the incredible heroism of a man like Picasso, whose moral loneliness was, at the time, quite horrifying, for none of his painter friends had followed him. Everyone found that picture crazy or monstrous." Picasso described what it is like at such moments when one is taking enormous risks: "Painting is freedom. If you jump, you might fall on the wrong side of the rope. But if you are not willing to take the risk of breaking your neck, what good is it? You don't jump at all. You have to wake people up. To revolutionize their way of identifying things. You've got to create images they won't accept." Stung, though not derailed by the mostly hostile reactions, Picasso quietly put away the canvas and did not display it publicly for several years.

Eventually, of course, the achievement of *Les demoiselles* and its unique status in the history of the period came to be widely recognized. (In this sense, its career is somewhat like that surrounding Stravinsky's *Le sacre du printemps*, which is discussed in chapter 6. Another work that tested its audience, *Le sacre* was first performed just six years later.) Or, speaking sociologically, the field elected to regard this work as a masterpiece.

One might have thought that, thus vindicated, Picasso would have shared the origins and evolution of the work with some pride. In fact, however, Picasso did not make public his many sketchbooks leading to *Les demoiselles*. Only sometime after his death was a complete pencil sketch of *Les demoiselles* found, folded inside one of the notebooks as if it were a shopping list, while the aforementioned oil sketch surfaced in the early 1990s. Picasso was always playing games with his public, and these are perhaps simply a set of posthumous surprises that he held in store for us. But Picasso was never one to spend much time looking backward; *Les demoiselles* may simply have been a way point for him, en route to the bigger game that was cubism.

THE PARTNERSHIP THAT MADE CUBISM

Almost an exact contemporary of Picasso's (he was born one year later), Georges Braque differed in most other respects from the Spanish artist. The short, compact Picasso was passionate, instinctively rebellious, and self-consciously prodigious; having no interest in music, he lived and breathed painting. The tall Frenchman Braque was unashamedly bourgeois, precise, and somewhat shy; he liked to box, dance, and play the accordion. In no way a prodigy, Braque was in fact not even a particularly good draftsman and was hesitant to draw the human figure. Drifting into painting, he was much influenced both by the fauvist school of painting and by Cézanne's work. Braque and Picasso apparently met in 1907, the year of *Les demoiselles*: The painting had a marked, if highly unsettling, effect on the young Braque. He commented later: "It made me feel as if someone was drinking gasoline and spitting fire." Within a few months, he had painted his *Large Nude*, unusual in its portrayal of traditional subject matter and reflective of the same kinds of artistic challenge Picasso was undertaking.

Picasso and Braque became friendly and started to work together. The collaboration began around 1908 and lasted for the next six years, until Braque was called to battle in the Great War. In some contemporary professions, such as science or corporate life, collaborations are common, but in the area of painting they remain rare. In this instance, over and above whatever personal attraction may have existed between these divergent personalities, the Picasso-Braque collaboration centered on artistic innovation. It represented, within the domain of painting, the same kind of intimate exchange Picasso had hitherto enjoyed across domains with his close friend Apollinaire.

In a word, Picasso and Braque, working together, invented and then probed the artistic style called cubism. Perhaps either one of them, working alone, might have continued the process begun by Cézanne, decomposing figurative painting into its component lines, forms, and planes, anticipating future, more abstract trends. But without question, the particular form cubism took and its speed in transforming the artistic world resulted from the unusually intense and productive collaboration between these two artists, still not thirty years of age.

For many months the two men were virtually inseparable. They painted during the day and then, at night, came together and studied one another's works. At times, their works were so similar that only an expert can tell them apart; as if to confirm the impersonality of their work, they refused to sign certain canvases: The process of cubist deformation, and not the identity of the specific artists, was primary. The two men appreciated one another's company; they jokingly referred to themselves as "Orville and Wilbur Wright"; and in an effort to please Braque, Picasso introduced him to women, including one whom Braque later married. There was good-natured competition as well as cooperation; sometimes, one man hid his work from the other; and especially after 1911, each man strove to outdo the other in inventiveness. Still, overall, the collaboration was intimate and symbiotic. As Braque was to recall nearly a half century later:

> We lived in Montmartre, we saw each other every day, we talked . . . Picasso and I said things to each other during those years that no one says anymore . . . things that would be incomprehensible and that gave us so much joy. . . . It was like being roped together on a mountain. . . . We were above all very absorbed.

An index of the importance assumed by the other individual is the fact that Picasso failed to keep regular notebooks only during this phase of his life. Ordinarily, Picasso used his notebooks as a pad and a goad for thinking—a place to try out ideas, reflect on them, chart his course. To be sure, this process did not cease completely during the cubist period, but just possibly, the role ordinarily assumed by the written journal was now instead taken by a live, interactive collaborator, critic, and reactor.

In the chapter on Freud, I discussed the important, perhaps indispensable, role that supportive individuals can play when creators have ventured into territory that is unfamiliar and, in others' eyes, beyond the pale. In Picasso's case we also see the shift from a general support system, as in the Barcelona and Parisian bohemian circles, to an intimate link to a single individual who can fulfill both cognitive and affective needs and who in this instance participates as a partner in the creation of a new symbol system.

Mary Gedo argues that Picasso had especially strong needs to be involved with other supportive individuals, needs that may well date back to unresolved attachment to and dependencies on his father and his mother. For the most part, these needs in the artistic domain were met by men of a literary stripe, such as members of the circles in Barcelona and in Montmartre. One can recount a string of such individuals, beginning with Sabartés in early life, Apollinaire and Jacob during the first years in Paris, and then the poet and artist Jean Cocteau, the poet Paul Éluard, or the sculptor Julio González* in later life. However, when it came to exploring new aesthetic dimensions, it may have been necessary for Picasso to work with someone with whom he could actually share the particularities and technicalities of his craft.

Cubism's Nature and Legacy

No story in the history of modern art has been told more often than that of the origins of cubism. The desire to create a convincing historical narrative occurs not only because cubism is so important a movement but also because a number of convincing stories can—and have—been told. I have already mentioned the crucial role played by the work and example of Cézanne, a role explicitly acknowledged by both of the founding cubists.

To this art historical account, one can add a complementary ensemble of stories about cubism's origins. Some see cubism as an accentuation of the primitive forms in African tribal masks of the sort Picasso saw in Matisse's home and examined at length in the Museum of the Trocadero, or in some of the flattened forms of classical Egyptian art; others see cubism as the elevation to high art of forms already widely used in caricature, poster art, and other kinds of "low" or "popular art." Some describe cubism as an attempt to draw as children do and to see as children do, or as a (possibly unconscious) return to the scribbled doodles, experiments, and marginalia with which young Picasso (like other schoolchildren) embroidered his journals. To others, cubism is a movement that drew on the insights and models of perception being discovered by psychologists around the turn of the century, or it is part of a larger artistic and scientific movement, where a single "right" or "privileged" perspective was eschewed in favor of multiple viewpoints or a more relativistic stance. With a touch of irritation, art historian Alfred Barr has commented:

Mathematics, trigonometry, chemistry, psychoanalysis, music and what not have been related to cubism to give it an easier interpretation. All this has

*Sometimes spelled Julio Gonzáles.

been pure literature, not to say nonsense, which brought bad results, blinding people with theory.

And Picasso echoes this sentiment in his remark: "When we invented cubism, we had no intention of inventing cubism, but simply of expressing what was in us."

Any epochal change or "paradigm shift" is most likely due to an aggregate of causes, no one of which is indispensable. One could envision cubism without Nigerian masks at the Trocadéro, optical illusions in William James's psychology textbook, the discoveries of Einstein, the writings of the symbolist poets, or the imagery of student notebooks, cabaret posters, or the master Cézanne. But the sheer number of emerging iconoclastic trends that pointed toward this shift in artistry is probably significant, raising the likelihood that the shift would occur and would gain some kind of acceptance.

Despite the "prepared" environment and the nigh inevitability of cubism, scholars do not yet agree on just how to characterize the movement's nature and significance. Is cubism pathbreaking in its own right, or does it chiefly signal the decline of figurative art and a paving of the way for completely abstract art? Does the fragmentation of the object occur to spur the viewer's fuller integration of mass and motion, or does the fragmentation instead underline not only the impossibility of our ever encompassing a full-blown object in two dimensions but also an insistence on the discontinuity of all experience? Is cubism an effort to convey the physical dimensions of objects, or does it call attention instead to the material elements of the paint and of the experience of making art? Are these works fundamentally serious—a study of the nature of reality and its perception—or have they been produced in a lighter vein, to amuse and provoke as much as to educate? Are the works designed as a political or social critique of contemporary society, or do they exhibit a studied neutrality on such controversial issues? Even the name has provoked controversy: Was it used dismissively, as a disparagement of artists who reduce everything to "the little cubes"? Or is it a mark of accomplishment that Picasso and Braque were able to capture the essence of objects by emphasizing their geometric configuration?

In my view, just as cubism did not grow out of a single factor, neither did it reflect a single perspective on aesthetic, philosophical, or practical issues. While the painters were deadly serious in their investigations, they also were witty and wanted to amuse themselves and others. They were eager to see how much of an object, or a set of related objects, could be conveyed on the two-dimensional (or, ultimately, two-dimensional-plus) surface, but they knew that every depictive decision involved costs as well as benefits. They wanted to challenge the primacy of a realistic, object-centered art, but they continued to feel the compensatory need to root their work in the world of objects. And unlike their counterparts in other expressive forms, neither ever made the decisive leap to a completely nonobjective art. Perhaps it was the sheer involvement in this highly novel—and almost dangerously anarchic—undertaking that fueled the pursuit: The answer that was ultimately embraced proved far less decisive than either the questions raised or the pursuit itself.

Some of cubism's paradoxical properties may have derived from Picasso and Braque's contrasting personalities and practices. As a more proficient depicter of the natural and the human worlds, Picasso may have been responsible for the stronger representational aspects, the focus on objects with their idiosyncratic peculiarities, whereas Braque pushed more toward abstraction. Picasso's virtuosity also contrasted with Braque's interest in, and contribution to, more technical aspects, particularly those having to do with the creation of purely spatial effects and experiments with composition.

But to assume that cubism simply reflected the strengths and weaknesses of each artist would be to oversimplify. Each artist spurred the other to try doing new things. Picasso might never have stilled his exuberant personality without Braque's

FIGURE 5.6.
Portrait of Uhde (1910), oil on canvas (31 7/8 x 23 5/8).
A geometric and monochromatic portrait characteristic of analytic cubism.
© 1993 ARS, N.Y./SPADEM, Paris.

austere example, whereas Braque might never have incorporated objects and graphic elements from diverse domains without Picasso's inventive example and goading.

The Phases of Cubism

Whatever their manifold interpretations, art historians agree that cubism passed through a set of phases from 1910 to 1916. The first phase, the period of early, high, or analytic cubism in 1910 and 1911, involved a working out of the method of decomposition. This phase was a time of austerity, of mundane subject matter rendered monochromatically (see figure 5.6). The goal was to analyze objects into their component parts, with natural forms being reduced to semigeometrical forms that in turn were further broken down, dislocated, or flattened. The actual identity of the objects became less important, as the act of depicting per se now exerted dominance over the objects being depicted; the apparition of space evident in the earlier paintings gave way to an increasingly shallow surface.

Indeed, as they looked at one another's works, the artists themselves often could not agree on the representational aspects. Once, around 1911, Picasso claimed to see a squirrel in one of Braque's canvases; then Braque spied it as well and spent the next week attempting, without success, to eradicate its traces. Picasso and Braque worked together in an unprecedentedly intimate way; each work was a response, in the only appropriate symbol system, to earlier works. Probably no one else could fully apprehend their personal and graphic interactions during that initial period as they felt their way toward a new graphic language.

A more informal, relaxed, and witty phase followed. During the middle period of 1912 and 1913, Picasso and Braque experimented with the use of objects, papers, letters, words, puns, disguises, sheet music, wallpaper, cigar labels, and other elements that could be incorporated into the works. Picasso assumed a greater role in the introduction of collage techniques, starting with the gluing of a piece of cane-patterned oilcloth to the surface of a canvas (see figure 5.7). Braque invented papier collé, in which pieces of wood-grained paper were incorporated into a picture, assuming a representational role while remaining recognizable as pieces of paper. Instead of carefully applying oil paint, the two artists now used much broader, harder daubs, sometimes mixing the paint with sand and sawdust; and they relied on these more "democratic" materials to express ideas. Parts of objects became larger and more recognizable. They also extensively used newspaper clippings, separate cutout words, and other artistic and numerical symbols, because of their intrinsic graphic effects and the political and social messages they conveyed.

This middle phase, in which the artists introduced external elements into the canvas, marked a shift from *analytic* to *synthetic* cubism. During the analytic phase, Picasso and Braque highlighted the analysis or breaking down of objects. In the synthetic phase, they emphasized the synthesis of objects from parts—for

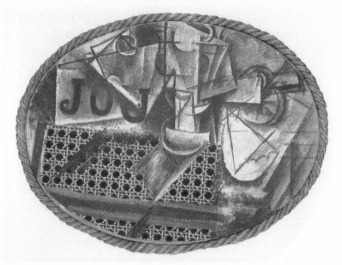

FIGURE 5.7.
Still-Life with Chair-Caning (1912), oil, pasted on oilcloth paper
on canvas, surrounded with rope (10 5/8 x 13 ¾).
A collage of objects characteristic of synthetic cubism.
© 1993 ARS, N.Y./SPADEM, Paris.

example, from the vestiges included in a collage. Pictures were made more boldly and rapidly, with little of the meditative quality of the earlier phase. One senses here that Picasso and Braque were comfortable with one another, but also that they were no longer collaborating so intensively. Competition became more salient, too, as Braque deliberately held back on introducing new ideas in Picasso's presence, in order to elaborate on them privately and then spring them on Picasso and the world. Synthetic cubism continued for perhaps a decade, but its course was not strongly influenced by Picasso and Braque after 1914.

Later phases of cubism also involve other artists, chief among them the younger Spanish artist Juan Gris (José V. González). In his history, Alfred Barr speaks of "surrealist," "rococo," and "curvilinear" cubism. Braque went to war, and Picasso, while continuing to use cubist techniques throughout his life, felt no compunction about reverting to elements of his precubist output, incorporating aspects of the history of art, particularly from the classical eras, and launching new, decidedly noncubist styles. By 1915 or 1916, the Picasso-Braque cubist era was effectively at an end.

Time Together, Time Apart

As remarkable as the fact that two artists could collaboratively launch a revolutionary movement is the fact that, at least in the earlier stages, two artists of such differing abilities and temperaments were able to progress together. Such a union required a suppression of the ego that, at least in Picasso's case, represented a considerable feat. However, it would be risky to assume that all of the dynamism occurred because of the two men's regular interaction. At least with Picasso, it was important at various times to get away from the action, to live and think in a new way for a while; these excursions also brought out a lightheartedness and love of life that were difficult to exhibit in the crowded, competitive, and often difficult circumstances of the city during the year.

The first of these sabbaticals occurred in 1898 when the adolescent Picasso, accompanied by a young painter named Manuel Pallarés, went to Horta de San Juan. There Picasso was introduced to peasant life, to farming, and to an environment rich in vegetation but surrounded by barren limestone mountains. According to Penrose, these surroundings provided inspiration a decade later for some of the first cubist landscapes. In the summer of 1906, accompanied by his mistress Fernande Olivier, Picasso journeyed to Gosol, a village on the southern slope of the Pyrenees. This peasant community, with its rugged scenery and Spanish folkways, made Picasso relaxed, happy, and productive. He was amazingly prolific, completing as many works in ten weeks as he had during the six months before. While many paintings in Gosol were pastoral, Picasso went beyond the sentimentality of his earlier work, confronting issues of classicism and primitivism; and he also worked intensively on some of the formal issues that had engaged him in the preceding months. Upon his return to Paris, he completed the *Portrait of Gertrude Stein*, which he had left unfinished at the end of the spring.

In the summer of 1908, Braque visited the town of L'Estaque in southern France and returned with Cézanne-style landscapes that were to be labeled cubist; at the same time Picasso returned from La Rue des Vois with paintings similar to Braque's. Then in the summer of 1909, after the first flourishes of cubism, Picasso returned to Horta de San Juan, but with a very different agenda. The still-familiar sights could be reexamined in light of the formal experiments he and Braque had been undertaking. Drawing on the inspiration of Cézanne, he produced landscapes that were recognizable but whose underlying geometric simplicity was now foregrounded. The insights he had forged with respect to the human figure were now confirmed with respect to the natural and artifactual environment of this remote village, as hills and houses were dissolved into ambiguously overlapping planes. Picasso returned to Paris with a whole sheaf of paintings that, more than any produced before, epitomized the development of the new style.

Even in the most intimate and successful relations between human beings, there is a need for some moments apart, for time and space of one's own, for opportunities to revisit old themes in light of a new perspective. Perhaps this is particularly true for those creative individuals who are entering uncharted territory. At such times, as I have argued, another human being can play a vital midwifery role, as clearly occurred in the unique relationship between Picasso and Braque, and, subsequently, in Picasso's relationship with several of his lovers. But a compensatory distancing may be equally essential. Both during the days of solitary work and during longer intervals of separate work, as in the summer, Picasso and Braque respected this rhythm of human relationships. (In retrospect, their decision to spend the summer of 1911 together in Ceret may have been a mistake.) Times of solitude proved as essential as times of intimacy. Eventually, tensions mounted, the need for distance magnified, and once Braque went off to war, the relationship was ended; as Picasso commented with metaphoric accuracy: "Thereafter I never saw Braque again." But perhaps it is not a coincidence that neither of them ever reestablished such a close relationship with another artist, and neither ever blazed comparably novel graphic trails in his own later work.

The Popular and Avant-garde Responses

Though art critics soon recognized the importance of cubism, the wider public hardly embraced this new approach. Even today, a century after its introduction, cubism continues to be emblematic of modern art and is treated by many as an enigmatic curiosity rather than a peak of artistic achievement. Still, while *Les demoiselles d'Avignon* was so problematic that Picasso kept it under wraps for years, and while Picasso rarely arranged for his cubist works to be exhibited publicly, the new cubist art became known from its early years. Indeed, in 1913 it was featured in a memorable show at the New York City Sixty-Ninth Regiment Armory, serving as the veritable introduction of modern art to the shores of North America.

Cubist art was controversial. Picasso tried to avoid this controversy by refusing to show in the salons and by declining the opportunity to answer its critics, but the controversy continued nonetheless. Many critics denounced the work as mediocre or as worthless—terms like *grotesque, barbarian, ultrarevolutionary, ridiculous,* and *deliberately shocking* were bandied about. Even those who saw merit conceded that the art was hermetic and elitist. Any number of spoofs and satirical works were mounted. The principal defender of the movement was Picasso's longtime friend Apollinaire, who, in a series of articles and in an influential book, laid out the rationale for cubist art and defended the importance of the movement and the contributions of its two founding members. As he put it, cubism had laid out a wholly new conception of beauty. Other sympathetic critics described cubism as bringing science into art, as being the first objective art, and as revealing the structure of objects as never before. And some, like John Middleton Murry, pronounced it great art even though they confessed that they did not understand it.

A distinction between popular taste and avant-garde taste hardly represents a new page in the history of the arts, and it certainly applies to most artistic movements from the middle of the nineteenth century on. But with the rise of cubism (and with other movements in other art forms of this period), one encounters a new phenomenon, where the artworks esteemed by the experts remain remote from the wider public indefinitely. Nowadays every culturally literate soul knows the names of Picasso and Braque, Stravinsky and Schönberg, Duncan and Graham, T. S. Eliot and Joyce; and yet, audiences gravitate far less to the works of these figures than to their more accessible predecessors Monet and van Gogh, Brahms and Wagner, Dickens and George Eliot, or to classical ballet troupes. The distance between the popular field and the avant-garde field has widened appreciably in the twentieth century; it is unclear at this time whether the gulf will continue to widen and which field is the more potent. To be sure, aspects of cubist art have become part of the landscape, from commercials to camouflage, but often without the audience's awareness of the sources of the imagery.

AFTER CUBISM: LIFE AS CELEBRITY

But if Picasso's most recent works were raising eyebrows, he was already becoming an international institution by the end of the cubist era. He was seen as a prodigy of great versatility, as a mature genius with unparalleled powers. While continuing to produce synthetic cubist works, he was also issuing realistic pencil portraits and drawings in a style more reminiscent of Ingres than of Cézanne. The paintings of the blue and rose periods, the beggars and the circus folk, were selling for sizeable amounts of money, and while cubist paintings were still regarded as curiosities, the paintings made after 1914 were again of interest to the art-consuming public. An entourage grew up around Picasso advertising his works, selling them to wealthy collectors, and describing him as the greatest artist of his era. And he had so outstripped his contemporaries that his principal identifications were—as they would remain—with the masters of the past.

The Works of a Grown-up Prodigy

One would have thought that the vaunted rebel Picasso would be either oblivious to, or distraught by, this notoriety, particularly since (as is so often the case) his enthusiasts understood little about what he was attempting to accomplish. In fact, however, following the departure of Braque for war, and the death from tuberculosis of his beloved mistress, Eva,* Picasso underwent a significant personal change.

By his mid-thirties, Picasso had lived several lives: as an unhappy schoolboy; a prodigious journeyman; a rebel at the academy; a young intellectual in the

---

*Her actual name was Marcelle Humbert, but she is always referred to by the name conferred on her by Picasso.

cafes of Barcelona and Paris; a struggling painter in Montmartre; an artist on the eve of fame, after his blue and rose periods; and a self-styled leader of the avant-garde during the various phases of cubism. Secure in the knowledge that he was a wealthy, successful, and acclaimed artist, Picasso opted for a more comfortable, bourgeois existence.

This life choice took several forms. Picasso moved to opulent quarters and began to associate with high society. He undertook collaborations with the leading creative spirits of his time, including the choreographer Serge Diaghilev and the composer Erik Satie, as well as Cocteau and Stravinsky. He traveled beyond France and Spain. Of greatest significance, after having a series of bohemian mistresses, Picasso married Olga Koklova, a Russian ballerina, the daughter of a general, and a lover of glamour and society. Completing the embracing of bourgeois life, Picasso and Olga Koklova had a son, Paolo, in the summer of 1921.

In the period after cubism, Picasso's work showed the earmarks of a prodigy who had grown up, perhaps too well. Having navigated through several stages in his own development, he drew heavily on his earlier work in the canvases produced in the succeeding ten years. He took note of, but did not partake significantly in, newer movements, such as surrealism and abstractionism. He began to experiment with other media of expression, particularly sculpture. This phase of Picasso's life is often labeled as a classical (or neoclassical) period, because Picasso deliberately embraced forms and models associated with Greek and Roman times. But as in the case of Stravinsky, whose neoclassical phase coincided with Picasso's, the neoclassical works bore the unmistakable stamp of Picasso's own style. Picasso himself captured the flavor of this work best:

> Don't expect me to repeat myself. My past does not interest me any more. Rather than recopy myself, I would prefer to recopy others. At least I would bring something new to them, I like discovery too much. . . . What after all is a painter? He is a collector who wants to make a collection by doing the paintings he saw in other's collections. That's how it starts but then it becomes something else.

Personal and Artistic Upheavals

While the period from 1916 to 1926 was one of relative stasis, the following decade proved a far more troubled time for Picasso. His personal life was becoming more complex and more stressful. Never easy with the notion—or the practice—of fidelity, Picasso grew more sexually adventurous. In addition to many occasional liaisons, the middle-aged artist began a serious involvement with the adolescent Marie-Thérèse Walter, often pitting her directly against his wife. He had a child, Maja, out of wedlock with Marie-Thérèse. (Maja's actual name was

Maria Concepcion, a mixture of the names of Picasso's mother and his late sister). The situation became even more strained when Olga sued for divorce and another mistress, the intellectual Dora Maar, entered the picture. Picasso enjoyed his children when they were young, but he soon tired of them and became estranged from Paolo.

Picasso's upheavals affected his painting. For perhaps the first time in his life, Picasso experienced significant periods where he did not paint. The output of pictures for certain years was far less than his previous norm. According to Gedo, Picasso typically produced about three hundred paintings and drawings in a year; but between 1926 and 1936 his output seldom reached one hundred works in a year. His various experiments with women and with artistic styles and media, and his mixing of bohemian and bourgeois life options, are seen by Timothy Hilton as Picasso's increasingly desperate ways of staving off a collapse of inspiration (and, perhaps as well, the ravages of age). Perhaps Picasso was unconsciously raising the ante in his life, as he had always done in his work, as a means of stimulating his imaginative powers.

Whether Picasso's works also declined in quality during this period is more difficult to say. Clearly, however, the themes became more tragic and the expressive forms more brutal. Picasso had always painted in a cool and clinical way; indeed, he had been called icy in the cubist period. But now the portraits of women were often brutal, filled with misshapen or dissociated parts, grotesque positions, and distorted facial features; and later versions of a female figure in the notebooks often were more distorted than the earlier ones of that figure had been. Picasso began to introduce mythic beasts into his work, particularly minotaurs (monsters with the head of a bull and the body of a human). The minotaurs were deliberately ambiguous figures sometimes depicted as evil rapists; at other times as calm, even innocent, observers; sometimes as triumphant figures; sometimes as victims of an evil monster. In the ever-abundant bullfights, injured bulls and dying horses were now featured. There were bacchanals, odd sexual couplings, contorted genitals, and warped crucifixions, plus a growing obsession with the voyeuristic and exploitative relationship between the artist and the model. Works of art had to shock. As Picasso said one day with conviction: "A work of art must not be something that leaves a man unmoved, something he passes by with a casual glance. . . . It has to make him react, feel strongly, start creating too, if only in his imagination. . . . He must be jerked out of his torpor."

Disturbing events were also occurring in Europe. Following the relative calm of the twenties, fascism had emerged with ferocity in Italy, Germany, and Spain. While Picasso was never a pointedly political person, his sympathies lay clearly with democracy, communism, and, above all, anarchy—and not in the least with fascism. The chaos that was occurring in Picasso's personal life appeared to be mirrored in the disintegration of traditional European civilization.

GUERNICA: THE SELF-DECLARED MASTERPIECE

For some individuals, a decline in fortune or spirits produces a crisis in work; Picasso was not immune from this plight. Yet one of Picasso's most engrained personality traits was his compulsion to paint "in opposition." A violent and destructive streak in Picasso motivated much of his work, particularly in his later years. "After all, you can only work against something," Picasso once commented. "I make paintings that bite. Violence, clanging cymbals . . . explosions . . . A good painting—any painting!—ought to bristle with razor blades."

By the mid-1930s, these various strains and conflicts in Picasso's life were coming together in his art. In individual works, and in series of paintings such as *The Dream and Lie of Franco* and *The Minotaur*, Picasso revealed a violent and tragic strain only hinted at during the earlier phases of his art. In the Franco series, the Caudillo is involved in comic, horrible, and obscene scenes; a bull acting a hero's part disembowels a Franco-like monster, consisting of a horse with a polyp's head. *Minotauromachie* (1935) is an incredibly complex composition, with a lone young girl holding a light and surrounded by a menacing minotaur, a disemboweled horse, a dying female matador, and various remote human figures peering at the strange scene, including a Christ-like artist figure escaping on a ladder at the left side of the canvas (see figure 5.8). (It has been suggested that *Minotauromachie* represents the sexual act as it looks to a

FIGURE 5.8.
Minotauromachie (1935), etching (19 ½ x 27 7/16).
A depiction of Picasso's violent fantasies.
© 1993 Ars, N.Y./Spadem, Paris.

FIGURE 5.9.
Weeping Woman (1937), oil on canvas (23 5/8 x 19 ¼).
The tragic pain of Picasso's *Weeping Woman*.
© 1993 ARS, N.Y./SPADEM, Paris.

child.) *Weeping Woman* (1937) shows a woman with an almost unbelievably fragmented visage and equally unendurable pain (see figure 5.9). We may assume that in his own imagination, Picasso was going through the most lurid and violent fantasies, and he himself often referred to this period as the most painful of his life, though a fair-minded observer would have to say that much of Picasso's pain was self-imposed.

When it came to the most brutal act of the period, however, the pain was in no way attributable to Picasso. On the 26th of April in 1937, German bombers in Franco's army wiped out Guernica, a small market town that had been the ancient capital of the Basque region. This wanton act killed thousands of people, who were thronging the streets on market day; it horrified the world and forever branded Franco and his military as inhuman. Terribly shaken by this event, Picasso resolved to capture it forever in a painting. He had been commissioned to prepare a mural for the Pavilion of the Spanish Loyalist to Government at the 1937 World's Fair; he decided almost immediately that this work would be his fateful contribution.

I have spoken of certain early paintings, like *La vie* and *Les demoiselles d'Avignon*, as defining works. Other works from the intervening periods, such as *Three Musicians* (1921), *Three Dancers* (1925—see figure 5.4 above), *Woman in an Armchair* (1929), or the aforementioned *Minotauromachie* might be included in such a list. Picasso himself said, "If it were possible . . . there would never be a 'finished' canvas but just different states of a single painting." But seldom in the history of human painting has a single work been so clearly destined to be "defining" as Picasso's *Guernica* (1937—see figure 5.10).

FIGURE 5.10.
Guernica (1937), oil on canvas (351 cm x 782 cm).
Picasso's self-declared masterpiece.
© 1993 ARS, N.Y./SPADEM, Paris.

Confident in his own mind that this work, done in his later eclectic cubist style but featuring a classically massive composition and theme, would be one for the ages, Picasso carefully documented his preparatory steps in exquisite detail. He declared: "[All of my paintings] are researches . . . there is a logical sequence in all this research. That is why I number them. It's an experiment in time. I number them and date them. Maybe one day someone will be grateful." Explaining further, he said: " . . . it's not sufficient to know an artist's work—it is necessary to know when he did them, why, under what circumstances. . . . Some day there will undoubtedly be a science—it may be called the science of man— which will seek to learn about man in general through the study of the creative man." Picasso in fact executed approximately forty-five sketches in preparation for *Guernica*, and with few exceptions, these are numbered and dated. Also, thanks to the photography of Dora Marr, Picasso's inamorata of the time, we also have a record of seven stages in the production of the final mural, as it was being prepared for display in the pavilion.

When the most famous artist in the world fashions the most illustrious painting of the century and leaves a detailed paper trail behind, one can be sure that the record will be carefully combed by experts of every description. From the curator Anthony Blunt, to the psychologist Rudolf Arnheim, dozens of scholars have had a go at the *Guernica* sketches. (Later I describe a similar pattern at work with the sketches for Stravinsky's *Le sacre du printemps* and Eliot's *The Waste Land*.)

These accounts contain strikingly similar conclusions. The compositional seeds of *Guernica*, say the scholars, are located in a number of works Picasso had recently completed, and especially in the aforementioned nine-work series *The*

Dream and Lie of Franco and in the large and complex etching *Minotauromachie.* The work is also influenced by other heroic canvases of the past, particularly those produced by Nicolas Poussin, Jean-Auguste Ingres, Matthias Grünewald, and Eugène Delacroix, and there is clear lineage to the classical theme of the "slaughter of the innocents." Picasso's lifelong fascination with the bullfight is echoed in both subject and tone. The work also harbors intimations of horrifying experiences from Picasso's own life, particularly his memory of the chaos when his family had fled the earthquake when he was three.

A hurried opening sketch (see figure 5.11) conveys to Arnheim the composition of the final work; Hilton calls attention to the first two sketches, seeing them as a re-creation of the basic *Minotauromachie* compositional pattern; and testimony by Picasso underscores that the earliest vision tends to be preserved across intermediate drafts. "Basically," he asserted, "a picture doesn't change, the first vision remains almost intact, in spite of appearances." I think it requires a certain wisdom in hindsight to see the final work in these rough sketches, but I am willing to concede, particularly in light of the organization of *Minotauromachie,* that Picasso had the general compositional matrix in mind from the start. Certainly, he knew that it would be a massive narrative canvas, even if he had not worked out the ratio of the sides, the precise cast of characters to be featured, or the ensemble of tones to be highlighted.

The Composing of Guernica

Picasso's working method, already familiar to us from his notebooks, comes across with full force in the *Guernica* sketches. As in a baroque concerto, there is a constant oscillation between parts and wholes, between chaotic limbs and twisted torsos, on the one hand, and a lightly sketched village panorama, on the

FIGURE 5.11.
First Sketch for Guernica "May 1, 1937 (1)," pencil on blue paper (10 5/8 x 8 ¼).
© 1993 ARS, N.Y./SPADEM, Paris.

FIGURE 5.12.
Composition Study "May 9, 1937 (2)," pencil on white paper (17 7/8 x 9 ½).
© 1993 ARS, N.Y./SPADEM, Paris.

other. Compulsive work on details alternates with more distanced overviews, in which the overall compositional pattern is manifest, and with lighthearted sketches, where Picasso gives reign to his wit. Six of the forty-odd sketches deal with the entire composition (see figure 5.12); the remaining sketches include experiments with individual animals, humans, facial features, and shapes, each in appropriate proportion to its importance in the final canvas. Featured in the ensemble is a careful experimentation with different eyes, a toying with the arrangement of the principal figures, a dozen different views of the bull (see figure 5.13), and a buoyant horse rendered in childlike fashion. During various sketches, Picasso himself and those close to him appear, and hints of them can even be discerned in the final version of the work.

A few features remain constant throughout (for example, the woman with the beacon of light); some (like the mother with the dead baby) travel to different loci (see figure 5.14); most others are altered extensively. Thus, as one example, the dying soldier comes to occupy an increasingly dominant role in the mural; as another

FIGURE 5.13.
Head of Bull with Study of Eyes "May 20, 1937,"
pencil and gray gouache on white paper (11 ½ x 9 ¼).
© 1993 ARS, N.Y./SPADEM, Paris.

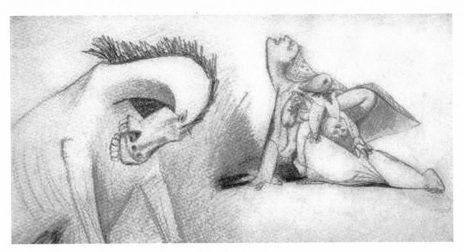

FIGURE 5.14.
Horse and Mother with Dead Child "May 8, 1937 (2),"
pencil on white paper (17 7/8 x 9 ½).
© 1993 ARS, N.Y./SPADEM, Paris.

FIGURE 5.15.
Final Mural of *Guernica, Initial State* (1937)
© 1993 ARS, N.Y./SPADEM, Paris.

example, the bull's positions, visage, and degree of detachment from the scene vary
markedly across the sketches and even over the seven stages of the final mural (see
figure 5.15). (Apparently, some decisions could be made only when the work was
assuming its final massive size.) Similarly, the relationship between the stolid bull
and the agonized woman undergoes continuous changes in the preliminary
sketches. Picasso seems unclear till the last on how benign, threatening, or stolid he
wishes the bull to be; and indeed, he suggested that the bull did not denote fascism

directly but said that the horse did represent "the people." These various changes represent Picasso's thinking via the symbol system of graphic depiction.

Guernica is of abiding interest to students of creativity, as well as to students of art, because of the way it, like other defining works, draws on the multiple strands of the creator's world. The work is at once of national and even global significance, even as it is intensely personal. A work of both classical composition and childlike imagery, *Guernica* conveys the vision of chaos as that might appear to an innocent child. Picasso captures aspects of all conflict, of the Spanish civil conflicts, and of the conflicts in himself about violence, sexuality, and artistic creation. He issues a powerful statement against war and a decisive indictment of Franco's fascism. As Picasso said: "I have always believed that artists who live and work with spiritual values cannot and should not remain indifferent to a conflict in which the highest values of humanity and civilization are at stake." And, more pointedly: "What do you think an artist is? An imbecile. . . . He's at the same time a political being, constantly alive to heart rending, fiery or happy events. . . . No, painting is not done to decorate apartments. It is an instrument of war for attack and defense against the enemy."

The work is an aesthetic *summum bonum*, with symbols that had punctuated Picasso's work for many years, now juxtaposed with a controlled chaos and cumulative power not previously achieved. As Blunt notes, "It combines the

FIGURE 5.16.
Nightfishing at Antibes (1939), oil on canvas (81 x 136).
A more playful and peaceful work from the *Guernica* period.
© 1993 ARS, N.Y./SPADEM, Paris.

emotional intensity of the Blue Period with the fantasy of the Metamorphic paintings, and on the other hand it makes use of the classical draughtsmanship of the early 'twenties as well as the strict formal discipline of Cubism." The final work, touching on nearly all of Picasso's obsessions, has an almost overwhelming density of information; and still, because of its enormous size and the epochal nature of the topic, *Guernica* succeeds.

Indeed, it is not hyperbolic to consider *Guernica* as the defining work among all of Picasso's defining works—his most confident and daring reach for immortal status. It captures the themes of love, life, and death of *La vie*; the sordidness and inhumanity of *Les demoiselles d'Avignon*; the unexpected frenzy and ecstatic violence of the surrealist *Three Dancers*; the deformations of *Woman in an Armchair* and *Weeping Woman*; and the tangle of themes and personages of *Minotauromachie*. Only the relative playfulness and peacefulness of *The Three Musicians* and *Nightfishing at Antibes* (see figure 5.16) are missing from the final version (though both of the latter works harbor signs of dark foreboding).

AN AGED PRODIGY

Picasso completed *Guernica*, clearly the culminating artistic work of his life, when he was fifty-five years old. Had he died at the time, the history of modern art would be approximately the same, and Picasso's reputation would have been secure. But Picasso lived for another thirty-six years, painting thousands of additional works, worth many millions of dollars, and creating in a range of media from commercially viable ceramics to surrealistically farcical plays. Meanwhile, his complex personal life became more so, complete with a tumultuous ten-year romantic relationship with the painter Françoise Gilot, legal disputes about publications and property, varying degrees of estrangement from his four children, and several forays into politics, including a direct embrace of the Communist party in the 1940s.

The relationship between the chaos of Picasso's personal life and his continuing artistic fecundity deserves comment. One can see in Picasso's life not only a steady sequence of new homes, mistresses, children, and summer escapes but also a steady stream of new styles and defining works. Some, like Gedo, have even claimed that each of Picasso's mistresses served as a catalyst for fresh artistic experimentation. Even if one resists the impulse to discern a one-to-one relationship between milestones in Picasso's personal life and developments in his artistic life, Picasso thrived in a certain sense from a life studded with involved imbroglios and sharp discontinuities. Working in a domain where development throughout the life course remained a possibility, Picasso may have helped to ensure his continuous evolution through the search for experiences that were novel, bracing, catalytic, and pregnant with flow.

Until well into his eighties, Picasso exhibited enormous drive and energy: He lived to work. He turned his attention to the art of the past and "reworked" many great artists, most notably Edouard Manet, Gustave Courbet, El Greco, Delacroix, and Velázquez, creating his own variations on *Las meninas*. In the 1940s he visited the Louvre with a handful of his own paintings and laid them alongside those of his masters, Delacroix, Ingres, and Zurbarán. In describing the incident later to friends, he was to comment, "It wasn't really that bad," thus signaling his belief in his artistic immortality.

Some regard this final artistic phase unfavorably. The biographer John Berger comments: "Picasso is only happy when working. Yet he has nothing of his own to work on. He takes up the themes of other painters' pictures. He decorates pots and plates that other men make for him. He is reduced to playing the child. He becomes again the child prodigy." Sometimes, it is said, Picasso himself did not know whether his later works were any good. Certainly, the works of Picasso's last years have had the most difficult time in achieving critical approbation. Many observers have not known what to make of works with grotesque sexuality or works that feature an almost clinical observation of the deterioration of his body and, possibly, of his mind as well (see figure 5.17). But it is worth noting that Picasso never ceased to be daring. People content to rest on their laurels would not have dared risk the opprobrium that was certain to be stimulated by the late, "decadent" works of Picasso.

The versatile and photogenic Picasso enjoyed an enviable press. He was the graphic magician of the time, endlessly productive and creative, ready to play to the crowd, in the manner of Charlie Chaplin or Einstein. A marvelous movie made in the mid-1950s, *Le mystère de Picasso*, shows the master painting for seventy-five minutes with hardly a comment. The film constitutes an eloquent brief in favor of the proposition that Picasso could create and destroy most anything he wished on a graphic surface and that he could be charming, witty, self-deprecating, and serious. In a typical riveting sequence he begins with a flower, transforms it into a fish, then into a chicken; switching from black and white to color, he then refashions the composition into a cat surrounded on the side by human beings. Biographers like Roland Penrose and photographers like David Douglas Duncan were assured continuous access to the master in implicit exchange for their flattering visual and verbal portraits. While Françoise Gilot's *Life with Picasso* documented that not all was as rosy as it appeared in the kingdom of the great man, plenty of acolytes were willing to question Gilot's descriptions and motives.

Since critical words about celebrities abound now, one cannot be surprised that many bitter and brutal descriptions of Picasso have appeared in the twenty years since his death. None of the other "heroes" in this book have been free of strong criticism, but the postmortem portraits of Picasso reveal a man marked by egotism and cruelty. At least for the last half century of his life, Picasso

FIGURE 5.17.
The Artist and His Model, VIII (July 4, 1970),
colored pencil on cardboard (9 ½ x 12 ½).
© 1993 ARS, N.Y./SPADEM, Paris.

insisted on being surrounded by individuals who would serve only him. To a greater extent even than Freud, he expected loyalty from his retainers and from his women, even as he reserved the right to treat them as he wished, to play them off ruthlessly against one another, and to fire or dismiss them at whim. He was sadistic and could physically abuse those who loved him, seeing, according to Marie-Thérèse Walter, a connection between rape and work. He refused to countenance criticism from any of his painter friends, not even Marc Chagall and Alberto Giacometti. And his treatment of his own children, in life and by virtue of decisions that were to be implemented after his death, has the makings of a tragic morality play.

Other creative individuals may have been responsible for the death or unhappiness of a small handful of individuals; those who remained involved with Picasso were likely to meet a bitter fate. This was especially true for women: Increasingly, Picasso identified with the minotaur, who required the sacrifice of women, body and soul, to him. His first wife, Olga, went crazy and died in 1955; his most carefree mistress, Marie-Thérèse Walter, hung herself in 1977; his most intellectual mistress, Dora Marr, suffered a nervous breakdown; his grandson committed suicide by drinking concentrated bleach when he was not allowed to attend Picasso's funeral service; his second wife (and widow), Jacqueline, whom he had married in 1961, shot herself to death the night after she had completed the plans for an exhibition of her personal collection of Picasso's works. Mary Gedo calls Picasso a "tragedy addict," maintaining that he was attracted to women who were fragile and that he remained in their lives until tragedy occurred. Picasso can hardly be considered innocent in this matter. "When I die," Picasso had prophesied, "it will be a shipwreck and as when a huge ship sinks, many people all around will be sucked down with it."

Picasso's male friends also suffered at his hand: Picasso pretended that he did not know Apollinaire when his chief promoter was accused unjustly of a crime; he refused to rally to the support of his friend Max Jacob when the writer he had known for forty years was sent to a concentration camp; he had affairs with the wives and lovers of several friends, including his early hapless comrade Casagemas; he intrigued to undermine the career of his young fellow countryman Juan Gris, made disparaging remarks about his old friend Braque, and abandoned his dealer Kahnweiler when the latter's money was seized from him during the Great War. His friend Sabartés said, "Picasso chooses friends as he chooses his colors when painting a picture, each one at its proper time and for a particular purpose." Richardson, a highly sympathetic biographer, describes the pattern in this way: "[Picasso] never outgrew the need for a devoted male friend whose loyalty, understanding, and patience would constantly be put to demanding tests, someone who would be prepared to set Picasso ahead of everyone else."

Only with respect to his old peer Matisse did Picasso maintain the civility and equality that should mark friendship. Already an established master when they met, Matisse was by far the more authoritative figure. Not that Picasso did not challenge Matisse: Indeed, *Les demoiselles d'Avignon* has been seen as an answer to Matisse's *Woman with a Hat* and *Le bonheur de vivre*. But Picasso did acknowledge throughout that Matisse was his equal, perhaps his superior, when it came to the production of line, the use of color, and the achievement of balance, purity, and serenity. He once said to Matisse: "I have mastered drawing and am looking for color; you've mastered color and are looking for drawing." He often commented that "in the end, there is only Matisse"; and when Matisse died, it was clear that he felt himself the sole surviving artistic master of the century.

My purpose in mentioning Picasso's negative side is neither to deny that he had virtues (he could be generous to friends, and he displayed courage during the Second World War), nor to detract from the quality of his artistry. One should not judge artists' (or scientists') achievements in terms of their human frailties, though in the end all of us must also be judged as human beings. Nor do I want to make a justifying kind of argument, maintaining that it is necessary (or even worthwhile) to endure an individual of Picasso's destructiveness in order to secure the fruits of his work. Though none of our modern creators were saints, none other seemed to have approached Picasso in an arrogant disregard for others.

I believe that we can attain some understanding of why Picasso was as he was. Picasso enjoyed the benefits and liabilities of a prodigious start. His gifts and energies meant that, with few exceptions, he was able to do whatever he wanted, whenever and wherever he wanted, throughout his life. His virtuosity was never seriously challenged, let alone vanquished, and he seldom met his equal, of either sex, in any sphere that he valued. (It is interesting that the two individuals who came closest to achieving this rank, Matisse and Gilot, befriended one another and that Picasso was extremely jealous of their relationship.)

But Picasso was not able to think beyond his gift. In many ways he remained childish, for example, in his relation to others and in his flouting of the "mature" world. As mentioned earlier, Picasso saw himself as having made a pact, a Faustian deal, possibly dating back to the time when his sister Conchita died, where he would do all and sacrifice all in the service of his talent or gift. A deeply superstitious and often highly fearful person, from a country and background that were still in many ways premodern, Picasso erected in his own mind a superstructure or narrative that was a thinly veiled pretext for behaving just as he wanted to: He convinced himself that he had to place his work and his survival above all other earthly concerns in order to realize his artistic mission. And so, building on his legitimate fears, he refused to talk about death and to acknowledge the death of others. Given this wholly egocentric behavioral pattern, he could be perfectly charming, gracious, and generous when he wanted to, but at the same time he stood prepared to sacrifice anyone and everyone who stood in the way of his work. He may have felt guilty at times about this behavior, but the guilt was simply absorbed into the larger passion surrounding the work.

In Picasso's strangely exploitative sequence of relationships I find only a handful of exceptions, perhaps four. In his early life, there was the relationship with his lover Eva, which seems to have been more meaningful for him than any other emotional bond. For many years there was the relationship with Matisse, which, however problematic, always contained a marrow of respect. Among his later lovers, there was his relationship with Gilot, whom he tried his best to destroy but toward whom he eventually seems to have acquired a grudging respect, because she had withstood his onslaughts. One could possibly add Stein to the list, though in later life he tended to dismiss her.

The final exception—and the one most important for art history and for our study—was his relationship, for something less than a decade, with Braque. The personal and work linkage was sufficiently powerful and enduring that it allowed the two men, working in tandem, to compose a new chapter in the history of Western art. Picasso suppressed his ego and his individuality as he had not done before nor did thereafter, and as a consequence, he was able to open new vistas. Later on, he was to call this the happiest period of his life. It must be left to clinicians to say whether the relationship to Braque replayed Picasso's relationship to members of his immediate family or had homosexual overtones or made it impossible for him ever after to pursue a path that was equally revolutionary. But it seems evident that, in order to step beyond *Les demoiselles d'Avignon*, Picasso needed to hold someone else's hand.

6

IGOR STRAVINSKY:
THE POETICS AND POLITICS OF MUSIC

Stravinsky, 1915

THE MOST FAMOUS sentence in Igor Stravinsky's autobiography reads: "Music is by its very nature powerless to express anything at all." When it appeared, this sentence surprised his audience. After all, Stravinsky had composed some of the most expressive music of the twentieth century, from the lyrical *Petrouchka* to the dramatic *Le sacre du printemps* (The Rite of Spring) to the elegaic *Symphony of Psalms*.* But ever the polemicist, Stravinsky was in actuality blasting those whom he regarded as his aesthetic opponents, such as the followers of Richard Wagner; such "impurists" were always marshaling music in the service of extra-musical ends, from national solidarity to religious freedom. Seeking to repair a perceived imbalance, Stravinsky portrayed the musician as a craftsman whose materials of pitch and rhythm in themselves harbor no more expression than the carpenter's beams or the jeweler's stone.

THE POLITICAL FACE OF CREATION

Stravinsky may have been correct that, in the absence of an externally im-posed "program," music is simply music. He spoke of the "poetics" of music, which in its literal sense refers to the making *(poiesis)* of music. Unintention-ally, however, Stravinsky vividly illustrated a different point through his own life: the extent to which the making of music is *not* possible without the exter-nally triggered factor of politics. All creative individuals—and especially all musicians—must deal with a set of associates who not only help the creators realize their vision but also eventually, with a wider public, determine the fate of the creators' works.

In comparison with the artistic and scientific pursuits we have surveyed so far, the making of music emerges as an intensely public activity. If merely scored and available for perusal, music has little effect. An ensemble of individ-uals (including performers, publishers, publicity agents, and ticket sellers) and a collection of materials (including instruments, a concert hall, billboards, and programs) are required if a musical idea is to achieve public expression. And when, as in Stravinsky's case, one elects to mount huge spectacles like a ballet or an opera, the number of individuals involved quickly reaches the hundreds.

When his friend and collaborator Robert Craft began to sift through Stravin-sky's correspondence, spanning nearly seventy years, he was astonished by what he found. It seemed that as much of Stravinsky's considerable energies had been directed toward the management of his musical life as toward actual composing and performing. Moreover, Stravinsky had thrown himself into this political arena with enormous gusto and determination. Craft comments:

---

*Hereafter, pieces are referred to by the name most commonly used in performance.

Whether or not Stravinsky's letters to bankers, brokers, lawyers, and estate agents provide sufficient documentation to diagnose a 'split personality,' the concentration, logic, and concern with minutiae that he devoted to business affairs are awesome, at any rate in a great musician. . . . Stravinsky's mind seems to divide almost equally into musical genius and moneylender. . . . After finishing *Le sacre du printemps* on a November morning in 1912, he apparently spent the afternoon writing letters about investment properties.

In introducing three extensive volumes of published letters, Craft remarks almost apologetically:

The correspondence does not include any of the extensive exchanges between Stravinsky and his banks in Russia from 1912 until the Revolution. It also omits the example of the numerous letters in which Stravinsky asks for advances from publishers, impresarios, patrons, and performance organizations—documents that contrast strikingly with the letters from people asking him to pay overdue bills. . . . This chapter does not examine any of Stravinsky's dozen or so ill-advised lawsuits.

With respect to his embroilment in personal and professional politics, Stravinsky represents an extreme, both within our sample of creative individuals and within the population of musical composers. (The analogy is perhaps best drawn with Picasso's increasingly entangled love life; quite possibly, both men derived pleasure from these conflict-laden affairs.) One need not engage in (typically futile) lawsuits to become a great physicist or an immortal composer. And yet, by throwing the political aspects of creation into sharp relief, Stravinsky reveals the extent to which an artist must work with the field that regulates his chosen domain. Only the rarest of individuals is fortunate enough to be embraced by the field without external prodding; only a few adult artists are blessed with another individual who is willing to run constant interference on their behalf, and, at least until the recent past, the need to justify publicly one's own creative output placed an even greater burden on women. Whether they do so well or poorly, eagerly or reluctantly, nearly all creative individuals must devote significant energies to the management of their careers. Such political activity by no means guarantees success; but in its absence, aspiring creative individuals risk permanent oblivion.

A RUSSIAN CHILDHOOD

Nearly all remembered childhoods are redolent of a gentler, simpler past; this seems particularly true for children reared in pre-Soviet Russia. From the writings of individuals like Vladimir Nabakov or Boris Pasternak, one receives the impression of cities in czarist Russia filled with delightful castles and scrumptious treasures, elegant hotels and clubs, countrysides dotted with snow-covered hills and splendid dachas, and a constant convivial atmosphere involving large, nurturant families, faithful servants, doting grandparents, and embracing nannies. In his nostalgia for the Russia of his youth, Stravinsky resembled others of the prewar generation, but his specific memories proved far less idyllic than those of his literary peers.

The third of four sons of landed gentry on both sides, Stravinsky was born in Oranienbaum, Russia, in 1882. He spent the winters of his youth in St. Petersburg, a city he especially prized; and he summered with the family in the country, at various estates owned by members of his extended family. The Stravinsky family's principal home was an intellectual center in St. Petersburg, frequented by individuals like the novelist Fyodor Dostoyevsky. Stravinsky's father was a well-regarded opera bass and a gifted actor at the Imperial Opera House. Young Igor heard much music at home and also attended concerts and operas where, in one of the most memorable experiences of his youth, the nine-year-old boy spied Peter Ilyich Tchaikovsky shortly before the great composer died.

Stravinsky seems always to have been interested in music, and some of his most vivid and faithful initial memories involve sound. He recalled a peasant who could not speak but who had a fascinating habit of clicking his tongue very noisily. The peasant would then sing a song of two sounds (the only ones he could pronounce) with great speed and dexterity. He accompanied these sounds by pressing the palm of his right hand under his left armpit and then made a series of noises that sounded (euphemistically) like resounding kisses. Stravinsky attempted to recreate this music at home. As a young child, Stravinsky also imitated the unison singing of women from the neighboring village as they wended their way home from work.

It is risky to overinterpret such childhood crystallizing experiences. After all, both the families of creative individuals and the individuals themselves are likely to search for early markers and, if necessary, to embroider memories until those prove "worthy" harbingers of the adult talent. Yet, it does seem reasonable to assume that individuals differ in the kinds of childhood experiences that attract them and that prove memorable, and in this spirit, we may think of Stravinsky's early aural experiences as analogous to Einstein's fascination with the compass (see chapter 4) or to Eliot's vivid visual and tactile sensations (see chapter 7). Stravinsky was also able to remember accurately the visual components of these scenes, a kind of embroidery that would have been unnecessary if one were merely trying to make the case that the "golden ear" had been present since early childhood. Ultimately, Stravinsky was also distinguished from other composers by his mastery of the visual components of dramatic performances.

Though immersed in music, Stravinsky was not a musical prodigy. Indeed, as a child, he seems to have been more interested in painting and in theater than in music per se. He began piano lessons at the relatively late age of nine and advanced quickly. He read opera scores in his father's library and attended concerts with keen interest. From early on in his musical education, he was interested in improvisation and persisted in creating his own melodies and variations, even though his family and teachers criticized these as a waste of time.

Stravinsky grew up in an atmosphere conducive to his musical and intellectual development; but unlike other White Russians of the period, he seems not to have had a happy childhood. His father, a lawyer and civil servant as well as an artist, was strict and cold. Young Igor had only dutiful feelings toward his mother, though he loved his German governess, Bertha, and was deeply shaken by her death in 1917. Among his siblings he liked only his older brother, Gury, who died on the Rumanian front during the First World War. Stravinsky remembers being quite lonely as a child: "I never came across anyone who had any real attraction for me," he recalled in his autobiography. In a manner reminiscent of Einstein, he did find some support from his uncle Alexandre Ielachich, a fervent music lover as well as a liberal intellectual, and from an older friend Ivan Pokrovsky, who introduced him to French composers.

By his own testimony, Stravinsky was not a good student and usually performed at or below the average level for his class. Unlike Picasso, however, who appears to have had genuine learning problems, Stravinsky was simply uninterested in formal schooling and preferred throughout his life to educate himself. Ignoring his son's antischolastic inclinations, Stravinsky's father insisted that Igor follow in his footsteps and receive legal training. Stravinsky did not like law school at all, and this alienation only exacerbated his tense relation with his father and his general disaffection with his current situation.

MUSIC AT THE CENTER

By the time of his entry to St. Petersburg University, Stravinsky had decided that music was his life's calling. Much of his education continued to be self-initiated. He disliked the study of harmony but liked counterpoint and was particularly excited by the opportunity to set and solve his own problems. He began to listen to new music; and like other promising young artists of the time, he soon found himself in a circle of intellectual and artistic peers, with a particular interest in the forms of contemporary expression in Russia and in Western Europe.

The most important event in Stravinsky's musical training was his 1902 meeting of Nikolay Rimsky-Korsakov, the dean of Russian composers. While responding unenthusiastically to Stravinsky's youthful compositions, Rimsky-Korsakov gave him shrewd advice about which studies he should undertake; moreover, and to young Igor's great surprise, Rimsky-Korsakov generously offered to supervise his composing.

For the next six years, until Rimsky-Korsakov's death in 1908, Stravinsky was the senior composer's pupil and, increasingly, his friend, confidant, and ersatz son. Much of the instruction was technical. Rimsky-Korsakov guided Stravinsky in orchestration, teaching him how to compose for each instrument; they would each orchestrate the same passages and then compare their versions. Stravinsky was an apt pupil, whose rapid advances pleased his mentor; and, perhaps for the first time in his life, Stravinsky found himself in a milieu that fully engaged him. The early crystallization in the musical domain was now transmuted into a lifelong course.

Stravinsky and Rimsky-Korsakov held similar philosophies about education, both favoring a strict disciplinary regime. As Stravinsky was to state later: "No matter what the subject may be, there is only one course for the beginner: He must at first accept a discipline from without, but only as the means of obtaining freedom for, and strengthening himself in, his personal methods of expression." And yet, their musical preferences were quite different. In many ways Stravinsky was more attracted to the music of Rimsky-Korsakov's Russian rivals such as Tchaikovsky than to his teacher's own programmatic music. For his part, Rimsky-Korsakov was suspicious of Stravinsky's interest in ancient Russian forms, his intoxication with current French music, and his curiosity about emerging hybrids of the Russian and European musical traditions. Displaying pride laced with ambivalence, he declared: "Igor Stravinsky may be my pupil but he will never be my or anyone else's follower, because his gift for music is uniquely great and original."

At this time the domain of music in Russia was in a state of flux, reminiscent in some ways of the "multiple options" discernible in the domain of physics in Germany or of painting in France. Exerting considerable influence was a group of five composers who had banded together around 1875 to promote a national school of Russian music. Included in their ranks, in addition to Rimsky-Korsakov, were Alexander Borodin, Cesar Cui, Mili Balakirev, and Modest Mussorgsky. These composers fashioned themselves to some extent in opposition to Tchaikovsky and to Mikhail Glinka, who were seen as more influenced by Western European music, and to Aleksandr Glazunov, who favored an academic style that was faithful to the classical orchestral forms. Of course, Stravinsky also drew on the works and traditions of many European composers of the past.

According to critics, Stravinsky's early work was unremarkable. Like workers in all creative fields, he was mastering the languages of his predecessors. Stravinsky drew on his contemporaries in a most catholic way. His initial compositions can be variously compared to those of Rimsky-Korsakov, Tchaikovsky, and other Russian composers in whose work he became interested and whom he often consciously imitated. There are intimations of Ludwig van Beethoven, Richard Wagner, Richard Strauss, and other favorite Germanic composers.

Eager to listen to new music, Stravinsky cofounded a Society of Contemporary Music in 1906; much of the tantalizing (but dangerous) new French music of Claude Debussy, Maurice Ravel, and others was performed there. While Stravinsky remained the eager pupil throughout most of the decade, his progress during that time was notable. Indeed, the critic Jeremy Noble claims that "the distance Stravinsky had already travelled in the four or five years since the sonata [of 1903 and 1904] is remarkable."

EARLY TRIUMPHS AND A FATEFUL ENCOUNTER

The first public performances of Stravinsky's works occurred in St. Petersburg in 1907 when his sonata was performed; his first symphony was performed in 1908. The composer was in his mid-twenties, not a youthful time for first performances. The pieces were not particularly well received by the audience or by Rimsky-Korsakov. More positive reactions greeted two brief pieces for large orchestra—the *Scherzo fantastique* and *Fireworks*—performed shortly thereafter. These pieces were explosive, brilliant, dynamic, and programmatic; the orchestration of simple motifs with rich harmonies was under firm control throughout. Perhaps more importantly, these brief compositions began to reveal Stravinsky's own artistic voice.

In attendance at a 1909 concert where the *Fireworks* was probably performed* was a young Russian lawyer-turned-impresario named Serge Diaghilev. Following an abortive career as a composer (Rimsky-Korsakov had discouraged him from pursuing this calling), Diaghilev had founded a publication called *Mir Iskusstva* (World of Art), which served as a rallying point for young artists, much as *Arte Joven* had done for Picasso and his Barcelona colleagues a few years earlier (see chapter 5) and as *Blast* was to accomplish for T. S. Eliot, Ezra Pound, and Wyndham Lewis in London a few years later (see chapter 7). The journal, which (again, like most of its counterparts) lasted but five years, helped situate contemporary avant-garde Russian art strategically between the academicians, on the one hand, and the political revolutionaries, on the other. It embraced art for art's sake, with a judicious blend of authentic Russian and contemporary European influences, and it gathered into Diaghilev's orbit the most talented young artists and writers of the time.

Diaghilev was a most remarkable individual. He was a grand and flamboyant seigneur, part gambler, part intellectual manqué, part artist manqué, part schemer, part dreamer. He loved intrigue, and for a charismatic homosexual living in the midst of a group of temperamental young artists and performers, such intrigue was never in short supply. He had an uncanny ability to pick out new talent and a virtually unerring sense of what was likely to shock (yet fascinate) an audience:

*No definitive documentation seems to have been found.

Sexuality and ecstasy, violence and death were his chosen themes. And while he sought the modern, he never lost sight of his audience and the box office.

Diaghilev knew himself. As a young man in his early twenties, he had written to his stepmother:

> I am firstly a great charlatan though *con brio;* secondly, a great *charmeur;* thirdly, I have any amount of cheek; fourthly, I am a man with a great quantity of logic, but with very few principles; fifthly, I think I have no real gifts. All the same, I think I have just found my true vocation—being a *Maecenas.** I have all that is necessary save the money—*mais ça viendra* [but that will come].

Stravinsky wrote of him: "He had a wonderful flair, a marvelous faculty for seizing at a glance the novelty and freshness of an idea, surrendering himself to it without pausing to reason it out." Fated not to be a great creative artist himself, Diaghilev takes his place within a tiny cohort of catalytic nurturers of talent in the twentieth century, among them the photographer Alfred Stieglitz, the teacher of composition Nadia Boulanger, the editor Maxwell Perkins, and the theater director Max Reinhardt—individuals who fostered the artistic history of the twentieth century.

Having conquered St. Petersburg through his influential publication, well-received performances, and powerful stable of talents, Diaghilev set his sights on Europe, and particularly, on Paris. First he organized an exhibit of Russian art at the Grand Palais in 1906, then five concerts at the Opéra the following year, then in 1908 a *Boris Gudonov*, which was a sensation.

In 1909, Diaghilev undertook perhaps his most daring step, as he launched the Ballets Russes dance company. Ballet had a mixed reputation as an art form, with many intellectuals considering it an ancient and somewhat passé activity. But Diaghilev felt that there were many great ballets and that the form retained tremendous potential, especially given his Russian troupe's special gifts. With performances of the Chopin-inspired *Les sylphides*, Borodin's *Polovtsina* dances, and other spectacles, Diaghilev's Ballets Russes took Paris by storm.

By 1909, Diaghilev had already assembled a wonderful group of dancers (including Vaslav Nijinsky), choreographers (Mikhail Fokine), and designers (Leon Bakst and Alexander Benois), but he lacked one thing: a composer who could work steadily with his troupe. Hearing the Stravinsky composition, he knew that he had found his man. As one who trusted his impulses completely, Diaghilev had no hesitation in asking Stravinsky point-blank to orchestrate the A-flat-major nocturne and the *valse brillante of Les sylphides*. He had also been toying with the idea of mounting a ballet around the story of *The Firebird*, and

---

* Gaius Cilnius Maecenas was a Roman patron of letters.

he soon commissioned Stravinsky, now in his late twenties, to prepare the score for that dramatic saga.

The meeting with Diaghilev and the invitation to join the Ballets Russes company changed Stravinsky's life overnight. From a pupil of the recently deceased Rimsky-Korsakov, a youthful composer with some talent but neither institutional affiliation nor guiding mission, Stravinsky became a valued member of what was possibly the most innovative performing artistic group in the world. Just as Stravinsky had taken immediately to the instructional discipline of the paternalistic Rimsky-Korsakov, he was attracted equally to the ensemble of quirky talent gathered around the indomitable Diaghilev.

Now, instead of working mostly alone, Stravinsky had almost daily intercourse with the ensemble—a new and heady experience for someone who had craved the companionship of individuals with whom he felt comfortable. Stravinsky turned out to be a willing pupil, one who learned quickly and reacted vividly to everything. He was sufficiently flexible, curious, and versatile to be able to work with the set designers, dancers, choreographers, and even those responsible for the business end of the enterprise. Benois remarked how unusual Stravinsky was among musicians by virtue of his deep interest in theater, architecture, and the visual arts. From Diaghilev young Igor learned two equally crucial lessons for ensemble work: how to meet a deadline and how to compromise on, or mediate amongst, deeply held but differing artistic visions.

THE BALLET MASTERED:
THE FIREBIRD AND *PETROUCHKA*

These disparate lessons came together when Stravinsky began work on the score to the ballet *The Firebird* in the winter of 1909 and the spring of 1910. As he later recalled: "I worked hard and this meant being in continual contact with Diaghilev and his collaborators. Fokine worked on the choreography of each number as I sent them to him. I was always at the company's rehearsal and that day used to finish with Diaghilev, Nijinsky [who was not in fact dancing in this ballet] and me sitting down to a large dinner washed down with a good claret."

The Firebird showcased Stravinsky's emerging gifts superbly. The story, in many ways a typical fairy tale, features the evil magical ogre king Katschei, the hero prince Ivan Tsaverich, the lovely princess Tsarevna, and the glittering "good fairy" Firebird. The fantastic creature is first imprisoned, then released, and finally helps Ivan save his love from the ogre.

This dramatic saga gave wide rein to Stravinsky's theatrical imagination. Seizing on devices to which Rimsky-Korsakov had introduced him, he found a specific register for each character realm—for example, using chromaticism to refer to the supernatural, a diatonic style for human characters, and Oriental strains to evoke legendary Russia. Stravinsky also had the opportunity to express in musical terms the characteristic physical gestures and movements of each of the protago-nists. Moreover, as the composition featured nineteen different scenes, he was able to mobilize his various orchestration techniques to full advantage. Though debts to French and Russian forbears were quite audible, the mastery of melody, harmonic progressions, and rhythmic movement marked the composer as one who had come into his own; he proved capable of creating vivid musical themes as well as clearly delineated sections and fragments that clashed energetically with one another. Not a few Hollywood films have been able to exploit devices that were handled to perfection in this, Stravinsky's first major work.

Diaghilev had confidence that Stravinsky would enter a new sphere as a result of his masterful work on *The Firebird*. The impresario declared on the eve of the first performance: "Take a good look at him. He is a man on the threshold of fame." And indeed, the reception to *The Firebird*, with Claude Debussy and other notables in the audience, was sufficiently enthusiastic to catapult Stravin-sky to celebrity status almost immediately thereafter. As the biographical entry in *The New Grove Dictionary of Music* notes:

> The success of *The Firebird* altered the course of Stravinsky's life. At that time Paris was the international centre of the world of art, the Ballets Russes one of its prime sensations; and Stravinsky's the most important original score in the ballet's repertory. This meant that overnight he became known as the most gifted of the younger generation of Russian composers, and during the next few years his music became better known and appreciated in western Europe than in his native Russia.

None of the other six creators I am describing enjoyed a more meteoric rise. The success of *The Firebird* also gave a cosmopolitan thrust to Diaghilev's com-pany and fused the fates of Stravinsky and Diaghilev for the next two decades.

Stravinsky was ambivalent about the success of *The Firebird*. It remained for the rest of his life the piece for which he was most famous and the piece that was most often performed and parodied (though generally not under copyright, which infuriated this instinctively litigious person). Perhaps underestimating its originality and its influence on his subsequent work, Stravinsky came to regard *The Firebird* as conventional in terms of conception and orchestration—a throwback to nineteenth-century narrative with its showstopping set pieces and its expressive excesses. Unhappy with some of the choreography, he seems to have been relieved when the suite began to be performed as part of an orchestral concert. As he commented sardonically: "It is more vigorous than most of the composed folk music of the period but it is also not very original. These are all good conditions for a success." But at the time, Stravinsky did not wallow in pride or in ruefulness; like other highly creative artists, he was too busy working on his next pieces.

The Innovations of Petrouchka

Visiting Stravinsky in the summer of 1910, Diaghilev found that the composer was working on an orchestral piece, "a picture of a puppet, suddenly endowed with life, exasperating the patience of the orchestra with diabolical cascades of arpeggios. The orchestra in turn retaliates with menacing trumpet blasts. The outcome is a terrible noise which reaches its climax and ends in the sorrowful and querulous collapse of the poor puppet." Fascinated, Diaghilev convinced Stravinsky to convert the piece into the ballet score *Petrouchka*. Stravinsky worked on the score during the fall and winter; it was performed at the Théâtre du Châtelet in Paris, in June 1911, and garnered an enthusiastic response.

The Firebird showed that Stravinsky could synthesize the lessons from his masters and fashion a piece that excited the field of his era. *Petrouchka* was a far more audacious work. The setting was both ancient and modern—a mix of traditional folk songs and popular urban songs against the background of a holiday festival. The mood shifts from the lyrical and the picaresque to the tragic, and unlike in *The Firebird*, the tragedy of the lonely puppet is genuine rather than formulaic.

The compositional techniques are innovative: Harmony alternates with polyphony, polytonality, and a touch of chromaticism; the predominant diatonic language is contrasted with a more dissonant idiom. Featured is the jarring Petrouchka chord in which a C-major triad (all white keys) and an F-sharp-major triad (all black keys) are superimposed. Stravinsky is able to create tiny episodes, some of them barely a phrase, that often sound quite discordant at first hearing, yet fit comfortably with one another; these recur in analogous contexts and combine to produce a larger, highly expressive, and satisfyingly integrated whole. There are also interesting characterizations: For instance, the poignant Petrouchka is portrayed through the seemingly contrasting vehicle of the carnival's wild abandon.

Probably the greatest innovations occurred in the rhythmic sphere. In the face of a seemingly inexhaustible invention of new meters, with binary and ternary rhythms superimposed, the overall sense is nonetheless of a completely integrated, almost mechanically precise score. Throughout, rhythm serves as the primordial organizing element, with a driving regularity punctuated by episodes of calculated asymmetry and syncopation. Perhaps not surprisingly, the composing of such an original score did not come easily to Stravinsky; in fact, he tried for a month to compose the poignant finale, seeking on the piano the last bars of the tableaus.

Far more so than *The Firebird*, *Petrouchka* has the feeling of a collage—a collection of individual pictures artfully integrated into a convincing larger tapestry. Unlike *The Firebird*, which follows the expected narrative sequence, *Petrouchka* is an effort to convey through suggestion the mood or feeling of the puppet and its world. Given that this work was created at precisely the same time as Picasso and Braque were experimenting with visual collages (see chapter 5), and Eliot was interspersing "overheard" conversational fragments in his poems (see chapter 7), one is tempted to envisage some artistic zeitgeist at work.

Once again, Stravinsky worked closely with the members of the Diaghilev troupe, with Benois serving in this instance as joint author of the libretto. But in contrast to the customary procedure, the musical score was composed first, and it therefore controlled the shape of the dance. While this approach was entirely to Stravinsky's liking, it alienated Fokine, who eventually left the ballet corps. Stravinsky also participated far more actively in the actual staging. As the biographer André Boucourechliev comments: "It is impossible to exaggerate the importance of the active role played by the composer in the stage presentation of the work, which finally confirmed his professional status as a man of the theater." As Stravinsky gained in knowledge and confidence, he also found himself engaged in strenuous disputes about characterization, choreography, and instrumentation. Ultimately, he and Benois were also to fight bitterly about control of the rights to the piece.

Like *The Firebird*, *Petrouchka*, ably conducted by Pierre Monteux, staged by Benois, and choreographed by Fokine, was a triumph. No doubt a considerable proportion of the great success of the premiere was due to Nijinsky's brilliant performance as the puppet. Stravinsky always paid tribute to his marvelous inventiveness: "As Petrouchka he was the most exciting human being I have ever seen on stage." The positive reaction was also important for Stravinsky himself: "The success of Petrouchka was good for me, in that it gave me the absolute conviction of my ear just as I was about to begin *Le sacre du printemps*."

A Telling Failure

Given Stravinsky's incredible productivity from 1910 to 1913, with three unchallengeable masterpieces completed during that brief interval, it is tempting to envision the young composer on an unprecedented roll, strutting from one success to another. Instead, Stravinsky actually devoted considerable energy during this period to *The King of the Stars*, a short cantata for male chorus and large orchestra, set to a text by the poet Konstantin Balmont. Stravinsky had great hopes for this composition, which he dedicated to Debussy, but the piece simply did not work. Indeed, due to the complexity of the choral writing and to other difficulties, the piece was not performed until 1939 and has rarely been heard publicly since then.

Acknowledgment of a singular failure, against this background of unprecedented triumphs, is important. It reminds us that even the most creative innovators can proceed down a false path and that they differ from others in the way they recover, rather than in their intrinsic infallibility. Indeed, as noted earlier, the student of creativity Dean Keith Simonton has collected evidence suggesting that the greatest creators simply produce more works, which includes more inferior as well as more superior works. One ought to think of *The King of the Stars* as a kind of failed *Les demoiselles d'Avignon,* a discarded early draft of *The Waste Land,* or Freud's "Project for a Scientific Psychology"—the creator's sincere, but still fumbling, search for a publicly accessible symbol system to capture an emerging, but still inarticulate, personal artistic vision. While unsuccessful by the usual public criteria, these particular searches may have harbored considerable significance for the creator himself: They helped him discover what he did, and did not, wish to achieve in his work and how best to pursue those goals in future works.

LE SACRE DU PRINTEMPS: COMPOSING SOUND FOR A NEW CENTURY

In the spring of 1910, while finishing the score for *The Firebird,* Stravinsky had a dream: "There arose a picture of a sacred pagan ritual: The wise elders are seated in a circle and are observing the dance before death of the girl whom they are offering as a sacrifice to the god of Spring in order to gain his benevolence. This became the subject of *The Rite of Spring.*" It is possible that the dream itself was inspired by a poem by a Russian modernist, Sergei Gorodetsky. Over the next three years, and particularly in the period following the completion of *Petrouchka,* Stravinsky worked on the score to this tableau. As is well known, the premiere of *Le sacre du printemps* was a major artistic scandal; but within a few years, the piece came to be considered a seminal work and, no less, a turning point in modern musical composition.

Writing about music or ballet is more difficult for me than writing about literature or poetry, but I shall try to re-create the composition of, and reactions to *Le sacre.* The events surrounding the actual composition constitute a complicated tale. Shortly after Stravinsky told Diaghilev about his vision, he was given a formal commission. Stravinsky realized that he would benefit from collaboration with someone knowledgeable about Russian pagan rituals, so he began to work intensively with Nicholas Roerich, a painter, archaeologist, and ethnographer. Though the most concentrated scoring was not to occur until two years later, Roerich declared as early as 1910 that "the new ballet presents a number of scenes from the celebration of a holy night among primitive slavs. The action begins during a summer night and finishes before actual sunrise, as the sun's first rays appear. The choreography consists of ritual dances, and the work will be the first attempt to reproduce life among a primitive people without using any definite dramatic story." Fokine was already committed to other projects, and so the choreography fell to Nijinsky; because Ballet Russes members were already involved in developing two new spectacles (the notable *Daphnis and Chloe* of Maurice Ravel and Debussy's *L'Après-midi d'un faune*), there was no prospect of a performance of *Le sacre* until 1913.

The composition of *Le sacre du printemps* did not proceed nearly as swiftly or smoothly as that of earlier works. The longer gestation period probably occurred because of the novelty and incredible complexity of the task Stravinsky had set for himself. In *The Firebird* he was working on a well-known form of narrative, using familiar musical techniques (if in a highly polished way), and collaborating intimately with the whole Diaghilev team. In *Petrouchka* he was using the relatively familiar story of a harlequin in a circus setting, and he had the good fortune of a principal dancer whose genius perfectly matched the part. But in *Le sacre* nearly all of the components were new—the theme, the folk material, Roerich as the collaborator, Nijinsky as the choreographer, and perhaps above all, an increasingly radical musical idiom that the composer was formulating for himself.

Drafts for the score of *Le sacre* exist, but in my view, there is less to them than meets the eye. Not exhaustive in any sense, they are particularly lacking in materials from the first period of composition: Stravinsky's "sketchbook" is more a logbook, or a record of critical points, in the evolution of the score. Still, some facts seem reasonably well established. The tides and scenarios were worked out with Roerich in the summer of 1911. Sketches for the parts "The Augurs of Spring," "Spring Rounds," and "Ritual of Rival Tribes" were prepared at about that time. Folk melodies—heard and remembered ones—were important ingredients in several of the sections. Stravinsky also conceived the chord whose rhythmic articulation has since become the signature of the work—the highly dissonant *sacre* chord, a combination of E-flat major with added minor seventh, and F-flat major. As he recalled, Stravinsky was unable to explain or justify the construction of the chord, but his ear "accepted it with joy." Interestingly, the opening sections depicting the awakening of nature, which lay the groundwork for the *sacre* chord, apparently were sketched later, and possibly even after the entire first part (of two) had been completed.

In both the initial vision and the early sketches, Stravinsky had in mind what the overall piece should sound like. (Here, the original vision resembles the early notions of *Guernica* and *The Waste Land*—very schematic but on the mark in terms of emotional tone and organizational structure.) "I had imagined the spectacular part of the performance as a series of rhythmic mass movements of the greatest simplicity which would have an instantaneous effect on the audience, with no superficial details or complications. The only solo was to be the sacrificial dance at the end of the piece."

Stravinsky generally composed a piece straight through, and, with some significant exceptions, *Le sacre* seems to have been drafted in much the same form as it is now heard; though the introduction may well have been penned at a relatively late stage (see figure 6.1). But, again, the composing involved problems: For example, there are no less than seven separate notations for the Khoborovod melody that frames the "Spring Rounds," and the slow chromatic sections in the

FIGURE 6.1.
Interim page, *The Rite of Spring*, various sketches for "Savage Dance," "The Ancestors," and "Sacrificial Dance."
© André Meyer/Bibliothèque Nationale, Paris.

opening movements of the second part clearly caused significant struggles. In the sketchbook, the pianistic parts appear in almost final form, while those without such a clear pianolike quality are most extensively worked through. Since Stravinsky always composed on the piano, it is scarcely surprising that the non-pianistic portions would have caused him the most problems.

One significant alteration occurred in the ordering of pieces. The "Abduction," which now comes close to the beginning, right after "The Augurs of Spring," had originally been scored to occur near the end of the first part, after "The Sage." Pierre van den Toorn, who has carried out the most thorough investigation of the composition of *Le sacre*, believes that this reordering was instituted to prevent the first part from being anticlimactic.

The sketches reveal another peculiar quality. While the work's greatest innovation is now considered to be in its rhythmic configuration, the most painstaking efforts seem devoted to orchestration rather than to the rhythm. Whether this is because Stravinsky had already conceived the rhythmic details or because he did not generally deal with them in his written sketches cannot be determined.

Without question, the composition of *Le sacre* was a long, complex, and arduous process that took its toll on Stravinsky. In one celebrated annotation on the final page of the sketchbook he declared: "Today, November 17 1912, Sunday, with an unbearable toothache I finished the music of the Sacre. I. Stravinsky, Clarens, Chatelard Hotel" (see figure 6.2). To compound matters, the rehearsal process did not go smoothly. Stravinsky fired the German pianist and began to play the piano part himself at rehearsals. Then, for unknown reasons, he stopped attending rehearsals regularly and left them in the hands of the capable conductor Monteux (though he made changes requested by Mon-

FIGURE 6.2.
Final page, *The Rite of Spring*, signed, I. Stravinsky, and dated, November 1912.
© André Meyer/Artephot-Ziolo, agence photgraphique, Paris.

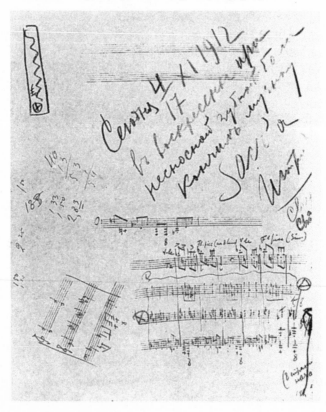

teux until close to the time of the first performance). There was time for only a very few stage rehearsals at the new Théâtre des Champs-Élysées in Paris, where the piece was scheduled to debut at the end of May 1913.

Still, despite the meager rehearsal time for so intricate and innovative a piece, there were few intimations of the tremendously hostile reaction at the premiere. Stravinsky had played the piece in four-hand version with Debussy in the spring of 1913. Debussy had been awestruck, "as though by a hurricane from the remote past, which had seized our lives by the roots," as an observer, Louis Laloy, had recalled. The dress rehearsal on May 29 had been attended by Debussy, Ravel, and the Paris press, none of whom appears to have had the slightest intimation of the turbulent reaction that would occur the next day.

LE SACRE DU PRINTEMPS: THE PERFORMANCE AND THE AFTERMATH

No other significant piece of classical music performed in modern times has been greeted with so overtly hostile a reaction as *Le sacre*. The audience at the Théâtre des Champs-Élysées was agitated from the opening bars. When the curtain rose to reveal dancers jumping up and down, hissing and howling followed. The din continued throughout the performance and included whistling, stamping of feet, honking of automobile horns, and shouting of insults. Apparently, the audience unrest reached such a pitch that it was not possible to hear the music: Choreographer Nijinsky had to stand in the wings and shout numerals to the dancers.

The extremity of the reactions is conveyed by oft-quoted eyewitness accounts. The artist Valentine Gross Hugo said: "It was as if the theater had been struck by an earthquake. It seemed to stagger in the uproar. Screams, insults, hoots, prolonged whistles drowned out the music, and then slaps and even boos." The author and photographer Carl van Vechten wrote: "Cat-calls and hisses succeeded the playing of the first bars and then ensued a battery of screams, countered by a foil of applause. . . . Some forty of the protestants were forced out of the theater but that did not quell the disturbance. The lights in the auditorium were fully turned out but the noise continued and I remember . . . the disjointed ravings of a mob of angry men and women."

Most of the initial written reviews were equally condemnatory. Opening season critics commented:

"Surely such stuff should be played on primeval instruments—or, better, not played at all."

"The music is ingenious since if the composer be more than two years of age, he must have suppressed all he knew in order to devise it."

"A crowd of savages, with knowledge or instinct enough to let them make the instruments speak, might have produced such noises."

"Practically it has no relation to music at all as most of us understand the word."

Ernest Newman, the dean of British critics, announced in the *Sunday Times* that "the work is dead," "the bluff is failed"; and he termed the event "the most farcical imposture in music of our time."

Why such a negative and hostile reaction to a work that had been appreciated in rehearsal and keenly anticipated by many of the Parisian cognoscenti? While the theme of a virgin who danced herself to death to propitiate the god of spring was provocative, it was certainly no more so than the bloodily erotic story recounted in Strauss's *Salomé*. The ballet was lengthy, but not significantly more so than other of Stravinsky's works and those of Tchaikovsky, Ravel, and other contemporary artists. Despite an air of surface chaos, the composition was highly structured and organized, in both its instrumentation and its rhythm.

One clue to the reaction may come from the undoubted technical skill of the composer and the widely acknowledged sophistication of the Ballet Russes. Audience members in Paris had been accustomed to attending the ballet and being provoked (as in Debussy's *L'Après-midi d'un faune*), but to remaining largely in command of the theatrical experience. Where other pieces had appeared outrageous, their authors had taken an ironic stance or had otherwise winked at the audience. *Le sacre*, however, seems to have fallen outside of the audience's customary categorical scheme, and the ensuing anomie was distinctly unsettling. All of the talent gathered on the stage of the theater seemed marshaled in an effort to shock, provoke, and challenge, and the audience simply decided not to collaborate in the effort. In particular, the early critics seem to have felt that they were being asked to accept too much, and they used their journalistic platforms to vent their anger.

Rather than there being any simple or single factor that caused the anger and alienation, I believe that the *combination* of factors engendered hostility. To begin with, the overt theme of a primitive sacrifice—a volitional self-annihilation—lacked any touch of pathos or moderation; it was unrelievedly amoral. The dissonant *sacre* chord was not played a few times: It was repeated for thirty-five solid bars and for a total of some 280 times in one section alone. Two- and three-note fragments were also reiterated many times in a monotonous and ceaseless alteration. There were not just frequent changes in rhythm: In some sections, nearly every bar differed from the previous one, with rhythms shifting abruptly from 9/8 to 5/8, 3/8, 2/4, 7/4, 3/4, 7/4, 3/8, 2/4, 7/8, and so on. The music was not just loud: It proceeded at unrelieved fortissimo for long percussive passages until suddenly stopping. Promising melodic passages appeared with tantalizing brevity, only to be dropped with unanticipated decisiveness. Stravinsky had thrived on juxtapositions since *The Firebird*; but now dissonant chords, irregular rhythms, exotic scales, and modified accent patterns virtually rained down on the listener. The method of melodic development—a process of breaking down, rearranging, and permuting simple four-note motifs based on Russian folks songs—shocked ears nurtured on nineteenth-century symphonic forms. The superimposition of simple diatonic thematic material and discordantly complex harmonic texture within a relatively plotless structure was also difficult to assimilate. Virtually every musical and balletic expectation had been violated most provocatively. What had been barely audible in *The Firebird* and tantalizing in *Petrouchka* transgressed the threshold of tolerability in *Le sacre*.

To add to this musical shock value, Nijinsky's choreography made little sense to the audiences of the time. Having dancers jump up and down or walk for no apparent purpose seemed just another flouting of convention: Symmetrical body movements were abandoned to shuffles, jerks, and stamps; instead of presenting pirouettes, arabesques, or pas de deux, the dancers simply mimed the jarring sounds and irregular rhythms.

Of all the commentators on the first *Le sacre*, the composer Ravel, one of Stravinsky's friends, may have had the deepest insight. Ravel declared that the piece's novelty lay not in the orchestration but in the musical entity itself. The orchestra had to be seen as a single multiregistered instrument seeking a single effect. Stravinsky himself was later to deny that the piece was revolutionary: "What I was trying to convey was the surge of spring, the magnificent upsurge of nature reborn," he declared. But it is the overall work that one must either accept or reject.

Apparently, the work alienated so many initial auditors for many of the same reasons it ultimately became accepted and even taken for granted. Of course, it is the field, rather than the work that changed. The rhythmic experiments came to be heard as exciting in themselves and as peculiarly appropriate to the first intimations of spring, the tensions among the boisterous young boys, the mysterious and severe sages, and the hapless virgin. The brief introduction and sudden abandonment of so many motifs in an ever-increasing cacophony conveyed the scattered contributions of nature to the primitive rite, even as it forecast the drive toward an inevitable destructive climax. The very introduction and dropping of sections called on the listener to carry out a creative, integrating function. In addition, the studied repetition of certain notes and phrases provided another kind of anchoring point for the listener. The playing of archaic folk themes by a full orchestra conveyed Stravinsky's sense of distance from the events being portrayed: It was as if a primitive rite were being performed with a full awareness of contemporary urban life, as happens with Eliot's *The Waste Land*. Debussy's comment on *Le sacre* is apt: "An extraordinary, ferocious thing. You might say it's primitive music with every modern convenience."

Paraphrasing Debussy, one might say that in composing *Le sacre*, Stravinsky used every gesture and trick he knew in order to communicate an original idea. The issue was less whether one liked the combination than whether one accepted it. Not surprisingly, older, more conventional, more traditional listeners were offended, if not insulted. Those who were younger, who enjoyed the spectacle, who shared the composer's impatience with the romanticism of the late nineteenth century and who sought to expand what was possible for the eye and the ear were invigorated. The very disconnections, disjunctions, repetitions, and abandonments that had so strained the early listeners became the essence of the work for a younger audience, which had its listening habits nurtured by repeated performances of *Le sacre*. The same lines of division determined the initial reactions to works like Joyce's *Ulysses*; Eliot's *The Waste Land*; or Picasso's *Portrait of Gertrude Stein, Les demoiselles d'Avignon,* and the early cubist works. And, as was the case with these works, initial distaste or noncomprehension gave way rather rapidly to a recognition—indeed to an insistence—that one was dealing with a novel work of power and, perhaps, a masterpiece. As the biographer Alexandre Tansman comments: "It is difficult to tell what is more admirable in *The Rite*— the boldness of the innovation or the total absence of the hesitation in its realization, combined as it is with the absolute certainty of an uncompromising convention that stops at nothing."

And what of Stravinsky's own reactions? Without doubt, Stravinsky was disappointed and dejected by the initial lack of comprehension of his efforts. The design and execution were clear in his own mind; he was satisfied with Monteux's conducting, though, with the passage of time, he became increasingly critical of Nijinsky's choreography. As with his earlier ballets, Stravinsky was pleased that *Le sacre* could so readily and effectively be presented by an orchestra alone. Whether he received a certain satisfaction from the scandal it caused is not clear; Diaghilev obviously derived some pleasure, and in later life Stravinsky became only too aware of the dividends of controversy.

Stravinsky continued to revise the work, more so than with any other of his works, and he did so mostly to clarify the design and harmony. He also revised his rationale for the work, downplaying narrative and imagistic elements and stressing the purely musical aspects. Also, the issuing of later, more definitive versions allowed Stravinsky to control the performances of the piece and to receive new royalties. The work was later rechoreographed, and Stravinsky was much more satisfied with Léonide Massine's version, which was performed in New York and Philadelphia in April 1930, with Martha Graham dancing the role of the Chosen One.

FROM POETICS TO POLITICS

On the eve of the performance of *Le sacre du printemps*, Stravinsky gave an interview to a journal called *Montjoie!* in which he described what he wished to express in his new composition. To a contemporary reader the description seems straightforward. Each of the approximately one dozen sections is sketched in terms of its purpose and orchestration. The composer concludes with words of gratitude to Nijinsky, the choreographer, and to Roerich, the scenarist.

Nonetheless, Stravinsky was infuriated by the publication of the interview, claiming to have been misrepresented. He seems to have been particularly incensed by the orotund opening, in which he declared: "In the *Prelude* before the curtain rises, I have confided to my orchestra the great fear which weighs on every sensitive soul confronted with the potentialities, the 'being in one's self which may increase and develop infinitely." To make matters worse, a Russian journal, *Muzyka*, published a translation of the interview, prompting Stravinsky to retort that the interview had been given "practically on the run," that the Russian translation was even less accurate than the French, and that the style of the piece was misleading. He declared to the editor of *Muzyka*: "It is highly inaccurate, full to overflowing with incorrect information, especially in the part concerning the subject of my work." But a version of the article revised by Stravinsky contains mostly grammatical changes. Finally, fully fifty-seven years after the original publication in *Montjoie!* Stravinsky declared in a communication to the *Nation* that the interview had been "concocted by a French journalist" and that he had disavowed it many times.

Stravinsky's concern about the way he is thought to have conceptualized his most famous work is not in itself surprising. What is anomalous are two further considerations. First, when given the opportunity to make corrections, Stravinsky made very few, and indeed in subsequent verbal accounts of *Le sacre*, he echoed many of the same remarks he apparently was rejecting in the *Montjoie!* account. Second, there is the oddity of a composer, whose music certainly could be expected to speak (or sing) for itself, caring so much about a chance interview published early in the century in an obscure French review.

A Legalistic Bent at Play

But as already noted, a concern with political minutiae seems to have characterized Stravinsky almost from the first. Like his father, Stravinsky had legal training. We may surmise that a legalistic (if not litigious) atmosphere pervaded the Stravinsky house and, perhaps, the intellectual and artistic circles in which his family traveled. Of course, Diaghilev was also trained as a lawyer, and Stravinsky had observed his mentor engaged in many negotiations throughout their twenty-year association; in some of them, Stravinsky and Diaghilev found themselves on the same side, but increasingly over the years, Stravinsky found himself at odds with his artistic mentor.

One source of information about the "political Stravinsky" can be found in his voluminous written legacy. Here, the Stravinsky-Diaghilev relationship does not come off very well. In addition to being terse, their telegrams to one another are devoid of any human touch and, with increasing frequency over the years, contain veiled or not-so-veiled threats. In letters, Stravinsky can be even more biting. For example, he writes to his friend conductor, Ernest Ansermet, about Diaghilev in 1919:

> His "moral integrity" about which he speaks incessantly is not worth much. . . . I was really ill when I learned of all this, not so much from his taking refuge in these "legal rights" as from his alluding to them, especially at a time when a friend finds himself in a difficult situation. A strange way to express friendship . . . I henceforth renounce all moneys that he might decide to send me without acknowledgement of my rights, moneys that I consider gifts and refuse to accept. And he should not complain about my behaving in this manner for he provoked it.

Stravinsky then enumerates his understandings of every property in dispute between Diaghilev and himself, in the process stipulating precise dates, lengths of commitments, and rights of ownership and performance as they apply in different parts of the world.

Some years later Stravinsky finds himself on equally tense terms with Ansermet. He writes: "Two words in response to your strange note of the 15th, *mon cher*. I am sorry but I cannot allow you to make any cuts in *Jeu de cartes*. The absurd one that you propose *cripples* my little March. . . . I repeat: either you play *Jeu de cartes* as it is, or you do not play it at all. You do not seem to have understood that my letter of October 14 was categorical on this point." Similarly barbed sets of exchanges take place with Monteux, who had given such exemplary performances of Stravinsky's early works, and with the conductor Serge Koussevitzky, whom Stravinsky labels as "the enemy."

Such charged relations extend even to the Swiss writer C. F. Ramuz, who idolized Stravinsky, and with whom the composer enjoyed a Braque-like artistic intimacy for some years, during which their families also became quite close. Their correspondence includes any number of tense communications about the ownership of different facets of works they coauthored. Throughout, Stravinsky seems determined to wreak out every advantage, no matter how small. He browbeats Ramuz: "I hold firmly to my argument, dear Ramuz, and it would deeply sadden me to learn that you were the one who composed that unfortunate page (of credits), knowingly, and with an ulterior motive." I am reminded of the quip that "the haggling is so bitter because the stakes are so small."

Toward nonartists Stravinsky can be even more peremptory and brutal. His correspondence is filled with litigious threats and unrelenting cajolings directed at the phalanx of agents, brokers, bankers, publishers, and publicity agents with whom he had to deal during his lengthy career. Much of the argument between Stravinsky and his collaborators was over tiny sums of money. Some of the concerns were no doubt genuine: Stravinsky had had to abandon his personal property when the Bolsheviks took over Russia at the end of the First World War; and in the 1920s, he had been financially responsible not only for his four children but also for an ever-expanding extended family of émigrés. Yet, even after Stravinsky's family had decreased in size and he had personally become quite wealthy, the penny-pinching and the litigating continued unabated. The ungenerous aspects of his personality come through all too regularly in correspondence with his children and with his hapless first wife.

Not that Stravinsky was incapable of wielding words in more flattering ways. When he wanted composers, performers, or agents to do his bidding, Stravinsky could turn on the charm. For example, Stravinsky induced the patron of the arts Werner Reinhart to pay him for performances that did not take place and even to make Stravinsky's negotiating victories look like concessions. And when he wanted someone to help him achieve a wish—for example, to travel to America during the First World War, to gain entrance into the French Academy, or to pledge to keep the story of a collaboration a secret—he could be unabashedly ingratiating. His egocentrism and focus on his own needs and desires were great, but not so great as to blind him to the "voice" he had to adopt to increase the likelihood of getting his way. For him, as for the famed war strategist Karl Marie von Clausewitz, threats and suits were just negotiations being carried on by other means. While other creative masters did not avoid legal hassles, Stravinsky appeared to revel in legal wrangling and to continue with it as long as he was alive.

To be sure, such inclinations to some extent reflect accidents of personality and upbringing. Certainly, an artist does not have to be as compulsive or embattled as Stravinsky, nor do most creative individuals live in the legal atmosphere that Stravinsky imbibed as a child. However, any artist involved in large-scale performances *does* have to enter the political arena, either directly, as Stravinsky did, or through the use of various representatives, agents, and patrons, which he also did in his later life.

Stravinsky's early career experiences mirror those of many other artists who do not have the option of hiring a representative and do not have a sponsor. To all intents and purposes, then, artists who wish to work with others must either fight for their own rights and beliefs or surrender those to people with greater power or more persuasive arguments. The most notable creators almost always are perfectionists, who have worked out every detail of their conception painstakingly and are unwilling to make further changes unless they can be convinced that such alterations are justified. Few intrepid creators are likely to cede any rights to others; and even if they are consciously tempted to do so, their unconscious sense of fidelity to an original conceptualization may prevent them from following through.

Work on a More Intimate Scale

Having composed three major works in a short span of time, and having participated actively in the roller-coaster-like atmosphere of mounting these productions, it is not surprising that Stravinsky did not attempt another large ballet for a period of time. Given his physical and mental exhaustion, the advent of the Great War, his settling in Switzerland, and the difficulty of initiating any large-scale cooperative performing enterprises at this time, it became virtually inevitable that Stravinsky would elect to work on a smaller scale.

Even before the war, Stravinsky had been attracted to Japanese art. As he put it: "The impression which [Japanese lyrics] made on me was exactly like that made by Japanese paintings and engravings. The graphic solution of problems of perspective and space shown by their art incited me to find something analogous in music." And so Stravinsky composed a set of miniatures, including several based on Russian folk songs, and set a number to music. His collaboration with Ramuz came to center on works that were both solid and small-scale, especially *Histoire du soldat*, which could be read, played, and danced by a small troupe of performers and a performing ensemble of limited size. Other works of the time, such as *Renard, Cat's Cradle*, and *Four Russian Songs*, were all of a much more confined, though not less original, scale than *Petrouchka* or *Le sacre*. It may have been at this time that Stravinsky arrived at his explicit philosophy of composition: that the setting of rigorous constraints on himself functioned as a liberating experience.

LES NOCES: A DIFFERENT KIND OF MASTERPIECE

As early as 1912, Stravinsky conceived of a choral work on the theme of a Russian peasant wedding. While his initial conception was to present an actual wedding spectacle, Stravinsky soon realized that he really wanted to present "wedding material through direct quotations of popular—i.e., non-literary—verse." As he explained it later, *Les noces* consisted of a suite of wedding episodes through which waft clichés and quotations of the sort overheard in *Ulysses*. Rather than a connecting thread of discourse, there is instead the creation of an atmosphere. And rather than individual personalities, there are roles that impersonate different types of character.

Actual composition of this masterpiece began in 1914. When Stravinsky played an early version for Diaghilev in 1915, the impresario was so touched that he wept; it was to become his favorite Stravinsky composition and the one dedicated to him. *Les noces* is said to have been Stravinsky's favorite composition as well. The music was composed in short-score form by 1917, but a complete score was completed only in 1923, just before the premiere.

None of Stravinsky's works underwent so much rescoring—"so many instrumental metamorphoses," as he put it. The initial version was scored for a large orchestra. Next Stravinsky divided the various instrumental groups into separate ensembles on stage; for instance, the strings were contrasted with the brass. In other versions Stravinsky contrasted the winds with the percussions, or combined pianolas (a kind of player piano) with bands of brass instruments. Later the brass were replaced by a harmonium and the strings by a pianola and two Gypsy cymbals. Finally, in 1921, Stravinsky arrived at a satisfying solution: "I suddenly realized that an orchestra of four pianos would fulfill all my conditions." He complemented the pianos with a collection of percussion instruments.

Despite the lengthy compositional period with its variety of contemplated orchestrations, *Les noces* emerges as a unified piece. It consists of three tableaux and four movements. Musical and literary references illustrate several components of the traditional wedding ceremony (referred to as matchmaking, separation of the couple, at the bride's house, at the bridegroom's house, the bride's departure, lament, parental blessing, sacrifice, ritual meal, funeral, the wedding bed, the burial of virginity, and so on).

The composition is typically intricate. Stravinsky carefully studied the phonetics of Russian folks songs and made sure that he captured the precise accents and stammering in the accompaniment. He also conferred a witty touch by means of syncopated rhythms and choral voices. The rhythm, which dominates the composition, is largely obsessive, synchronous pulsation; there is a fundamental motif of a fourth divided into a minor third and a major second. The melodies are largely folk songs, and the timbre features a contrast between percussion, on the one hand, and the continuity of the singing voice, on the other. With the instrumentation restricted to percussion instruments, the piece features simple combinations of piano, xylophone, and triangle.

Les noces can be instructively contrasted with *Le sacre*. It has been described as a kind of civilized, "cultural" answer to the explosive "pagan" ritual of *Le sacre*. Absent are the harsh violence, abrupt shifts, or ear-blasting passages of *Le sacre*; the piece is austere, concise, concentrated, and intellectually controlled while still spirited and humane. The recurring "vertical" chromaticism of *Le sacre* is replaced by music that is largely diatonic. Instead of an extravagant spectacle, there is a formal tableau; instead of a splendid orchestra, there is a compact and rigorously functional ensemble of players, with voices as important accompaniments. Both compositions do create their own form, with thematic material shifting in the face of a pulsating rhythmic drive; but the thematic material in *Les noces* is far more closely related and integrated with the rhythm. As in *Le sacre*, the music's direction comes from coupling, rotation, and metric transformation of a small number of melodic scraps. Like *Ulysses* or *The Waste Land*, both completed at almost the same historical moment, the text is a montage of related, but deliberately unorganized, popular sayings; the consciousness of the audience member must provide the integration, which occurs at the level of intuition rather than formal analysis.

The extensive experimentation that Stravinsky went through in creating *Les noces* gives insight into his approach to composition. Stravinsky generally had a clear conception of the shape of the piece that he was creating; and with the help of the piano, he was able early on to identify its basic themes and rhythms. Not an inspired melodist, he relied as much on the scraps of the classical and folk musical cultures as on his own experience with the optimal instruments and ensembles in fulfilling his musical ideas and in determining how to juxtapose various fragments and sections to achieve the musical and expressive effects that he sought.

Stravinsky was engaged in a complex endeavor in which he had to balance literary themes, dramatic personalities, and dominant moods against the available instrumental and musical resources. We might say that the primary symbol system in which he worked was tonal music, but that the music had to be reworked constantly in light of linguistic, personal, visual-scenic, bodily-kinesthetic, and metrical considerations. Various drafts represent his changing efforts to mediate among these elements.

In my view *Le sacre* and *Les noces* are the two most important compositions by Stravinsky, comparable to *Ulysses* and *Finnegan's Wake* by Joyce, *Les demoiselles d'Avignon* and *Guernica* by Picasso, and, if one can cross the art-science chasm, to the two theories of relativity formulated by Einstein. We see at work what I have dubbed the ten-year rule, with significant innovations or reorientations occurring at approximate decade-long intervals after an initial decade in which the skills of one's trade have been mastered. In Stravinsky's case, the situation is complicated by the fact that the two compositions were begun at almost the same time, with *Les noces* having an extraordinarily lengthy gestation period. The reactions to *Les noces* were initially mixed, but its genius was gradually recognized; nowadays many find it a more satisfying work than the grander, but less elegantly shaped, *Le sacre*.

Trying to label the stylistic provenance of these pieces serves little purpose. Yet in both cases, one observes Stravinsky struggling to reconcile the different influences upon, and pressures within, himself. The three great works of the immediate prewar period constitute Stravinsky's distancing himself from the Russian balletic tradition, as both the themes and the instrumentational means become increasingly more radical: One can speculate that Rimsky-Korsakov would have been proud of *The Firebird*, ambivalent about *Petrouchka*, and personally offended by *Le sacre du printemps*. The latter piece belongs properly to, and helped to constitute, the Parisian avant-garde; it is remote from Russian national or Russian-European music.

In contrast, *Les noces* may be thought of as a kind of return to, and confirmation of, Stravinsky's Russian origins. The piece denotes an actual pivotal life event—the peasant wedding. Both the language and the music draw widely and deeply on folk materials, in a manner that reminds one of Béla Bartók; accordingly, the piece appears remote from current Western European concerns. The piece represents a further development in an increasingly personal idiom of a master in his prime.

A FRESH RELATION TO THE MUSIC OF THE PAST

With *Les noces*, Stravinsky climactically spanned the Russian past and the modern era. In some ways, the effort to mediate between the two strands within him never ceased, he remained simultaneously Russian and modern throughout his career.

But from well before the completion of *Les noces*, Stravinsky had embarked on a new enterprise—a rediscovery of the classical music of the past and its re-creation through the embracing of a neoclassical style. As a neoclassicist, he paid homage to both the melodic sensibility and the forms favored by composers from the classical era. As had always been the case, Stravinsky was his own best teacher, and so he now pored over work from the seventeenth and eighteenth centuries with the same discipline he had displayed two decades before in tackling modern masters. As Boucourechliev comments: "He was determined to make the whole of history his own, to use it for whatever attracted or inspired him at that moment, whatever the occasion or circumstance, and to use it to create a new work by Stravinsky."

While walking through the Place de la Concorde after the end of the Great War, Diaghilev suggested to Stravinsky that he study some music written by the eighteenth-century composer Giovanni Pergolesi. Stravinsky liked the music and decided to create a Pergolesi-inspired piece based on the figure of Pulcinella. Diaghilev arranged for Picasso, whom Stravinsky had met a few years earlier, to create the decor for the piece. Thus, the two indispensable creators of modern art became close collaborators for the only time in their lives. According to Stravinsky, "Picasso accepted the commission to design the decor of Pulcinella for the same reason that I agreed to arrange the music—for the fun of it." Stravinsky added: "[Picasso] worked miracles and I find it difficult to decide what was more enhancing—the coloring, the design, or the amazing inventiveness of this remarkable man."

The composition of *Pulcinella* occurred at a critical juncture for Stravinsky. In 1920 he moved from Switzerland, a neutral no-man's land, to France, thus casting his lot with the West and identifying more explicitly than before with the Western classical tradition. Stravinsky was aware of this pivotal moment: "Pulcinella was my discovery of the past, the epiphany through which the whole of my late work became possible. It was a backward look of course—the first of many love affairs in that direction—but it was a look in the mirror too."

The parallels with Picasso have often been remarked on. The two men were born a year apart, both of them somewhat outside the orbit of mainstream Western European culture. Both gravitated to and made their first major splashes in Paris in the early 1900s, with Picasso more precocious than Stravinsky. Their most determinedly avant-garde works were produced in the years just before the Great War, with Picasso working alongside Braque, and Stravinsky immersed in the world of the Ballets Russes. During the war both tread water to a certain extent, with Picasso also meeting his first wife, who, interestingly enough, turned out to be a member of the Ballets Russes. Then, around the end of the war, both men embraced a middle-class life in Paris and moved into a neoclassical phase of creation, during which each was quite cognizant of what the other was doing. This postwar period also engendered an ingratiating playfulness in their work, as well as a preoccupation with more intimate compositional forms.

An interest in the work of the past is certainly an understandable step for a master, and particularly for one steeped in his particular art form and conscious of its origins and his own niche in its evolution. Such a historical bent may also be a normal reaction to an early career, in which one has quite explicitly rejected the canons of the past and one's own roots. What one absorbed intuitively as a young student can now be revisited in a more conscious and detached way; and because one has already made a decisive break with the past, it is no longer perceived as a crushing weight. Frequently one goes back to more remote times, as Stravinsky noted: "It is in the nature of things that epochs which immediately precede us are temporarily further away from us than others which are more remote in time."

I submit that for Stravinsky and Picasso the opportunity to engage in a stimulating and sustaining dialectic with the past was one of the prime reasons each could contribute creatively for so long. Reworking and learning from the past, they discovered further dimensions of their own voices. In this way they exploited an option not available to scientists or mathematicians. Had they lacked this playground of the past, they might have had little choice but to become yet more individualistic and radical, a tack that might have proved troublesome and counterproductive.

While Stravinsky's early work had been frankly and productively collaborative, his work in the 1920s and 1930s was more individual. Stravinsky did not stop collaborating, but he was more likely to initiate works and to do so with one or two other select collaborators of equal status, rather than as part of an established ensemble like the Ballets Russes.

In addition to several fruitful collaborations with Ramuz and the one with Picasso, Stravinsky also worked on *Oedipus Rex* with the French poet Jean Cocteau, and on *Perséphone*, with the French novelist and dramatist André Gide. He considered a collaboration with Berthold Brecht but found himself unable to work in revolutionary political theater. He began his longest, most fruitful, and most important association with the Russian-born dancer and choreographer George Balanchine; such rapport and mutual respect existed between the men over a forty-year period that there seems to have been little of the tension that characterized Stravinsky's other collaborations. I suggest that, of all Stravinsky's collaborators, Balanchine was closest to being his equal in terms of background, talent, and aspirations. Their tastes and their views of the relation between dance and music were cut from the same cloth, and both men were products of the same social and artistic tradition, with Balanchine exactly one generation younger than his father-figure and mentor, Stravinsky.

Throughout this period Stravinsky deliberately strove to relate contemporary and earlier works. Like Eliot in *The Waste Land*, he pointedly used materials from other eras. As far as he was concerned, it was not necessary for an audience to appreciate a quotation directly; sensing at an unconscious level the use of themes with some substantial history and allusiveness sufficed. (This idea was quite similar to what Eliot termed the "objective correlative," as discussed in chapter 7.) Like his English counterpart, Stravinsky also spurned work that wallowed in individual self-consciousness; he wanted to confirm and sustain a tradition, not create an idiosyncratic style. He saw all of European music as a single, indissoluble whole to which one could contribute. As he once expressed it: "Did not Eliot and I set out to refit old ships? And refitting old ships is the real task of the artist. He can say again, in his way, only what has already been said."

As Stravinsky became more established, he found himself in a position to dictate the terms for each of these works; ever the perfectionist, it was very important for him to maintain control over as many facets as possible. Increasingly, he stipulated the most stringent criteria for performance of his pieces, often insisting on conducting or playing his own works. Like a dictatorial military leader, he allowed conductors and interpreters virtually no leeway. He composed piano music for himself and commissioned violin music for a young Russian violinist, Samuel Dushkin, who willingly and wholly bent to the master's whims. The Stravinsky of the 1920s and 1930s became a small industry, all focused around his own career.

THE MATURE THINKER AND PERSON

While working out his relation to the musical past, Stravinsky was also defining other aspects of his mature personality. In 1926, while attending the seven-hundredth anniversary of the Celebration of Saint Anthony in Padua, Stravinsky underwent a profound religious experience. Shortly thereafter, Stravinsky rejoined the same church he had abandoned in his youth. By a curious coincidence, Stravinsky's return to the Russian Catholic fold occurred almost simultaneously with Eliot's conversion to the Anglo-Catholic faith (see chapter 7). One cannot help wondering whether, in addition to the men's need for membership in a traditional church at a time of personal and worldwide turmoil, the respective conversions of these two exiles represented an atonement for past "sins" of aesthetic iconoclasm. It may also have been part of a bargain they struck with God, in an effort to sustain their creative powers. Even if his motives may have been mixed, Stravinsky experienced very powerful religious feelings, which remained with him throughout his life, affecting his daily activities. He declared: "I regard my talent as God-given and I pray to him daily for the strength to use it. When I discovered that I had been made custodian of this gift, in my earliest childhood, I pledged myself to God to be worthy of it. . . . First ideas are very important—they come from God." And as he once told Robert Craft, in order to compose religious works he had to "not only believe in the symbolic sense, but in the person of the Devil and the miracle of the Church."

For most of the rest of the world, irrespective of the personal demons with which he was wrestling, Stravinsky represented the quintessential cosmopolitan artist: well connected, well groomed, and living the good life in Europe; married, but with an attractive and artistic mistress named Vera de Basset (whom he married in 1940 following the death of his invalid first wife); sailing and later jetting all over the world to promote his own music and to confer his blessing—or to pronounce his curse—on others' music. As he grew older, Stravinsky contributed actively to this legend by his pungent writings and by his participation in various efforts to dramatize his intriguing persona. He was unquestionably a witty, charming, articulate, and literate individual, whose companionship delighted those charmed few admitted to—and retained in—his circle. Indeed, though I would clearly have been eager to know all of the individuals chronicled in this book, I believe I would have most enjoyed eavesdropping at the Stravinsky dining table.

But Stravinsky preferred to view himself as a workman in a long tradition:

> I was born out of due time in the sense that by temperament and talent I should have been more suited for the life of a small Bach, living in anonymity and composing regularly for an established service and for God. I did weather the world I was born to, weather it well you might say, and I have survived—though not uncorrupted—the histericism of publishers, musical festivals, recording companies, and publicity—including my own.

Clearly, the flamboyant, controversial, public Stravinsky was balanced by a cerebral, hardworking, private craftsman. He saw himself as embodying an

Apollonian principle of order and balance, with only occasional forays into the turbulent Dionysian realm.

Stravinsky worked at least ten hours a day for many years. Beginning by playing a Bach fugue on the piano, he would compose for four to five hours in the morning and then, after lunch, orchestrate and transcribe for the rest of the day. His approach was very orderly; as his biographer Mikhail Druskin notes: "Stravinsky's work table resembled that of a surgeon rather than that of a composer. The neatness and precision of his scores recalled those of a map, with every syllable, every note, and every rest perfectly drawn." He had available all conceivable writing implements and scoring paraphernalia he might need, and he used these like the most highly skilled craftsman.

Stravinsky introspectively described his own composing activity: "For me as a creative musician, composition is a daily function that I feel compelled to discharge. I compose because I am made for that and cannot do otherwise. . . . I am far from saying that there is no such thing as inspiration. . . . Work brings inspiration if inspiration is not discernible in the beginning." (I am reminded of Freud's similar plaint: "When inspiration does not come to me, I go half way to meet it.") Stravinsky remarked on the opportunistic aspects of composing: "I stumble upon something unexpected. This unexpected element strikes me. I make a note of it. At the proper time, I put it to profitable use."

In describing the composing of *Petrouchka*, Stravinsky paid tribute to the role of his own bodily intelligence: "What fascinated me most of all in the work was that the different rhythmic episodes were dictated by the fingers themselves. . . . Fingers are not to be despised; they are great inspirers and in contact with a musical instrument, often give birth to unconscious ideas which might otherwise never come to life." Noting his tendencies to obsessiveness, he commented: "I would go on eternally revising my music were I not too busy composing more of it." And he added: "They think I write like Verdi! Such nonsense! They don't listen right. These people always want to nail me down. But I won't let them! On the next occasion I do something quite different; and that bewilders them." These words echo those of Picasso, Graham, and other introspective creators.

Always a reader and an intellectual, in a way that Picasso never was, Stravinsky hewed out a coherent musical philosophy during his middle years. While he did not enjoy literary composing per se, he was articulate. Working with gifted ghostwriters like Pierre Suvchinsky, and Alexis Manuel Lévy (who wrote under the pen name of Roland-Manuel), he voiced this philosophy in two seminal works: his autobiography of 1936 (*Chronique de ma vie*) and his Charles Eliot Norton Lectures at Harvard, delivered in 1939 and 1940 and published in 1942 as *The Poetics of Music*.

In these writings Stravinsky develops his positive views of music, while also seizing the occasion to castigate his opponents. Stravinsky's annoyance at the pretentiousness of Wagner's music, through which that composer sought to combine all art forms and to elevate his works to the status of a religion, motivated Stravinsky to assert, memorably, that music in itself is powerless to express anything. He wanted to replace unending melody with discrete order, syncretic and synthetic forms with self-contained ones, and emotional self-expression with strictly musical statements.

Suppressing whatever revolutionary impulses may have existed in his own person and animated his earlier music, ignoring the rich emotional associations of his early masterpieces, Stravinsky stressed the importance of conventions and traditions, and the utility of self-imposed constraints. He loathed disorder, randomness, arbitrariness, the Circean lure of chaos. Music was akin to mathematical thinking and relationships, and one could discern powerful, inexorable laws at work. In the paradox-packed closing lines of *The Poetics of Music*, Stravinsky declared: "My freedom will be so much the greater and more meaningful, the more narrowly I limit my field of action and the more I surround myself with obstacles. Whatever diminishes constraints, diminishes strength. The more constraints one imposes, the more one frees one's self of the chains that shackle the spirit."

Stravinsky's philosophical statements about music and composing have taken on a considerable importance, analogous in spirit, if not equal in potency, to those put forth around the same time by Eliot with respect to literature. In fashioning coherent (and surprisingly congruent) philosophies, these men differ from Picasso, whose intellectual aspirations were less well honed. Eliot and Stravinsky also shared a conservative political orientation, one tinged with anti-Semitism and a sympathy for fascism; in one letter to his German manager, Stravinsky declared: "I loathe all communism, Marxism, the execrable Soviet monster, and also all liberalism, democratism, atheism, etc. I detest them to such a degree and so unreservedly." Stravinsky's political conservatism did not blind him to the value of radical musical innovations. The same person who praised tradition and constraints had once declared: "I am the first to recognize that daring is the motive force of the finest and greatest artist. I approve of daring; I set no limits to it."

However, the two artists differed in important ways. Stravinsky was not concerned about political matters, except to the extent that they impinged on his own artistic labors or pertained to the fate of his beloved Russia. And while much of Eliot's poetry now seems to be directly autobiographical, chronicling the agonies of his personal life, Stravinsky's music appears to have evolved intrinsically. Perhaps in the deepest sense, this lack of association between musical and extramusical events confirms Stravinsky's conviction that music cannot express anything by itself.

FINAL MASTERY

In 1947, after the Second World War, Stravinsky was leading the life of the expatriate in southern California. Already in his mid-sixties, he had transcended his initial revolutionary period and had worked through an entire neoclassical agenda. Both the Russian and the European worlds had fallen apart. His parents, his first wife, and one of his children were dead; his other children were grown; and it would have been easy either to retire or to succumb to the allure of Hollywood. Indeed, a number of attempts were made to get him to compose for the popular American screen and theater. Stravinsky's attitude about those opportunities is captured wonderfully in the story of his encounter with the well-known American impresario Billy Rose. Rose had heard *Scenes de ballet* by Stravinsky and liked it, but he felt it could be improved by an arranger. Rose wired:

> your music great success stop could be sensational success if you would authorise robert russell bennett retouch orchestration

Stravinsky immediately wired back the deflating response:

> satisfied with great success

In later life, Stravinsky spoke about two crises he had had to deal with: the loss of Russia and his native language, after 1920; and the need to adjust after the Second World War to a new form of music, the austere serial style that Schönberg had developed in the early 1900s and that was being widely adopted in élite musical circles. In each case, he was able to make an adjustment and thereby to prolong his musical life.

Stravinsky was fortunate enough to create a third career, one that enabled him to compose with originality until the last years of his life. The energy and inspiration came from him, of course, but these were reinforced by pivotal contacts with two younger artists. In 1947, on a visit to the Chicago Art Institute, he had been greatly impressed by William Hogarth's engravings that depicted *A Rake's Progress*. He talked with his friend Aldous Huxley, the writer, about his plan to compose an opera based on this theme; soon thereafter Huxley introduced him to the young British poet W. H. Auden, who had also immigrated to America.

At Stravinsky's invitation, Auden joined the renowned musician in creating a full-length opera, *The Rake's Progress*. The two men worked for three years on the piece, spending approximately a year on each of three acts. From all evidence their collaboration delighted both men, who enjoyed obsessing over the details of versification as well as gossiping about the present and historical great figures of Europe and the United States. The opera was performed to considerable critical acclaim in Europe and in the United States in the early 1950s. Often considered the culmination of Stravinsky's neoclassical period, *The Rake's Progress* demonstrated that he was able to execute a major work in the English language and to reach new audiences without compromising his artistic integrity.

At about the same time that he met Auden, Stravinsky also made initial contact with the gifted young American conductor Robert Craft. Craft was intrigued by the compositional innovations associated with the Viennese school of twelve-tone, or serial, music, which Schönberg had instituted several decades earlier. Stravinsky was of course aware of these experiments; and earlier in his career, he had listened with sympathy to some of Schönberg's work, calling *Pierrot Lunaire* "this brilliant instrumental masterpiece." He had declared in his 1940 lectures: "Whatever opinion one may have of Arnold Schoenberg's music, it is impossible for a self-respecting mind equipped with genuine musical culture not to feel that the composer of *Pierrot Lunaire* is fully aware of what he is doing and that he is not trying to deceive anyone."

Nonetheless, Stravinsky had kept his distance from the serialists for several reasons, ranging from his personal antipathy to Schönberg, to his dislike of a priori compositional schemes, to the understandable uneasiness induced by a major competitor, one who was arrogant about his work, sarcastic about the ballet, and dismissive of Stravinsky's own efforts. (Schönberg had declared: "I have made a discovery that will assure the preponderance of German music for a hundred years.") Two somewhat paranoid personalities inevitably clashed. The gulf between the composers, while understandable, was unfortunate, particularly since they lived near one another in Los Angeles and shared many of the same interests and acquaintances.

Craft was not to be undone by these Old World tribal feuds. He gently pressured Stravinsky to listen to the music of the Schönberg circle, and Stravinsky found it far more stimulating than he had anticipated. He was particularly attracted to the work of Schönberg's younger associate Anton Webern, whose pointillistic, intervalic approach proved more congenial to his ear than Schönberg's grander and more harmonically oriented style. When Schönberg died in 1951 (roughly coincident with the conclusion of the *Rake* project), Stravinsky felt licensed to begin his own experimentation with serial techniques.

Even as the contact with Diaghilev had inspired Stravinsky in his twenties, and the revisiting of the classical repertoire had invigorated Stravinsky in midlife, the exposure to serial (twelve-tone) music fueled Stravinsky's compositional powers in late life. At a life stage when most creators have ceased to work altogether (like Eliot) or are susceptible to repeating themselves (as Picasso sometimes did), Stravinsky embarked on a set of compositions that, while never widely popular, are considered by some critics to be as important and innovative as his earlier works. In compositions like *Canticum sacrum* (1956), *Agon* (1957), *Threni* (1958), and *Movements for Piano and Orchestra* (1958–1959), Stravinsky created works in the language of serial music that retained his personal voice and reflected his lifelong aesthetic vision. Rather than being archetypical serial music, these works combined the tonal and the serial. Melodic invention and emotional immediacy were perhaps less evident than in earlier music; but his thematic and contrapuntal skills continued to develop, and the minted Stravinskian tonality, rhythmic organization, and sharp juxtapositions remained.

We may think of this synthesis as what might have happened if Picasso had crossed the Rubicon into purely abstract art while adhering to key compositional principles that had animated his earlier periods, or, more metaphorically, if Einstein had succeeded in fusing the relativistic and quantum-mechanical approaches. To be sure, Stravinsky's work in the serial mode proved less accessible (and less frequently performed) than some of the earlier works. Stravinsky appeared to accept this with resignation tinged with a defiance; he stated in his autobiography: "The general public no longer gives my music the enthusiastic reception of early days. . . . Their attitude certainly cannot make me deviate from my path."

Stravinsky's strength began to ebb in his eighties, and he suffered a series of debilitating illnesses, which gradually reduced his composing and performing activities. Still, he remained a vivid presence in the international artistic scene through a curious activity—the issuing of a long series of books and articles he and Craft penned. Craft is a gifted writer, a knowledgeable musician, and a sharp observer of the contemporary artistic scene; and in the twenty-odd years during which he lived with the Stravinsky family, virtually becoming one of its members, he came to know the mind of the master extremely well. He continued to invigorate his mentor by introducing Stravinsky to new music and encouraging him to listen again to some of the Germanic music he had earlier castigated.

What began with a series of questions and answers executed at Stravinsky's seventy-fifth birthday in 1957 culminated in writings where, as *The New Grove Dictionary of Music* entry indicates, "the two authors were beginning to sink their individual identities in a new character which was distinguished by some of the salient characteristics of both." Controversy has arisen about the extent to which Craft placed words in Stravinsky's mouth (just as the same question could be raised about earlier collaborators, Suvchinsky, in his *Autobiography*, and Claude Roland-Manuel, in his *Poetics*). But we clearly know far more about Stravinsky's views and sensibilities than we could ever have known, were it not for Craft's tireless conversing and chronicling of this oddly evocative friendship. It is as if Boswell and Johnson, or Goethe and Eckermann, had collaborated on a set of writings over fifteen years, or as if Françoise Gilot had remained on good terms with Picasso and thereby served as a continuing catalyst for the expression of his views. As Stravinsky once put it: "It is not a question of simple ghostwriting but of somebody who is to a large extent creating me."

We have seen that, earlier in his life, Stravinsky received needed cognitive and affective support from Diaghilev and Roerich, as well as from the members of his tight-knit ensemble. In the absence of such support, Stravinsky might well have been unable to break away from the Rimsky-Korsakov mode of *Fireworks* and *The Firebird* and develop the more innovative languages of *Le sacre* and *Les noces*. During his middle years, Stravinsky enjoyed the support of a wide circle of friends and followers; but, like Picasso, he seems to have conducted his neoclassical experimentation in conversation with his redoubtable predecessors as much as with his illustrious contemporaries. In old age, however, Stravinsky may have felt the need for someone of greater vitality who could again play a nurturing role, this time providing parentlike guidance as well as intellectual sustenance. It is perhaps because Craft met the ensemble of needs so perfectly that Stravinsky sustained his creativity in old age more fully than did our other creators and that he remained active among the living musical creators of his era. I see Craft as the last, and in many ways the most influential, of the series of collaborators who provided cognitive and affective support to the master throughout his long life.

Like Picasso, Stravinsky lived through much of the twentieth century and helped place his distinctive mark on it. He was able to absorb an enormous set of influences and yet retain his own highly distinctive voice—or set of voices. He may have lacked Picasso's indefatigable energy and protean facility, but he surpassed Picasso in the coherence of his work, the consistency of his personal philosophy, and the ability to articulate his vision in words as well as in his chosen artistic medium.

Having elected to compose music, and to do so largely in formats that required the participation of many other individuals, Stravinsky was consigned to collaboration in a way less necessary for an individual working in a relatively solitary pursuit like painting or poetry. From Diaghilev he received a primordial model of how to collaborate; and he internalized much of the most positive, as well as many of the least attractive features, of that dominating personality. During the height of his career, Stravinsky as a collaborator could be quite unpleasant, and as noted, Craft himself was shocked by the inflammatory paper trail left by Stravinsky during the early decades of the twentieth century.

In later life, Stravinsky appears to have become increasingly at peace with himself and with those around him. While remaining a stickler for details and a perennial skinflint, he seems to have been able to enjoy life, with access to the friends, the travels, and the publicity he needed, as well as the privacy he prized. Drawing on his understanding of the theater, he became a dramatic personality of his time. He was fortunate that he was able to continue composing until close to the end of his life, and to do so in an idiom that made sense to him and took advantage of the century's progress in his domain. He had the shrewdness to initiate collaborations with younger men like Auden and Craft that kept him in touch and engaged with the environment of the day. In this respect he was much more fortunate than Picasso, whose quest to remain young was more fervent but less well guided, secured chiefly through a never-ending search for young lovers and done with scant effort to remain in contact with the most innovative and fertile artistic streams. More so than our other creators, Stravinsky seems to have been able to preserve what was important from childhood while enjoying as well some of the fruits of later life.

I must mention one discordant note. Stravinsky's relations with his three surviving children were bumpy and, toward the end of his life, increasingly disrupted by legal entanglements over the ownership of rights. The children had never fully accepted Vera de Basset, Stravinsky's long-time mistress and second wife, and by the end of his life, his wife did not want to have anything to do with them; even their attendance at his funeral and at various memorial services became an issue. As with several of our other creators, a connection to the wider world seems to have been purchased at the cost of smooth and loving family relationships.

The waning of one's powers provides no pleasure for anyone, and this is perhaps an especially bitter experience for the creative titans of a century. But Stravinsky dealt with aging as well as any other master of our era, continuing to compose, being personally happy with his wife and his "adopted" son, Craft, and able to relinquish some of the most combative aspects of a creative life carried on amidst other creative individuals. As a final gesture of peace, he was buried, at his wish, in his beloved Venice, near to Diaghilev, with whom he had quarreled a half century earlier, but with whose founding and catalytic genius he wished to be reconciled in the end.

7

T. S. ELIOT: THE MARGINAL MASTER

Eliot, 1914

THE WASTE LAND RECOVERED

IN 1968 A MANUSCRIPT long thought to be lost was located in the Berg Collection of the New York Public Library. The packet consisted of fifty-four pages, most of them typed, but some handwritten on one or both sides. Some of the pages had few marks on them, while others were heavily edited by various hands, and several had been completely crossed out. The typed language itself was equally varied, with much of it in colloquial English, much of it in elegant or abstruse English, and scattered lines in a range of European languages as well as Sanskrit.

No ordinary manuscript, the recovered sheets were interim drafts of *The Waste Land*, arguably the most famous poem, and in all likelihood the most influential poem, in English of this century. T. S. (Thomas Stearns) Eliot, a St. Louis–born poet who had settled in England, had begun to write the poem, or verses that became a section of the poem, around 1914, and had finished a complete draft of nearly a thousand lines by the end of 1921. He had then given this draft to both his wife, Vivien,* and a close friend, Ezra Pound, an American-born poet who had also settled in Europe. Along with Eliot, these "friendly critics" had made substantial changes to the poem, with Pound suggesting changes that reduced the poem to approximately half its length. In the words of the Eliot scholar Helen Gardner, "Pound turned a jumble of good and bad passages into a poem."

Eliot appreciated immediately the significance of Pound's help. Confident that the poem would be regarded as a significant work, Eliot presented the manuscript as a gift to an American agent, John Quinn, who had ably (and without charge) represented Eliot's publishing interests in the United States. Quinn died in the year following the receipt of the manuscript, and during his estate's disposition, the manuscript was misplaced; indeed, Eliot came to believe that it had been lost. Its discovery forty-five years later not only solved a literary mystery; it has also provided unique insights into the genesis of a major literary work, revealing the role that a sympathetic but candid critic can play. And it has also crystallized the question of why a pair of young American expatriates would be writing about the decline of civilization from their vantage point following the Great War.

THE MANY STRANDS IN ELIOT'S BACKGROUND

Although born and raised in St. Louis, along the Mississippi River, Eliot also grew up in the shadow of New England. His ancestors on both sides had come from the Boston area, where they had been leaders in religion and education

---

*Sometimes spelled Vivienne.

since the seventeenth century. His paternal grandfather, William Greenleaf Eliot, had moved to the St. Louis area where, as an influential nondoctrinaire religious leader and a gifted financier, he had founded Washington University. T. S. Eliot's father, Henry Ware Eliot, an affluent, achievement-oriented businessman, was the president of a hydraulic press company; his mother, Charlotte Champe Eliot, was a poet of considerable ambition and some talent who considered herself a failure because she had not completed her formal education. Both parents were highly moral, if not moralistic, individuals who believed in a lifetime of "good works"; they were also perfectionists. A distant cousin, Charles William Eliot, had become the president of Harvard University almost twenty years before T. S. Eliot's birth; such achievement was not considered atypical in the larger Eliot clan.

Eliot was a somewhat sickly child whom his mother shielded. He was surrounded by women—not only his mothers and sisters but more distant relatives, as well as a beloved Irish nurse, Annie Dunne. Recognized as extremely intelligent and talented from an early age, he was expected to adhere to very high academic and moral standards. Eliot felt that he carried within him the expectations of his Puritan ancestors and that these strictures placed great pressures on him. His greatest pleasures came from summers spent along the Massachusetts shore, where he would read and sail. He also cherished the experience of living along the mighty Mississippi River. "I feel that there is something in having passed one's childhood besides the big river which is incommunicable to those who have not," he wrote in later life.

From all reports, young Tom Eliot was an extremely sensitive child. According to his sister, Eliot, when he was still learning to talk, used to produce the rhythm of sentences without shaping words. He was entranced by sensory impressions—smells, noises, sights—and drawn to effigies, candles, and incense. Decades later he recalled as well the steamboat horns blowing in the new year, the river in flood with its cargo of human bodies, animal carcasses, and assorted debris, as well as the prayers of his Irish nurse and her discussions about the existence of God. To be so powerfully attracted to such sensory impressions is in itself unusual. Even more unusual were Eliot's capacity to remember them vividly even decades later and his strong inclination to capture these impressions in lines of poetry.

Written language had been an important vehicle of communication on both sides of his family for many generations. Charlotte Eliot had written a great deal of religious poetry, much of it didactic. Young Eliot had a very capacious linguistic memory. He issued his own newspaper at school, stocking it with jokes. He also wrote nautical stories, doggerel, and more serious verse by the time he was a teenager. He indulged his fantasy life by spinning tales about the South Pacific and Hawaii. While not strikingly precocious, Eliot's poetry exhibited an instinctive sense of form and a well-defined tone. And his considerable parodic gifts were showcased in a decent imitation of Ben Jonson's work, which Eliot completed at the age of sixteen.

Eliot was also a notable student. He performed very well at Smith Academy (also founded by his grandfather) in St. Louis and excelled during an extra year at Massachusetts's Milton Academy. He read extremely widely in English, Latin, Greek, and French, and remembered a good deal of what he had read. The only area of academic difficulty for Eliot was physics; throughout his life, he displayed little interest or ability in the sciences, situating himself squarely within the humanities, the more ancient of the two academic cultures described by the scientist-humanist C. P. Snow.

OUT OF JOINT AT HARVARD

On virtually every ground one might have anticipated a natural fit between Eliot and the Harvard of the early twentieth century. Eliot's family had maintained a long relationship with Harvard, and had spent their summers in the Shore area north of Boston. Eliot was bookish, literary, and witty—all desirable features in a Harvard man. (Interestingly, Eliot saw himself as priggish, whereas contemporaries described him variously as being bright, reserved, mischievous, attractive, and graceful.) As an added feature, Eliot was a member of perhaps the most distinguished Harvard class ever—the class of 1910, which included the future essayist Walter Lippmann, the politician Hamilton Fish, and the political revolutionary John Reed—amidst a cohort from adjoining classes that included the poet Conrad Aiken, Eliot's lifelong friend, and the literary critic Van Wyck Brooks.

Reflecting the leadership of President Eliot and the times of President Theodore Roosevelt (himself a Harvard graduate), the atmosphere of the school was generally considered to be liberal, democratic, progressive, materialistic, individualistic, and pluralistic. Most surprisingly, these features did not appeal to Eliot. Harvard struck Eliot as cold, and he was troubled by what he perceived to be the inferior status afforded humanistic studies.

Eliot was discovering in himself a contrasting set of impulses. He considered American literary work of the era to be largely empty. He found himself attracted to arcane literature of the past and of foreign lands, rather than to the facile and jingoistic verses of Cambridge's favored poet, Henry Wadsworth Longfellow; he gravitated to teachers like Irving Babbitt, who identified with conservative and Roman Catholic causes; he preferred Oriental mysticism to the straightforward and plebeian Unitarian creed of his forbears; he disdained Charles Eliot's renowned elective system, seeking instead the security of an established canon and a systematic, cumulative course of study.

Continuing to write, Eliot contributed verse to the undergraduate periodical, the *Advocate*; already the themes of urban squalor and withering nature, as well as an incipient Oriental sensibility, can be discerned in what are in other ways unremarkable juvenilia. As in much youthful poetry, the feelings are strong, but they have not been clearly differentiated and captured in words by the author, and, perforce, remain far from self-evident to the reader.

A crucial event in Eliot's Harvard education was his discovery of Arthur Symons's book *The Symbolist Movement in Literature*. Eliot responded enthusiastically to Symons's rejection of prosaic realism, his treatment of art as a variety of religion, his quest for a spiritual vision where poetic symbols could capture the essence of things, and his conviction that a poem could create a world of its own, referring to nothing beyond its own elements. Symons drew Eliot's attention to French poetry, and particularly to the work of Jules Laforgue, a little-known late–nineteenth-century poet.

Eliot immediately found an "elective affinity" with the works of Laforgue, who had experimented with the use of different voices to represent suffering, ordinary discourse, and magisterial commentary. Embracing a view of the world as decadent and depressing, Laforgue was able to juxtapose youthful boredom and aging withdrawal; he treated serious ideas through irony, humor, and emotional distance. Eliot was to say later: "I do feel more grateful to him than to anyone else. I do not think that I have come across any other writer since who has meant as much to me as he did at that particular moment."

No doubt abetted by the French symbolist poetry that he had recently discovered, as well as the earlier example of the Parisian poet Charles Baudelaire, Eliot was also developing a more highly personalized sensibility. He found himself walking through the decaying areas of Boston, alternately attracted to and repelled by what he saw. At the same time, he retained access to the drawing rooms of the Boston Brahmins, and to the fashionable clubs and drinking spots of Harvard's élite. He felt that Boston, once vivid and dynamic, was now inert, its citizens tense and alienated. The consistent, if dour, Puritan life had been replaced by crass commercialism and urban disrepair. The clash between these realms—the world of the slum and the world of Cantabrigian gentility, the suffering of the impoverished and the hypocrisy of comfortable socialites—made a deep and deeply dissettling impression on the sensitive young student. In poems of 1908 and 1909 he sought to capture his newly emerging ensemble of feelings.

Sometimes, crystallizing experiences are quite specific, as in Eliot's discovery of the second edition of Symons's book in the Harvard Union in December 1908, or his first excited reading of Jules Laforgue the next spring. The following year, Eliot had an experience that greatly influenced him but that he never could characterize precisely. About the time that he graduated from Harvard, while wandering through Boston, he perceived the streets shrinking and dividing, and he felt engulfed in a great silence. "You may call it communion with the Divine or you may call it temporary crystallization of the mind," Eliot later commented on what seems to have been a hallucinatory experience.

In my view, the experience itself symbolized a change that was taking place in the young Eliot—a shift of sentiment that he could most adequately grasp through a poetic image. Until this point in his life, he had pretty much followed the script his family had envisioned for him. He had studied well, performed well, written well, and gone with seeming effortlessness through the scholastic and social paces that were expected of him. Because he had been attracted to the discipline of philosophy, it was assumed that he would continue his studies at Harvard and eventually become a professor of philosophy.

Within his own consciousness, however, Eliot had come to feel increasingly alienated. He did not like the Boston, the St. Louis, the America of his era. He felt equally distanced from the clubby students, the Boston Brahmins, and the sordid underclass of urban life. Coming from a family circle that was dominated by older women, he did not know how to relate to women of his own age; he felt threatened by them and frustrated by his inability to deal with his sexual urges.

Eliot was attracted to another world and a road less travelled. There was something alluring about France and Britain, with their far longer histories, greater literary sophistication, heightened valuing of religious and spiritual matters, and deeper sense of irony. While intrigued by the study of philosophy, he was also searching for a poetic voice, one that could synthesize the concrete sensations, the powerfully felt emotions, and the ideas about life and civilization pressing upon him.

The "crystallizing moment" in Boston, with its hint of divine inspiration, held out a possibility of dealing with these contrary emotional stirrings, of capturing his alienation through poetry. Eliot stressed the importance of the unattended moment—as he wrote in *The Waste Land*, "the awful daring of a moment's surrender . . . by this, and this only, we have existed." The diverging street served as a vivid image, and streets were to recur as pregnant symbols throughout Eliot's writings. But the vision, the unattended moment, was but a fleeting intimation; in the years ahead, Eliot had to find his way of realizing this dimly felt synthesis.

NEW ATTEMPTS AT LIFE

Given Eliot's strong feelings of alienation and his search for a more comfortable balance, a decision to travel to Europe following his graduation was virtually inevitable. Eliot went first to France, where he sought to capture the atmosphere of Laforgue: "The kind of poetry that I needed to teach me the use of my own voice did not exist in English at all; it was only to be found in French." Eliot attended lectures of French savants like Henri Bergson and Émile Durkheim and found himself particularly attracted to the conservative views of Charles Maurras, an Irving Babbitt–like figure who styled himself in catholic, classical, and monarchic terms. Eliot befriended a number of young Frenchmen, most notably the future novelist Alain-Fournier (Henri-Alban Fournier) and a medical student-cum-writer named Jean Verdenal, with whom Eliot became personally close and whose death in the First World War proved singularly traumatic for Eliot.

Eliot seems to have been no happier as a visitor living in Paris than he had been as a student in Boston; again he was struck by the contrast between the intellectual salons, the artistic avant-garde, and the sordid poverty of the streets. But his letters back home indicate that, like other talented young artists, he was soaking up information and impressions at a remarkable rate. He also made at least one voyage to London, visited Germany, and, in the summer of 1911, completed a draft of his first important poem, "The Love Song of J. Alfred Prufrock."

While Eliot considered remaining abroad and attempting to survive as a poet, he ultimately decided to return to Harvard in 1911 and to begin doctoral study in philosophy. Equipped with hindsight, we may look at this period as a respite, or even as a regression, as Eliot was once again fulfilling the expectations of his ambitious, conservative, and professionally oriented family. Indeed, Eliot spoke of the frightening "nightly panics" he had experienced while sojourning in Europe. But this was also a time when Eliot continued to play out in his mind various life choices, personal philosophies, and modes of expressing himself. He took courses with Bertrand Russell, who saw him as a silent young dandy with little enthusiasm or *joie de vivre*. He studied Sanskrit, read Hindu and Buddhist sacred texts, and continued to write poetry; in the terminology of the psychoanalyst Erik Erikson, Eliot was experimenting with multiple identities and disparate voices. As if to signify this uncertainty, his handwriting actually changed several times during this period.

As a graduate student, Eliot immersed himself in the writings of a contemporary British philosopher, F. H. Bradley, who explored the relation between appearance and reality, experience and truth, subjective experience and objective truth. These perennial philosophical questions intrigued Eliot, and he eventually sought to treat them in his own poetry and in other writings. It is not clear that Bradley's actual solutions impressed Eliot, but their sensibilities coincided. Eliot shared with Bradley an interest in rituals and in the role of order in one's own existence, an attraction to subjective experience, an interest in unifying discordant beliefs, and a distrust of "conceptual" intelligence. His dissertation topic—"Experience and the Objects of Knowledge in the Philosophy of F. H. Bradley"—gave Eliot the opportunity to examine an issue of singular importance to him: "a maddeningly brief visionary moment and its contradictory interpretations." Bradley is perhaps the only philosopher who continues to be known outside the circle of experts because he happened to be the subject of a major poet's doctoral dissertation.

Eliot seems to have been somewhat happier as a graduate student than as an undergraduate, and he was valued by the Harvard philosophy faculty, whose members urged him for several years to join their ranks. But shortly before the outbreak of the First World War, Eliot returned to London and ultimately remained in Europe for almost twenty years. Renouncing the pastoral university existence, he declared: "How much more self-conscious one is in a big city." By this time, he was content to lead the life of the marginal man: He had cast his lot as a writer and had committed himself to succeeding on foreign shores.

TWO POETS JOIN FORCES

It took hubris for young Eliot to abandon the life of a comfortable American and a promising philosopher in order to attempt a career as an artist abroad. Especially as viewed in Paris and London, the capitals of civilization, the United States remained a backwater, with little significant accomplishment in the arts. Only a most extraordinary individual, like the writer Henry James, had succeeded in transplanting himself to Europe; and after a half century there, he still felt like an outsider.

But Eliot had been preceded in 1908 by a gifted young American poet from Idaho named Ezra Pound. A forceful and controversial character, Pound had impressed himself upon the English literary circles by his personality as well as his five published volumes of poetry. Meeting the yet-unpublished Eliot in September 1914, Pound immediately became enthusiastic about his younger Harvard-educated counterpart. He wrote to Harriet Monroe, the editor of *Poetry*, that "Prufrock" was the best poem he had yet seen from an American, and he told the writer H. L. Mencken that Eliot was "the last intelligent man I've found." As Eliot recalled: "In 1914 my meeting with Ezra Pound changed my life. He was enthusiastic about my poems and gave me such praise and encouragement as I had long since ceased to hope for." The two men, similar in background though different in temperament, fast became close friends.

For the next several years Eliot, Pound, and the British writer and painter Wyndham Lewis joined forces to promote new forms of expression in English. In Eliot's view such innovation was desperately needed; the art of the time was hopelessly romantic and propagandistic. He later commented that "the situation of poetry in 1909 or 1910 was stagnant to a degree difficult for any young poet of today to imagine." The young Turks learned technique from one another; for example, Eliot was influenced by Pound's indirectness, fragmentation, and startling juxtapositions.

They also buoyed one another psychologically. Pound's support for Eliot was especially important. Much more aggressive than Eliot, Pound introduced him and his writings to a spectrum of individuals both in England and in the United States. He served as a kind of agent and impresario, advancing Eliot's cause to all who would listen, even defending his expatriate status to the Eliot family. It is quite possible that, without Pound, Eliot would not have remained in England, effected the transition from philosophy to poetry, married his first wife, met the agent Quinn, or published verses in America.

One may legitimately wonder why the poetic writings of an American who had been influenced chiefly by French symbolist poetry should prove of interest to the British literary public. In my view, certain changes taking place in the Western world at the time were less visible within insular England than elsewhere. The great empires of the nineteenth century were breaking up; the liberal consensus, according to which an enlightened aristocracy worked collaboratively for the greater benefit of society, had evaporated; the disparity between social classes was becoming more jarring and more problematic; the squalor of urban life, far from being a temporary dislocation of industrialization, was now seen as permanent; and accepted religious forms and ideological systems were being undermined, even though a need for them was felt more urgently than ever.

Intimations of these and other disruptive trends had begun to emerge in the art forms around the turn of the century. Novelists like Émile Zola in France and Theodore Dreiser in the United States sought to capture the upheavals of their time; painters like Picasso and Braque in France had intermingled the high and low forms of art; composers like Strauss, Schönberg, and Stravinsky had in their respective ways rejected the comforting nineteenth-century romantic forms. But in Great Britain, where the aforementioned trends were somewhat less evident, the poetry of Alfred, Lord Tennyson and Rudyard Kipling, the novels of George Eliot and Thomas Hardy, and the works of academically oriented composers, like Edward Elgar, and painters, like Walter Sickert, still held sway.

Early twentieth-century literary life in Britain came to be influenced—even dominated—by non-British individuals: James Joyce, William Butler Yeats, and George Bernard Shaw were all from Ireland; Joseph Conrad was from Poland; and Eliot and Pound were from the United States. These individuals had mastered the language, but perhaps because they were essentially foreigners, they discerned trends that were less salient to those who had come to take the nineteenth-century Victorian assumptions for granted. Doubtless, even without such marginal individuals, English artistic forms would have changed; the revolutionary Virginia Woolf, after all, was a British blueblood (though, perhaps significantly, not an Englishman). But the non-natives, the marginal writers, helped transform artistic forms more rapidly and also placed an international stamp on them.

Evidence of the Poet's Gifts

Eliot's early poem "The Love Song of J. Alfred Prufrock" represents an important contribution to this modernization of English letters. The absurd name of the protagonist, coupled with the romantic label of a love song, immediately marks the poem as one that traffics in incongruities. An Italian epigraph, a refrain about individuals "talking of Michelangelo," and references to Hamlet, Lazarus, an eternal Footman, and the Fool reveal the speaker as one with a knowledgeable, high-artistic sensibility. Slang references to one-night cheap hotels, sawdust restaurants, a bald spot in the middle of one's hair, thin arms and legs, skirts that trail along the floor, and rolled-up overall hems signal an author equally at ease (or equally ill at ease) with the life and talk of the street, the music hall, and the pub.

> *Let us go then, you and I,*
> *When the evening is spread out against the sky*
> *Like a patient etherized upon the table*

When reading the arresting opening line, one immediately encounters a poetic voice that can comfortably juxtapose the most disparate elements: a romantic evening with an anesthetized patient. These startling couplings recur throughout the poem: visits made amidst overwhelming questions; time to meet and to murder others; comings and goings and talk of Michelangelo; the measuring out of life in coffee spoons; the act of daring . . . to eat a peach, to disturb the universe; squeezing of the universe into a ball; a slightly bald head brought in on a platter. Of course, the couplings are not only revealing about bizarre juxtapositions in the world; they also suggest the idiosyncratic sensibilities of an individual who would be struck by such incompatibilities.

The poem also deals with incongruities at a more conceptual level: The protagonist is at once in the thick of the poem and removed from its mundane world; the voices are, varyingly, young, middle-aged, and aging; the identification occurs equally with the masculine and the feminine elements, as well as with a life of action and with a pose of paralysis. The verse varies from the regularly metric and rhymed to the occasional ellipsis and the unexpected short line. Momentary mental states are featured, as a means of conveying the fragmentation of the world, indeed, the several worlds that had so unsettled the youthful poet. A Laforguian sense of self-mockery and irony pervades and unifies the poem.

"Prufrock" and a second early poem of the same genre, "Portrait of a Lady," announced Eliot as a gifted young poet. Most poets write poetry when they are very young, and there is still-unpublished Eliot juvenilia dating back to his preteen years. Usually this early poetry is of little intrinsic interest, though in Eliot's case, as noted, it does reveal a commanding vocabulary, some sense of tone and form, an engaging humor, and a concern with themes of death, loss, and the passage of time. During adolescence, or shortly thereafter, however, there is generally a rapid spurt: Within a few short years the poet discovers a distinctive voice—or, in Eliot's case, a distinctive ensemble of voices. As occurred most remarkably with John Keats, the Eliot voice combined the philosophical and the prosaic; he was able to deal with philosophical themes that engaged him while wrapping them in the language and imagery of ordinary life.

Eliot was by no means certain that he wanted to remain the mouthpiece of Prufrock, or of the observer of the "Lady." Part of him was exploring authorship in French, capturing Parisian scenes in tight French quatrains. His early works also contain experiments with Elizabethan drama, esoteric imagery, and straight satire, as well as a ponderous exploration of religious themes. In letters to friends he lays out entire poetic dramas as well as frankly vulgar doggerel. (Versions of these lighthearted epistolary experiments continued throughout his life.) From early on, Eliot exhibited his wonderful mimetic flair, and he found it easy to parody different literary styles. He was also writing serious philosophy, both scholarly articles and a lengthy dissertation, and for significant periods, there were no poems at all.

Indeed, while recognizing that "Prufrock" had merit, Eliot was in no hurry to publish it and might not have done so at all, or at least for a while, had it not been for the urgings of Aiken and Pound; he confessed that in some ways it felt like his swan song. However accomplished he may have appeared to others, Eliot was still uncertain about what (or who) he wished to become. Ever the marginal individual, he rarely exhibited the striking self-confidence of the young Freud, Einstein, and Picasso.

SETTLING DOWN IN EUROPE

As the Great War began, Eliot felt more than ever that his future lay in Europe. Renouncing the second year of a Harvard traveling fellowship, he elected to try to survive as a teacher and writer in the London area.

Until this time, Eliot had had virtually no sexual experience or love life at all. As he confessed in a letter to Conrad Aiken: "I should be better off, I sometimes think, if I had disposed of my virginity and shyness several years ago; and indeed I still think sometimes that it would be well to do so before marriage." But in 1915, some months after encountering the energizing Pound, he met a bright, spirited, dramatic, candid, and sensitive young English woman named Vivien Haigh-Wood. Tom and Vivien fell in love and were soon married.

There has been much speculation (even in *Tom and Viv*, a recent London play) about Eliot's first marriage, which was by all accounts singularly, even tragically, unhappy. Clearly, the initial attraction was between opposites: The thoughtful, reclusive, scholarly, and extremely introverted American virgin with the spunky, fun-loving, wild-dressing, sexually more experienced Englishwoman. They must have deemed their traits and aspirations complementary, viewing each other as a solvent for personal fears and failings. Although they struggled for almost two decades to make the marriage work, it apparently never got off the ground. Bertrand Russell, who had known and liked Eliot for some years and who apparently had had a brief affair with Vivien during the early years of the marriage, offered one interpretation: He believed that Eliot had married Vivien in the hopes of being stimulated but had soon discovered that such arousal was simply not possible. Clearly, Eliot was uncomfortable in his sexual identity, though no convincing evidence exists that he had significant homosexual leanings.

Shortly after the marriage it emerged that Vivien was given to frequent debilitating illnesses. Some of these were clearly physical (swollen face, pituitary troubles, pleurisy, spinal injury), but she also experienced considerable nervous disorder and, quite possibly, some psychosomatic and hysterical elements as well. She seems to have been very weak, yet capable of wounding the hypersensitive Eliot by her remarks. At least on the surface, Eliot was a good nurse to his wife, attending to her troubles, expressing considerable concern about her condition, and remaining faithful to her. At the same time, he suffered deeply over their incompatibility, confessing shortly after the marriage, "I have lived enough material for a score of long poems in the last six months." Eliot himself became increasingly disturbed, ultimately suffering a breakdown in the months directly preceding the composition of *The Waste Land*. What cannot be determined is the extent to which the neurotic strands in each member of the couple contributed to the other's difficulties—whether each drove the other crazy by what each said and did.

For a while, Eliot sought to lead the typical life of the struggling young artist, teaching during the day and writing at night. By all indications, he was a dutiful and effective teacher, much appreciated by his students. Once again, however, Eliot deviated from type. He decided to work instead at a bank and, to his and others' surprise, found that he liked this distinctly nonartistic vocation. Eliot enjoyed playing with numbers, adhering to the office routine, and dressing up and filling the role of a conventional professional manager. Perhaps not surprisingly, given his lineage, Eliot proved adept at this task and was eventually given the formidable responsibility of handling the war debts.

Expanded Works and a Growing Reputation

Though Eliot was struggling to make ends meet, writing relatively little poetry, and going through an unusually painful marriage with very few rewards, his literary career was beginning to flourish. With considerable assistance from Pound, Eliot was able to get much of his work published. The collection *Prufrock and Other Observations* was published in June 1917 in England and three years later in the United States. Eliot appears as a writer, critic, and poet in leading English and American journals of the time such as *Poetry*, the *Dial*, and the *Nation*. An influential book of criticism, *The Sacred Wood*, was published in 1920. By the early 1920s, barely into his thirties, Eliot was already recognized as one of the major literary figures on both sides of the Atlantic, respected (and sometimes feared) because of his formidable literary talents and his growing power in the publishing world.

No doubt Pound's (and then Eliot's) membership in leading literary circles furthered Eliot's career. Eliot was at the very least a fellow traveler in the Bloomsbury circle, the influential literary and intellectual group that included writers such as E. M. Forster, Lytton Strachey, and Virginia Woolf. In particular, he discovered a personal and professional affinity to Woolf. The fact that virtually a generation of young Englishmen were away at war, and that many of the most gifted never returned, created a vacuum in cultural matters, which foreigners like Pound and Eliot were available—and no doubt eager—to fill.

However, Eliot's success certainly required effort on his part. His recently published correspondence reveals clearly that Eliot was concerned from the first about ensuring recognition for his work. The once-shy Eliot was gaining in boldness and authority. He carefully cultivated influential sponsors in the United States, such as Schofield Thayer of the *Dial*, the wealthy patron Isabella Stewart Gardner of Boston, the important New York publisher Alfred Knopf, and John Quinn, his loyal, unpaid agent. In Britain he acted gingerly to remain on good terms with Richard Aldington, Bruce Richmond, and other leading literary figures. He adopted the appropriate degree of deference or authority in relating to each of these individuals, making adjustments as needed from one letter, encounter, and year to the next.

In one lengthy, amazingly directive letter to his brother, he instructs Henry Eliot on how to contact influential people in Boston and New York, which publications to visit, introductions to make, and the like. Eliot counsels his brother "to secure introductions to editors from people of better social positions than themselves." He expresses a desire "to have these magazines mentioned know my name in some personal way . . . and to get some steady connection such as the writing of an 'English letter' or discussion of current French stuff."

Eliot could also perceptively analyze his own position; in a letter to his old philosophy teacher J. H. Woods, he declared:

> There are only two ways in which a writer can become important—to write a great deal, and have his writing appear everywhere, or to write very little. . . . I write very little and I should not become more powerful by increasing my output. . . . My reputation in London is built upon one small volume of verse. . . . The only thing that matters is that each of these should be perfect in their kind, so that each should be an event. As to America: I am a much more important person here than I should be at home. . . . If one has to earn a living, therefore, the safest occupation is that most remote from the arts.

And apologizing to agent Quinn for a volume that combines prose and criticism, he says, "It is time I had a volume in America and this is the only way to do it."

Family members and influential mentors typically have shown young creators which techniques can help them influence the field. Eliot clearly picked up tactics from his mother, who had long since attempted to manipulate the scene on behalf of her gifted son, and from Pound, who knew the Anglo-American literary terrain intimately. Still, Eliot was a master planner of his own career. Indeed, perhaps because he learned from others' mistakes, he was less aggressive in public than Pound and less unyielding in personal relations than his mother. From the time of his permanent move to Britain in 1914, until he had become an established international literary celebrity in the 1930s, Eliot was extremely planful and reflective about every career move he made.

While still somewhat shy, stiff, and subjectively uncomfortable with himself, Eliot made an excellent impression on most of the powerful and influential individuals he met: His graciousness, reserve, encyclopedic knowledge (which he wore lightly), delicate touch of humor, and perennial hint of marginality all served to ingratiate him with others. It does not hurt to have an appealing persona; and while not frankly charismatic, Eliot had perhaps the more important property of making powerful individuals want to come to his aid. Even when his books received less-than-positive reviews and did not sell well, they strongly impressed the intellectual leaders of Britain; in the end this positive penumbra proved pivotal.

THE WASTE LAND:
BACKGROUND AND COMPOSITION

As early as 1914, Eliot had begun work on a "long poem," and he continued to amass fragments for it over the next several years. It is not clear that Eliot had the theme or structure worked out, but he no doubt wanted it to be an important "defining" statement about his era, as well as a description of the kind of spiritual journey that he and his mother had long found fascinating. There were many false starts and abandonments, as well as early, more effective intimations in a French poem called "Dans le restaurant." By early 1921 Eliot was able to show four parts of a poem, entitled "He Do the Police in Different Voices,"* to his friend and associate Wyndham Lewis. He wrote to his agent Quinn that he had wanted to complete the poem by June, but that family matters had intervened.

Two factors helped, in their respective ways, to propel the composition of the poem that eventually became *The Waste Land*. The first, positive factor was Eliot's reading of Jessie L. Weston's book on the Grail legend, *From Ritual to*

---

*A line quoted from Charles Dickens's *Our Mutual Friend*.

Romance. The notion of a quest beginning in doubt and devastation had long been on Eliot's mind, but Weston's rich description of mythical quests and themes from diverse cultures provided an attractive and capacious vehicle for Eliot's still disparate notions. The second, initially disruptive factor was Eliot's rapidly declining health.

By September 1921, a combination of personal and professional pressures had brought Eliot to the brink of a severe nervous breakdown. A doctor counseled complete rest, and Eliot went away, first to Margate on the Kent coast, and then to Lausanne, Switzerland. Over the course of three months, spent largely alone, Eliot produced a complete draft of the poem, now about a thousand lines long.

While scholars still dispute the complete story on the process of composition of *The Waste Land*, one point is clear-cut: Eliot gave the manuscript to Pound at some time around the end of 1921, and Pound edited it severely. This heavily annotated manuscript is what Eliot offered as a token of gratitude to Quinn and is what was rediscovered several decades later. Eliot scholars have now pored over Pound's proposed revisions, as well as those made by Vivien Eliot and by Eliot himself, and the poem's compositional process is now as well understood as that of any other twentieth-century literary work.

Over the years, and especially in the final months of 1921, Eliot had produced a collection of sections or episodes covering a wide range of situations: lower-class life in contemporary London; classical scenes involving mythical characters; descriptions of natural scenes featuring winter, bones, deserts, and other evocative elements; conversations conducted in several languages; lines and fragments from high literature (Shakespeare, Dante, Baudelaire); parodies of esteemed authors (Pope); hymns of praise; fiery sermons; maritime narratives; and Sanskrit phrases. The episodic quality of "Prufrock" and "Portrait" had been greatly elaborated. One was now confronting language not merely from a single somewhat bewildered middle-aged man, but also from a welter of voices reflecting the consciousnesses of actors and objects drawn from the broadest sweep of time and space.

While highly suggestive and full of sections with undeniable power, the original manuscript was bloated. Instead of conveying a feeling for these different universes, the first-draft version rubbed the reader's nose in them and brought the reader back to each one over and over again. There was much indecisiveness, repetitiveness, and monotony: too many voices and too little sense of overall direction, control, and locale. Pound's feat was in carving away the overstated sections that pulled the poem in diffuse directions and in both sharpening the remaining verses by crossing out unnecessary or misleading words or phrases and eliminating many hedges and ambivalent tones. He was merciless with Eliot's deliberate or unwitting attempts at parody, finding most of them inferior and unnecessary, and he also deleted or challenged phrases that struck him as misogynist, anti-Semitic, or otherwise personally idiosyncratic. He counseled that the embedded and the appended poems be published separately. He helped to bring out the work's musical qualities, both in the meter of individual lines and in its broader themes and rhythms. At the same time, he left the major themes and overall subject matter unaffected, largely limiting his efforts to the clarification, reinforcement, and precise communication of Eliot's almost completely despairing message.

In his own words, Pound was the "midwife"; in Eliot's words, Pound was *il miglior fabbro*—the better craftsman. The result is a far sharper, more compact, and telling poem. Each of the sections is more directly stated, while the actual links between the five sections of the final version are now left to the reader's construction.

Vivien Eliot's proposed changes, while far less extensive, nicely complemented Pound's. Having an excellent ear for specific lines, as well as an enviable skill at parodying the utterances of ordinary Londoners, Vivien praised the best of the verses, improved some of the existing ones, and suggested some pithy new ones—for example, "If you don't like it you can get on with it" and "What you get married for if you don't want to have children."

It would be simplifying, but misleading, to declare that Pound provided the literary advice, while Vivien Eliot supplied the affective support. Indeed, both individuals seem to have served both roles. Thanks to the one-two punch of his two talented editors, Eliot received help both on the overall sweep of the poem and on the word choice for specific lines.

Eliot himself did considerable editing of the poem, addressing both structural issues and ones involving individual lines. He did not always follow Pound's suggestions, and some commentators feel that he would have been better off had he followed even fewer of them. Yet few dispute that the final poem as a whole is far superior to the original draft; faced with a stark choice, Eliot would have been better off acceding to all of Pound's suggestions rather than ignoring them all.

The compositional process for *The Waste Land* provides an excellent example of the indispensable role others can play in the birth of a creative masterpiece. At the time of the writing, Eliot was in desperate circumstances. His personal life was unhappy, and he still lacked confidence about his place in the world of letters. He produced a sprawling draft of great promise, but one that might have been quite difficult to assimilate. Eliot was fortunate that two individuals to whom he felt close were able to work with him, and he was equally fortunate that he could take their criticism constructively.

Like other creators involved in a major breakthrough, Eliot was striving to hammer out a new symbolic system or language, a poetic voice that could capture his feelings of personal despair, on the one hand, and the decline of European civilization, on the other. The various fragments he had formulated over the years resembled the individual sketches prior to Picasso's *Les demoiselles* and *Guernica*, as well as the snippets and folk songs that became absorbed into

Stravinsky's *Le sacre* and *Les noces*. Eliot's two hand-picked critics realized that he had found the proper tone and the proper means for achieving his goal: Neither tried to second-guess him on the design and the message of *The Waste Land*. Rather, like parents who seek to help their young child communicate effectively what is on his or her mind, Eliot's midwives pared away excess verbiage so that the widest possible audience could grasp his dominant tone and sentiments. Once again, as we saw with Einstein, Freud, Picasso, and Stravinsky, a creative individual on the threshold of his most dramatic achievement has benefited from close, almost parent- or sibling-like ties to respected intimates.

The Significance of the Achievement

Few poems in history have been so quickly recognized as important, both as a revolutionary contribution to letters and as a spiritual signature of a generation. Though an older, and overtly religious Eliot was later to diminish the pretension of the poem, calling it "a personal and wholly insignificant grouse against life . . . a piece of rhythmical grumbling," it signified far more to a generation of European and American readers.

Eliot captured in words and in form the widespread malaise, the apocalypse of a European civilization that had once seemed integral and integrated, but now seemed increasingly fragmented, irresolute, and even pointless. Eliot achieved this message, which had been put forth explicitly in Oswald Spengler's *Decline of the West* just a few years before, almost entirely by indirection. Nowhere in the poem are there overt references to Western civilization, to human disintegration, or to a decline or absence of values. (Pound talked Eliot out of using a Joseph Conrad epigraph in which a protagonist voiced his desperate feelings.) Rather, this sensibility is conveyed by vivid portraits—for example, of the listless love-making between the typist and the house agent's clerk; the use of images from classical literature, such as Phlebas, who was once tall and handsome, and the androgynous Tiresias, who has suffered all; the implicit contrast of the fallen, no longer fertile West with a wiser and more pacific East, as exemplified by the final mysterious trio of "Shantihs" from the Upanishad, intimating the "peace which passeth understanding."

Eliot's achievement was striking in another respect. The poem was difficult and forbidding, filled with lines that only the learned could understand and allusions that even the extensive footnotes did not fully elucidate. Yet, rather than mystifying or turning off readers—and here I refer specifically to younger readers—the enigmatic and abstruse quality of *The Waste Land* seems to have contributed to the poem's effectiveness and to have carried readers beyond the evident snob appeal entailed in reading a document of such apparent profundity. Eliot somehow succeeded in conveying the messages of the poem even when the individual lines and disparate parts made little sense. As one read and reread the poem—and, like other modern works, it demands and rewards rereading—individual sections may not have become pellucid, but Eliot's elegaic mood came across with ever clearer conviction and power. Here, again, the analogies to *Les demoiselles* and *Guernica*, to *Le sacre* and *Les noces* immediately suggest themselves.

REACTIONS TO *THE WASTE LAND*

The early reviews of *The Waste Land* were far more positive than those received by comparable epochal works, for instance, Stravinsky's *Le sacre* or Picasso's *Les demoiselles*. To be sure, there were negative reactions. The poet Amy Lowell called it "a piece of tripe," the poet William Carlos Williams complained that "I felt at once it had set me back twenty years," and the *Manchester Guardian* spoke of "so much waste paper." But most reviewers were either wholly or guardedly positive. Conrad Aiken referred to "a series of brilliant, brief, unrelated, or dimly related pictures by which a consciousness empties itself of its characteristic contents"; Edmund Wilson claimed that it "enhanced and devastated a whole generation"; Karl Schapiro called it "the most important poem of the twentieth century." And the *Times Literary Supplement* declared on October 26, 1922, "We have here range, depth, and beautiful expression. What more is necessary to a great poem?"

From the time of publication, *The Waste Land* has been scrutinized as much as any other twentieth-century literary work. Its polyphony has engendered multiple readings and disputes. Controversy within the field centered around a number of points. Some saw the poem as chiefly a personal statement of despair, while others felt that it strove for universal expression. Most regarded it as highly experimental and avant-garde, but a few felt that it was deliberately or inadvertently regressive. A recurring point of debate concerned one of the Aristotelian literary virtues—whether the poem held together. Those more critical of the poem complained that it was incoherent, a random or meaningless pastiche with neither continuity nor direction. In response, one group of defenders claimed that the incoherence was indeed the point of the poem: Since modern life and sensibility *are* chaotic, a poem that would faithfully convey this air must itself exemplify disorder. Other defenders claimed to find order, nonetheless. The critic I. A. Richards, one of the most thoughtful early readers, described the poem as unified emotionally and not logically, achieving its effects by the "music of ideas"; he spoke of the contrast and interaction of emotional effects as the unifying thread. In Cleanth Brooks's view, the pervading stance of irony created a sense of oneness across the experiences and sections of the piece. There was equally heated argument about voice: Were there many voices or one? If the latter, was it Eliot, a proxy for Eliot, the sage Tiresias, or the modern European mind? With time and further research, many writers came to focus on the autobiographical elements that reflected a depressed, impotent, and marginal Eliot.

Whatever the reaction, Eliot had become an important poet, and *The Waste Land* clearly was a formidable work that proclaimed—and perhaps realized—its author's ambition for it to be great. Even the diffident Eliot dared to say it was "a good one," "the best I have ever done." As with *Le sacre*, a still youthful artist had managed to convince the field that his most recent work marked a culmination, a defining moment in his own development, in that of his contemporaries, and, perhaps, in the evolution of the domain. It was not just that the poem occupied importance in Eliot's mind and that he thought well of it; it was that he succeeded in conveying this feeling of specialness to others. Over and above any intrinsic merits, the length, the framing, and even the pretentiousness and portentousness of the presentation all conspire to render this a defining work in the eyes of the author and his public.

The Poet's Special Accomplishment

In my view, *The Waste Land*, more than any other poetic work of its era, conveyed the tones and the themes that occupied the consciousness of literate contemporaries. In less than five hundred lines, Eliot succeeded in touching on and entering an amazing number of worlds. Each of the lines, and certainly each of the stanzas, is pregnant with meaning and could itself launch a separate poem on a separate topic. This prodigiousness not only conveyed to the reader a universe (or, indeed, several universes) of poetry but also provided an enormous number of entry points for the audience. Some sections featured colloquial language, vivid caricature, a sustained natural description, mystical images, witty repartee, lugubrious urban scenes, narrative vignettes, a pure play of sound, lurid snapshots, and much more. As in other crowning masterpieces of the modern era, such as *Ulysses* or *Remembrance of Things Past*, the cluster of overlapping themes, which appear initially and then are developed in one way or another later in the work, also contributed powerfully to the work's effectiveness. Among the themes in *The Waste Land* are those of fertility and other vegetative rites, the Fisher King, the deck of tarot cards, the story of the Grail, the griminess of London with its ancient bridges and churches, the light banter among bar guests, the overheard conversation among members of high society, the prospects of lust without love, the possibility of redemption, and the beguiling echoes of Eastern philosophy and religion. These significant motifs appear in a poem of strong lines and unadorned statements, conveyed largely in classic pentameter form. *The Waste Land* emerges as a dense and yet powerful portrait of the range of thoughts that occupy a deeply troubled mind, indeed, the modern mind. And though far from a straight narrative, it parallels ancient quests closely enough to give the reader a feeling of a completed, rounded experience.

Finally, I suggest that the poem's mood fits exactly the feelings of the European population at the end of a long and largely fruitless war—a vision Eliot had anticipated on his walks through Boston over a decade before. Among the young and the intellectual—for the purposes of this poem, the field—there was consensus that the war had achieved little and that the prospects for a vital, progressive civilization were slender and diminishing. Religion appeared irrelevant, amorality carried the day, and once-lively cities were now decrepit and disintegrating; many of these individuals were searching for expressive artifacts that could capture these desolate moods. *The Waste Land* succeeded better than any other work in doing this; thus, it served as an emblem for an entire generation of the dispossessed, Gertrude Stein's "lost generation." As the poet C. Day Lewis put it: "[*The Waste Land*] gives an authentic impression of the mentality of educated people in the psychological slump that took place immediately after the war." Whatever literary critics said, ordinary readers became the field that made *The Waste Land* famous.

A Decline in Effective Work

Eliot had written "Prufrock" in his early twenties and *The Waste Land* in his early thirties. The ten-year rule I have described can be seen at work here, with major milestones in the development of Eliot's work and thinking having occurred approximately one decade apart. Eliot lived for another four decades following *The Waste Land*, and he created more estimable verse, one additional poetic masterpiece, and a significant outpouring of writing in other genres; yet his output in his preferred genre had achieved its high point with the publication of *The Waste Land*, a productivity record unlike that of other artists of his stature. No further significant innovations occurred in Eliot's verse thereafter, and if he had stopped writing poetry after 1923, Eliot would probably still occupy roughly the same position in the history of poetry, though not in the history of letters.

One can propose both personal and domain-related reasons for this decline in effectiveness following the poet's early success. On the personal side, Eliot's own life took an increasingly conservative turn after the early 1920s, and this conservatism seems to have contributed to a decline in both the amount of poetry written and in its adjudged interest to those individuals who read contemporary poetry. As mentioned before, poetry in general, and lyric (personal) poetry in particular, is a craft in which artists tend to make their contributions at a young age. Most of the great poets of recent centuries created their defining work in their twenties or thirties, and many either died, stopped writing poetry in the succeeding decades, or continued to write in the same vein, without appreciable growth or change. Poets who achieve breakthroughs in middle age or later, like William Butler Yeats or Robert Penn Warren, are more exceptional than their peers in novel writing, musical composition, or visual art. As the novelist Marcia Davenport recently expressed it:

"All the great poets died young. Fiction is the art of middle age. And essays are the art of old age."

ELIOT AS A PUBLIC PERSON

Despite the publication of *The Waste Land* and the recognition of Eliot as the voice of his generation, Eliot's domestic life remained a source of great anguish to him. Vivien's illnesses and public outbursts continued, and Eliot, while clinging to physical health, remained in fragile psychological condition. Immediately after completing *The Waste Land*, Eliot declared: "[I] am about ready to chuck up literature altogether and retire; I don't see why I should go on forever fighting a rear guard action against time, fatigue, and illness and complete lack of recognition of these three facts."

Domestic conditions did not improve. Finally, in 1933, when on a trip to the United States, Eliot decided that he could no longer remain in a relationship that had caused him such pain and had only worsened with time. He arranged through a solicitor to secure a separation from Vivien, and after returning to England, he hardly ever saw his estranged wife again. Vivien never recovered from this brutal and cowardly rejection; she eventually was placed in a mental institution, where she died in 1948.

Eliot sought relief from his personal miseries through immersion in a number of demanding endeavors. He was actively involved in the publication of so-called little magazines (collations of poetry, prose, and opinion) on both sides of the Atlantic. He worked twelve to fifteen hours a day, waking early in the morning, putting in his time at the office, writing letters and reviewing books until late at night, and penning articles, of which he produced several hundred— some anonymous, many still uncollected. Part of this frenzy of activities no doubt stemmed from desire and ambition, but at least part of it stemmed from a horror of having free time on his hands and a fear that he could no longer create poetry.

Most important was his role in the *Criterion*, which he launched at the same time as *The Waste Land* (the long poem was actually published in its first issue) and over which he presided until the late 1930s. Eliot's stated goal in this magazine was "bringing together the best in new thinking and new writing in its time." Because of Eliot's tireless efforts and because of his increasing influence in the literary and social worlds, he achieved considerable success, though (as is inevitably the case with little magazines) the success was realized more in esteem than in financial terms.

Reading over the correspondence generated in conjunction with Eliot's editorial role with the *Criterion* and with similar publications, one encounters nearly as much intrigue and controversy as characterized Stravinsky's musical world. Eliot made a great fuss over a single word or punctuation mark; cajoled and pressured laggard writers; became involved in financial intrigue and in struggles for power. Such an entrapment in the field seems to come with the territory of becoming a central figure in the world of art and ideas, especially when one does not have personal financial resources on which to draw. Yet, unlike Stravinsky and Picasso, Eliot seems to have gotten little (even unconscious) pleasure out of the contretemps associated with his position. Regarding these as the necessary evils of the literary life, he seems to have been relieved when he was able to abandon such bickering and assume the position of elder statesman.

After remaining in the banking business for some years, Eliot was given the opportunity in 1925 to join the publishing firm headed by Geoffrey Faber. For the rest of his life, Eliot remained an important figure in the publishing firm of Faber and Faber, where he was able to encourage promising young poets, among them W. H. Auden, Ted Hughes, Louis MacNiece, and Stephen Spender. Eliot proved to be a first-rate editor and an excellent member of the board room. Though he did not enjoy teaching in person, he provided highly useful and precise written feedback to aspiring writers, in the process repaying the debt that he owed to Pound. Perhaps most surprisingly, even though his personal philosophy took an increasingly conservative turn, he (like Stravinsky) retained an expansive view of literature and encouraged writers of diverse dispositions and styles. And, in contrast to his supplicant role as editor of a little magazine with great aspirations but few resources, he had here the opportunity to be both judge and benefactor, roles far more congenial to him as the years went on.

The Reverberations of Religious Conversion

Eliot had long been critical of most American brands of Protestantism; in his view, they were insufficiently rigorous in thought and action. He had been attracted to conservative strains of political thought, as articulated by his undergraduate teacher Irving Babbitt and by the French intellectual Charles Maurras. And he had become a faithful reader of, and a proponent of, writings of earlier eras, particularly the metaphysical poets and the dramatists of the seventeenth century.

Still, it was a shock to most of his literary circle when, following his religious conversion, Eliot announced in 1928 that he was a "classicist in literature, an Anglo-Catholic in religion, and a royalist in politics." In the years that followed, Eliot developed an entire literary, political, and social philosophy that was deeply, if idiosyncratically, conservative. What had been latent in early writings now became the overt themes of his poetry and drama. Eliot probably needed to endorse these strict beliefs to maintain his fragile mental state; but his increasingly distanced pose made him a less attractive figure to many of those with whom he had grown up.

I have spoken in earlier chapters of the proclivity of creative individuals to enter into certain kinds of Faustian bargains with the gods, or with themselves, as a means of sustaining their creativity. Like Stravinsky, Eliot had tortured relations with his peers and made peace with his own God, in his case by becoming a faithful member of the Anglican church. Eliot, like Freud, also embodied the ascetic role; he would not even allow himself a piece of candy until he stopped smoking when well into his sixties, and he would not even shave in front of his wife. One could argue that his remaining for so long in an unhappy marriage reflects a masochistic tendency.

ELIOT AS A LITERARY FIGURE IN MIDDLE LIFE

As dazzling as Eliot's rise as *the* poet of his generation was his swift ascent to the position as *the* chief literary commentator in English of his time. Eliot wrote a great deal about English literature of the period and of the past; he also presided over literature in other European tongues and traditions with which he was familiar. He wrote much and he wrote well. But his stature, again, seems to have derived in significant measure from the *way* that Eliot wrote.

To begin with, Eliot was able to speak from the pulpit of the poet, a most respected role in European letters. He maintained that poets were best able to write about poetry, even as he questioned the credentials of nonpoets who would presume to judge poetry. He specialized in issuing judgments about works or authors, conclusions put forth as confident epigrams, much as if they were lapidary lines from one of his poems. Whether in conversation, correspondence, or published writings, he could render judgments quickly and confidently—a necessary skill for one who would engage in literary journalism. These judgments were not always supported by logical or systematic argument, but they were carefully organized in a rhetorically persuasive pattern. Above all, he had learned to offer his opinions in a temperate, impartial way. There was the air of the Olympian judge, raising the reputation of certain hitherto neglected or underrated individuals or bodies of work and calling received favorites into question. Whatever his personal interest and stake in the reputations of one or another writer, Eliot maintained the stance of a disinterested observer, whose job it was to protect literature.

Eliot's literary redistricting proved surprisingly successful. Just as many had felt a shock of recognition when reading his poetry, so also did many who read his often surprising literary and cultural judgments. His defense of Dante as an equally gifted and more integrated person than Shakespeare was taken seriously. His criticisms of Milton as overly intellectual were noted; his discomfort with urban intellectuals, women, liberals, Jews, and "impure races" struck some responsive chords. His attention to the metaphysical poets, to the French symbolists,

233

and to the sermons of Lancelot Andrewes affected the reading habits and sensibilities of more than one generation of students of literature. And while his conservative politics probably caused few to change their allegiances, it conferred a certain respectability on views that might otherwise simply have been considered abhorrent.

Of course, it is ironic that an individual who felt so marginal could have presumed to speak for an English, and even a European, mainstream. Here was Eliot—a New Englander living in the Midwest, a conservative at progressive Harvard, an American in England, an effeminate urban intellectual with stereotypically negative views of women and intellectuals—presuming to prescribe attitudes for those who were decidedly less marginal. This irony was not lost on Eliot, who continued throughout his life to think of himself as an outsider. But Eliot presumed to make a virtue of necessity. He claimed that only foreigners could undertake the sort of summative and evaluative exercise in which he engaged; English natives would experience too much difficulty in seeing the terrain with a clear, unprejudiced eye. And, echoing Henry James, he maintained that an American can become a European in a way that no Englishman can.

While Eliot's pronouncements in many fields now seem curiosities, his views about poetry—many enunciated when he was still quite young—have continued to compel (and to merit) attention. Eliot saw poetry not as an unleashing of emotion or of personality but rather as an escape from these components; but he shrewdly pointed out that only someone who *possesses* personality and emotion knows what it means to want to escape from them. As he put it, "The more perfect the artist, the more completely separate in him will be the man who suffers and the mind who creates." Echoing almost exactly the sentiments of fellow modernists Picasso, Stravinsky, and Graham, he noted that immature poets imitate, but that mature poets steal, in the process making the plagiarized content into something personal and, not infrequently, something better.

Eliot reflected a great deal on the poetic process. As he saw it, the poet has a sensibility that can devour any kind of experience. The poet's mind is a receptacle for seizing and storing up innumerable feelings, phrases, images, and the like; these remain in unconscious and inchoate form until they can coalesce as a new compound in which they are presented together. He spoke of a "logic of the imagination," which is, in its distinctive way, as powerful as the logic of concepts or ideas. Further, he argued that the reading of poetry is an emotional experience, like listening to music, that can be impeded by the exercise of one's powers of reasoning. The best poetry, in his view, is unconsciously memorable; it arises from, builds on, and relates to a rhythm in the unconscious.

In what is probably his most well-known contribution to literary analysis, Eliot introduced the idea of the objective correlative. Poets do not directly communicate emotions, he said. Rather, they create a situation or image such that the emotion in question, when apprehended, will be successfully and effectively communicated. One needs, he said, "a set of objects, a situation, a chain of events which shall be the formulate of that *particular* emotion, such that when the external facets, which must terminate in sensory experience are given, the emotion is evoked." The most notable poets were those who possessed the gifts of creating such objective correlatives. He concluded that "unless we have those few men who combine an exceptional sensibility with an exceptional power over words, our own ability, not merely to express, but even feel any but the crudest emotions, will degenerate."

In writing of the objective correlative, Eliot may well have had in mind his own experiences, including his efforts to capture in words his early sensuous preoccupations and his uncanny walk through the streets of Boston. Yet it should be apparent that Eliot's views of literature, and particularly of poetry, went beyond his personal experiences and prejudices. He achieved genuine insights into the processes of poetry, and particularly those of his own era, including his own work. In using his critical and philosophical training to illuminate the art of his time, Eliot was able to combine roles that are usually filled by different individuals having different training and contrasting sensibilities. Whatever success and acclaim he achieved as a poet, on the one hand, and as a critic, on the other, was multiplied by the fact that he could readily fill both of these roles as well as any other individual in his time.

Eliot's reflections on poetry help illuminate his own intellectual gifts. Of course, as a poet, he trafficked principally in linguistic symbols—in the meanings, nuances, and combinations of the words in each of the languages in which he wrote. But in writing his poems, he drew on materials from other realms—on concepts of a logical-philosophical sort, on historical and literary sources from many cultures, on understandings of the world of other individuals, and, perhaps above all, on sensitivity to his own emotional life. Except in his literary and philosophical writings, Eliot did not deal directly with these materials. Rather, he explored them by means of the literary images and expressions that constituted his poems. The numerous changes he made in his own drafts, as well as his excellent suggestions to other poets, reflected efforts to capture these meanings precisely and directly within the medium of poetic language.

In contrast to Eliot's enduring contributions to literary discussions, his pronouncements in nonliterary areas ranged from the mundane to the bizarre. He cultivated the mindset of the early Christians and looked forward perversely to a ruined civilization; he distrusted any ideas of progress; he praised traditional figures and forms, spurning anything that smacked of the contemporary. While not overtly fascist or totalitarian, he refused to recognize the value of ideas of individualism and democracy, he spoke mystically about "blood ties" among those in the same tradition, and he spurned every opportunity to denounce the Holocaust. Altogether he seemed increasingly of another millennium, mired in Eurocentrism and obscurantism. When someone commented to Eliot that modern people know so much more than did the ancients, Eliot agreed, but added with asperity, "and they are what we know."

From all indications, Eliot was not an easy person to get to know well, and few individuals felt close to him. Especially in later life, he seems to have been most comfortable maintaining a distance from other individuals; and more than one time, in a manner that I have noted with each of our previous creators, he shocked and severely hurt a person to whom he had apparently been close by abruptly cutting off all ties. The question therefore arises: How did Eliot relate to other individuals earlier in his career, at a time when he might have needed their support politically, and their friendship personally?

In carrying out this study, I have repeatedly been astounded by the speed with which talented young individuals, like members of a rare species, immediately spot those of their peers who are appropriate members of the same cohort. The young Picasso, barely able to speak French, met and aligned himself early on with Max Jacob, Gertrude Stein, Guillaume Apollinaire, Henri Matisse, and Georges Braque; within a year or two after focusing on composing, Igor Stravinsky is supping nightly with Serge Diaghilev and Vaslav Nijinsky and exchanging compositions and compliments with Claude Debussy and Maurice Ravel. Despite his later remoteness, the young Eliot fits into this precocious pattern. He complemented his early friendships with Americans Conrad Aiken and Ezra Pound with significant new relationships with Wyndham Lewis, Richard Aldington, Aldous Huxley, Julian Huxley, and many other leading members of the younger British artistic establishment.

Cynics might conclude that it is not the "future greats" ferreting out other "future greats," but rather those bent on success who are using one another and feigning interest in one another's careers in order to enhance their own. Unquestionably, those who cannot or will not engage in any such social contacts are at a disadvantage, even as those with a genius for self-promotion or scouting talent may achieve at least a temporary edge on others.

But in my view, a different phenomenon is at work. The most talented young individuals have accomplished the feat of mastering the existing work in their domain, or, in the rarest of cases, in more than one domain. Already at the cutting edge of contemporary work, they wish ardently to advance further. Such individuals immediately detect the scent of others at the same crucial point in their development. Of course, they realize that these peers are competitors, and some cannot transcend this parochial vantage point. But at least in many cases, these aspiring innovators realize as well that what benefits the cause of their group, such as the Pound-Eliot-Wyndham coterie that gathered around a self-styled avant-garde publication called *Blast* in 1914, will work to their own advantage as well. And so, the competitive edge is in part moderated by involvement in a larger cause.

In attempting to piece together the enigma of Eliot's relation with others, I discern the following picture. In terms of personality traits, Eliot was clearly an insecure introvert, and he risked leading a life in isolation from other individuals. Still, his personality was reasonably appealing, and others were quite happy to meet him and even to help him. In adolescence and early adulthood, when needs for peer friendships are most pronounced, Eliot developed his most important and enduring friendships. The fact that these friendships could help to advance his career was certainly germane, but there is no reason to assume that Eliot's contacts with Pound or Aiken were purely cynical, any more than his marriage to Vivien represented anything but a sincere effort to forge a life together with an attractive young woman. However, as Eliot aged, and as his need for political support from others was less acute, his natural tendencies toward isolation and distance returned with a vengeance. Only with a few longtime friends and a few appealing youngsters was he able to relax and reveal himself.

Eliot continued to admire talent of the highest rank. He stood in awe of James Joyce, whom he regarded as the outstanding master of his time. Eliot viewed *Ulysses*—published virtually simultaneously with *The Waste Land*—as the most important work of the age; Joyce had accomplished with a quarter of a million words the same kind of portrait of past and present (one more triumphant in tone) that Eliot had sought to suggest in a mere 433 lines. Eliot also pursued the publication of *Finnegans Wake*. Perhaps more revealingly than he intended, Eliot said, "Joyce I admire as a person who seems to be independent of outside stimulus and who therefore is likely to go on producing first-rate work until he dies."

Another individual whom he regarded as a peer was Virginia Woolf. Eliot admired her literary innovations and was on friendly terms with her for many years. At one time he indicated that he valued her positive opinion over those of anyone else on the literary scene. And the author of *The Waste Land* identified strongly with what Stravinsky was attempting to achieve in *Le sacre du printemps*:

> Whether Stravinsky's music be permanent or ephemeral I do not know; but it did seem to transform the rhythm of the steppes into the scream of the motor horn, the rattle of the machinery, the grind of the wheels, the beating of iron and steel, the roar of the underground railway, and the other barbaric cries of modern life; and to transform these despairing noises into music.

ELIOT IN LATER LIFE

It has often been noted that precocious children seem in certain respects very much like older persons. Even among the precocious, Eliot stands out as someone who seemed old before his time. In the manner one tends to associate with first-born children, Eliot (actually the youngest of six children) identified from early childhood with his parents, his forebears, and other representatives of tradition. Finding the example of the Eliot family not quite compatible with his own idiosyncratic personality, Eliot searched for alternative father figures—like Babbitt, Maurras, and Pound—and, like them, eventually came to identify with the most conservative political and religious causes.

As if in compensation, Eliot retained an endearing childlike quality throughout his life. His energetic letters to colleagues, as well as his delightfully lively letters to the children of his friends, are more likely than his formal writings to convey his humor and his informality; but his return to doggerel, his famous *Old Possum's Book of Practical Cats*, and his recurring self-parodies show that even at his most sober, Eliot could appreciate the incongruities in life and in himself. In later life he became quite friendly with Groucho Marx, and the two genuinely enjoyed sharing their love of cats, cigars, puns, music-hall banter, and bawdy lyrics. In his stubbornness, compulsiveness, and obsessiveness, Eliot also preserved some of the less attractive features of the young child.

Most of us yearn for our lost youth—and none more so than lyric poets. But from all indications, Eliot enjoyed the appurtenances and encumbrances of maturity. He was comfortable in the role of banker, businessman, and distinguished editor at Faber and Faber. He was a faithful and regular parishioner in his church, attending mass nearly every day. During the Second World War, he valued his role as an air-raid warden. He traveled widely to receive accolades and occupied readily the niche as an elder literary statesman. After living for eleven years with an invalid named John Hayward, for whom he helped to care, he abruptly left Hayward and married his devoted secretary, Valerie Fletcher. Eliot made it quite clear to everyone that this was the best decision of his life and that he owed his years of final happiness to his second wife.

Between 1925 and 1940, Eliot continued to write poetry, ever at a deliberate rate. He published *The Hollow Men* in 1925, *Ash Wednesday* in 1930, and *The Four Quartets*, his longest poetic work, in 1942. These poems were well received and maintained Eliot's reputation at a respectably high level, but few believed that they represented significant growth; and some were uncomfortable with, or at least uninspired by, their overtly liturgical, purgatorial, and redemptive themes.

In my own view, *The Four Quartets* represents a major poetic accomplishment, one worthy of a place in the ten- or twenty-year creative sequence of events. In a manner of other later, more comprehensive works, like *Les noces* or *Guernica*, *The Four Quartets* retains the language and sensibility of a modern style but is deeply rooted in both an artistic and a personal tradition. Eliot comes to grips with his own personal background and his enduring spiritual and philosophical concerns about time, place, memory, and self in a more straightforward and less ironical way. Unlike later works of great creators, however, *The Four Quartets* does not amount to any kind of a breakthrough; indeed, it is more like a final work, reminiscent of composer Richard Strauss's *Four Last Songs*. Always strongly influenced by the past, Eliot now seemed to be living there as well. Perhaps in the absence of a powerful stimulus like Laforgue or a trenchant critic like Pound, Eliot was condemned to create poetry that looked chiefly backward.

Eliot had always been interested in drama, particularly in the possibility of verse drama. He had once commented that in a play "we may perceive a pattern behind the pattern into which the characters deliberately involve themselves; the kind of pattern which we perceive in our lives only at rare moments of inattention and detachment, drowning in sunlight." As if to compensate for the decline of his poetic output in the 1930s, he composed a series of plays over the next two decades: *Murder in the Cathedral* (1935), *The Family Reunion* (1939), *The Cocktail Party* (1950), *The Confidential Clerk* (1954), and *The Elder Statesman* (1958). These later works are distinctive in their religious and spiritual themes, their virtually mathematical mode of organization and composition, their perennial blend of the comic and the tragic, and their ever-ironical stance.

It had long been noted that Eliot's verse had a strongly dramatic quality, with its vivid personifications and sense of voice; now it was observed in complementary fashion that Eliot's dramas were distinguished by their excellent poetry and that they could be read as effectively as they could be performed. While each of these works is notable, and perhaps a few—most probably *The Cocktail Party* and *Murder in the Cathedral*—are likely to remain in the repertoire, much of their significance comes from the fact of Eliot's authorship. Indeed, had these plays not been written by the premier literary figure of his age, some of them probably would never have been performed on the London or New York stages.

In daring to write for the stage, Eliot displayed the personal courage that we have come to expect from titanic creative figures. It could not have been easy for the shy Eliot, already revered as a poet and critic, to risk his reputation in so public and perilous a medium as the London or New York stages. Yet his works of middle age do not have the power of the later works produced by Picasso, Stravinsky, and Graham; they are more reminiscent of the shift of focus found in the less successful efforts of Einstein in his later years.

To coincide with the 101st anniversary of Eliot's birth, the American novelist and critic Cynthia Ozick scathingly attacked Eliot in the somewhat unlikely pages of the *New Yorker*. Ozick begins by noting, quite appropriately, the unparalleled power that Eliot exerted over his peers and the younger reading public during the height of his influence, from about 1930 to 1960. She then questions his authority on nearly every dimension: the quality of his poetry (only forty-three poems in all, of which only "Prufrock" is still regularly anthologized); the insipidness of his plays (strictly period pieces); the unwarranted lèse-majesté of his criticism; his anachronistic belief in the value of a canon, which it was his peculiar responsibility to define; his despicable views about women, Jews, liberals, and others he characterized as low-lifes, as observed from his haughty stance as a member-in-good-standing of the British literary, social, and political establishments.

Ozick charges that Eliot was not even what he thought he was. He saw himself as a classicist but was actually subjective and mystical; he identified with Donne and Dante but more closely resembled Tennyson and Whitman; far from being the first of the modernists, he is better seen as the last of the romantics. Ozick concludes: "Not since Dr. Johnson has a man of letters writing in English been received with so much adulation and seemed so formidable. . . . It may be embarrassing for us now to look back at that nearly universal obeisance to an autocratic, inhibited, depressed, rather narrow minded, and considerably bigoted fake Englishman."

Obviously, if I agreed in substantial measure with Ozick, I would not have included Eliot in this study. Eliot is scarcely alone among modern masters in being attacked today; and, indeed, for reasons of their personal behavior as well as their contributions to the domain, his peers Stravinsky, Picasso, and Graham have been subjected to equally potent denunciations. Certainly, Eliot's attitudes, while deplorable, do not suffice to reduce his poetry to insignificance, any more than do Yeats's fascist or Picasso's Communist or Einstein's pacifistic tendencies; and, notably, although Eliot's published remarks may have been virulent, he seems to have been kindly in personal relations, even to those individuals whom, on demographic grounds, he might be expected to disdain. Just because we now understand better how *The Waste Land* documented a personal hell, we need not disparage the poem; indeed, in some ways, this new knowledge can enhance our appreciation of its achievement.

As for Eliot's poems, plays, and literary criticism, they constitute an important contribution to the corpus of the modern era, the first three quarters of the twentieth century. The best works—"Prufrock"; *The Waste Land*; *The Four Quartets*; *Murder in the Cathedral*; and his collected lectures, *The Use of Poetry and the Use of Criticism*—are likely to remain as definitive of our era, with *The Waste Land* as this age's signature. Subsequent poetry would have been different had Eliot never lived. What Joyce and Woolf contributed to the novel, what Richards presented to the philosophy of literature, Eliot gave in equal measure to poetry and to nontechnical literary criticism.

Each of our modern masters can be viewed as marginal in various ways: Martha Graham as a woman in a world of male innovators and impresarios, Gandhi as an Indian dealing with the English and the South Africans, Picasso as a Spaniard in France, Stravinsky as a Russian in France and the United States, and Freud and Einstein as Jews in an increasingly anti-Semitic Europe. But none of these masters remained as centrally and determinedly marginal as Eliot.

Eliot's marginality is rooted in paradox because, coming from an established family and living in the country of greatest influence in the world, he had the option of remaining decidedly mainstream, in the manner exemplified by so many other male members of his family. Rather than being marginal by necessity, he was clearly marginal by choice. Having perhaps felt somewhat marginal in a psychological sense, he deliberately made decisions that would underscore his marginality: the midwesterner at Harvard, the conservative among liberal students, the American perenially abroad, the intellectual at the bank, the reactionary among the artists, the celibate among the profligate, the believer among the atheists, the poet among the philosophers and the philosopher among the poets, and the dramatic poet and the dramatist who created poetic verse. Evidently, Eliot had a strong drive to feel marginal; and when this marginality was not obtained naturally, he took actions that would ensure his marginal status.

Yet, a feeling of marginality implies a need for community as well. (If one felt ensconced in a community, one would not feel marginal.) There are clear indications that Eliot felt a need to belong to a community, and for this reason he eventually joined the Church of England, espoused a specific political line, and strove to be a good, "compleat" Englishman. The critic Donald Davie maintains, however, that Eliot remained very much an outsider in England, knowing only a certain hermetic London circle. Just as clearly, Eliot felt a need to define himself at least in part as an outsider in a community, a congenital alien. He used this language, frequently calling himself "a metic, a foreigner." And echoing a Faustian bargain, he once declared: "The arts insist that a man shall dispose of all he has, even of his family tree, and follow art alone; for they demand that a man be not a member of a family or a caste or of a party or of a coterie, but simply and solely himself." Perhaps not surprisingly, his greatest breakthroughs occurred when he was able to position himself in a sharply marginal way: the young aesthete observing the ways of Boston in "Prufrock" and "Portrait"; the individual of early middle age distressed by the dissolution of civilization as it had been known in *The Waste Land* and *The Hollow Men*; the disinterested literary critic in later life who delighted in taking subversive positions, in favoring neglected artists and in disparaging sacred icons. Eliot's literary voice—or as he might have said, his different voices—epitomized marginality, which was purchased at considerable cost. Probably none of our other modern masters had a more painful personal life than did Eliot, with his abundant illnesses and traumas amidst rare moments of pleasure; his existence was so grim that he frequently spoke of himself as being imprisoned, though he also felt licensed to poke poetic fun at his feeling of victimization.

Eliot thus epitomizes a strand I have encountered in all of the modern masters being described: a feeling of marginality and the capacity to exploit that marginality in the service of one's life mission. The case of Eliot also suggests that when such marginality does not arise of its own accord, the would-be creator may attempt to raise his or her ante of marginality to a level of fruitful asynchrony. Since marginality only makes sense within a possibility of community, one will observe an oscillation in the creator's life between feelings and moments of marginality, and feelings and moments of community. Perhaps this is another way of recognizing that highly creative individuals belong in part to the whole world, and in part to themselves alone, with the oscillation between these poles providing much of the positive and negative asynchronies in the life of the creator.

When he finally separated from Vivien, consolidated world fame, and married Valerie, Eliot seems to have achieved a measure of peace and satisfaction. These achievements coincided with a clear decline in the power of his literary work, a decline occurring at a time of life when artists like Graham, Yeats, Picasso, and Stravinsky were still in their prime. Even as Eliot's singular attainments emerged when his marginality was at its peak, a further accumulation of triumphs might have necessitated a continuing marginality that he neither wanted nor could sustain.

INTERLUDE TWO

The three creators considered in chapters 5, 6, and 7 have often been linked together in the artistic history of the twentieth century. Each born during the 1880s, each rising to the top of his chosen craft by the early 1900s, each occupying a commanding role in his domain for close to half a century, each producing frankly innovative works while at the same time identifying so closely with tradition that they have also earned the epithet neoclassicist, Picasso, Stravinsky, and Eliot lay equal claim in their respective domains to the tide of artistic masters of the modern era.

In addition to these widely acknowledged similarities, we may also note continuities in light of the themes of this book. First, these artists came from bourgeois families that valued the particular art form in which they worked; in this respect, they differ from Einstein and Freud, who chose professions that their families could admire but not understand. Each found himself gravitating at an early age to the center of artistry—Paris in the case of Picasso and Stravinsky, London in the case of Eliot; and each soon found himself at the center of the most innovative group of artists and intellectuals. Their most radical works benefited directly from the aid of individuals to whom they felt personally and professionally close: Georges Braque for Picasso; the Diaghilev circle for Stravinsky; Ezra Pound and Vivien Eliot for Eliot. Nor should one ignore the succumbing to the bourgeois life in midcareer, or the role of younger persons at the close of their lives: the rejuvenation provided by Jacqueline Picasso, Robert Craft, and Valerie Eliot. Finally, in an effort to ensure the survival of their creative juices and their capacity to pursue work, they each found themselves striking bargains of a sort: Picasso pursued youth and self-preservation with sadistic contempt for those closest to him, Stravinsky fueled the engine of litigation at the cost of personal closeness, and Eliot chose a path as ascetic as Freud's while also displaying a disregard for the feeling of others that bordered on the sadistic.

Our artists can also be arrayed on other important dimensions. They range in degrees of intellectuality, with Eliot capable of being a professor, Stravinsky keenly interested in certain esthetic and philosophical issues, and Picasso, the least scholarly, marked by a cat-and-mouse relationship with the ideologues and intellectuals of his era. With reference to religious matters, there are also notable differences: Stravinsky and Eliot found it necessary to embrace a variety of Catholicism at almost the same moment in their lives, while Picasso's stance was more superstitious than formally religious. And on political grounds, there is an enormous range, with Eliot the instinctive conservative and Picasso the reflexive anarchist.

It is important to note here a decisive difference from creation in the sciences or mathematics. Individuals in these latter areas begin to be productive at an early age and certainly have the option of making numerous innovations during their early years. However, unlike the arts, these domains progress and accumulate at a rapid rate, stimulated by the discoveries of the most creative individuals; tools fashioned earlier in life may become irrelevant or dysfunctional. The creators must keep moving or, like Einstein, they will fall forever behind their peers and be increasingly ignored by the field. And unlike individuals in the arts, scientists and mathematicians cannot revisit the past of their disciplines and set up dialogues with earlier masters. If they attempt to do this, they have in the process become historians, philosophers, poets, or (in the case of Einstein) wise old-timers rather than contributors to the contemporary scientific or mathematical domain.

Artists maintain the realistic option of continuing to create in a novel way so long as they retain their health. Like Martha Graham, Picasso and Stravinsky took advantage of this opportunity, going through a full course of phases and continuing to create until well into their old age. They exploited the opportunity of several ten-year cycles of renewal. Eliot occupies a more anomalous place in this respect. Never as protean in output as Stravinsky or Picasso, he also ceased writing poetry, for all intents and purposes, when in his forties. To be sure, Eliot continued to write, embarking on the risky path of playwriting and serving as a major literary critic and a virtual gatekeeper for several domains of literature. Whether he stopped writing poetry because of the nature of the domain, often described as a young person's medium, or because of factors in his personal life, we cannot say.

And what of the nature of the creative activity in which these three men engaged? Unlike Freud and Einstein, it is not appropriate to think of Picasso, Stravinsky, and Eliot as solving problems or as creating, except incidentally, a new conceptual scheme. Their work could not be translated into textbooks or discussed propositionally and built on or discarded by others working in their paradigm. Rather, their mission was to create new works of art within a genre and to accumulate a corpus of creations that reflected their evolving vision.

If these men had lived at a different time or had had different personalities, they might have become ordinary workers or contributors to some long-established tradition. However, they lived at a time when the traditions in their art form were spent and avant-garde work was at a premium. Eliot, Picasso, and Stravinsky rose to the challenge. Having mastered the symbol system of their domain at a young age, they went on to stretch the symbol system in new ways, in ways that had not been explored before but that, when explored, made sense to themselves and, ultimately, to their contemporaries.

Picasso, with Braque, broke apart the figures of previous Western art and brought the same kind of aesthetic attention to forms and shapes that had previously been directed toward representations themselves; and, in works like Les demoiselles d'Avignon *and* Guernica *he succeeded in wedding the new language to the most powerful human themes. Stravinsky experimented with rhythms and dissonances as never before, creating works like* Le sacre du printemps *and* Les noces *that were genres unto themselves. Eliot introduced themes and patterns of verse that sounded odd to the English ear, in the process giving voice in* The Waste Land *to the feelings of an alienated generation. As we shall see in the next chapter, Martha Graham carried out much the same kind of distancing from traditional dance forms as she explored the forms of the body and the themes of her land.*

From a greater distance, one can discern continuities with earlier work in the domain: Picasso with Cézanne, Stravinsky with Debussy, Eliot with Laforgue. One can as well see the striking parallels across these innovators: the interest in fragments, in form for its own sake, the tension among the mundane trifles of daily life, the lure of the primitive, and the ponderous themes of the past; the oscillation between tiny, mundane details and an exquisitely planned whole.

But these historical considerations should not obscure for us the precise nature of the work carried out by these individuals. They were, to begin with, craftspersons: They returned each day to their desks, to work mostly in isolation on products that might take months or even years to complete. Such pieces were worked out through the exploration of media and genres that, in a sense, had existed for centuries but that these men were stretching in new, unfamiliar, and perhaps even frightening directions. These products remained in their own hands until such time as they chose to part with them, and they remained expressed in the symbol system (painting, music, language) in which they were created.

But then these aesthetic objects entered a wider world, as paintings were displayed, plays and musical works were performed, and poems were published. These latter activities thrust the individuals directly into the world of other individuals, including the power brokers of the vaunted field. The extent to which such contact had to be initiated personally varied across individual, domain, and field, with the demands clearly being the strongest in the case of Stravinsky's large-scale works; in each case, some intercourse with other persons became an essential part of the creator's life.

Still, all of these individuals had the option of withdrawing into privacy and, particularly in their later years, of leaving much of the negotiation to others. They were not called upon to perform themselves. To be sure, Picasso's performances on a canvas at a given moment had more of an aesthetic significance than the marks made by Eliot or Stravinsky who, had they lived later, could have created all of their works on a computer terminal; but Picasso, too, had the option of throwing away or painting over a work, if that was his wont. In this respect, our artists, like the scientists of previous chapters, are creatures of the study or the studio. We encounter here the most fundamental difference from the activities of our last two creators, Martha Graham and Mahatma Gandhi. For even though their creative work emanated from wholly different domains, both were ultimately performers, judged by others in terms of the effectiveness of their performances at a given moment.

8

MARTHA GRAHAM: DISCOVERING THE DANCE OF AMERICA

Graham, 1935

DANCE LAYS FAIR claim to being the original art form, the one most likely to have been practiced by the first human beings. In the West it is also the art that most recently underwent a major transformation: Painting, music, and literature had already turned modern when dancers took the decisive steps in the 1920s and 1930s. Of the art forms, only dance assumed its modern form primarily in the United States, in large part as a reaction to European, Asian, and African forms of bodily expression. Though many individuals contributed to this historic invention, the dominant figure was Martha Graham—among the creators being discussed, the only woman and the only one to make her life in America.

THE DOMAIN OF DANCE AT
THE TURN OF THE CENTURY

Martha Graham's breakthrough entailed a reaction to two major strands of dance. On the one hand, classical ballet was a form that dated back several hundred years. With its five basic positions of the feet, prescribed positions of the body, and articulated geometrical relationships among the dancers, ballet epitomized precision in an art form. To be sure, ballet had received an infusion of energy early in the century, courtesy of the innovations associated with Diaghilev's Ballets Russes, and particularly the choreography of Mikhail* Fokine. The Ballets Russes explored anew the formal properties of the dance and began to move away from plot. Nonetheless, particularly from the perspective of impatient, adventurous, and irreverent Americans, ballet was still considered a constrained European form.

The other major strand consisted of dances associated with non-European peoples, particularly the folk dances of Asia, Africa, and native American populations. These dances, often based on rituals, reflected the practices, values, and emotional palette of the populations in which they had evolved over the centuries. American and European audiences found these exotic spectacles intriguing but typically thought of them as popular art or craft forms, rather than as exemplars of an evolving high art.

Innovators Emerge

By the turn of the century, those with artistic inclinations evinced dissatisfaction with the ballet, exotic folk dances, and other popular forms of bodily entertainment, such as burlesque, ballroom dancing, and acrobatics. What passed for

---

*Sometimes known as Michel.

dance seemed obsessed with the mimicking of plants and animals, while bereft of genuine human emotions and remote from contemporary life. Against this background, a series of remarkable American women began to experiment with more authentic dance forms.

The person generally recognized as having first led dance into the modern era is Isadora Duncan, who was born in California, but who spent her adult life in Europe. Universally referred to as Isadora, she saw the body as being, above all, a vehicle for expressing emotional content. The dance ought to be considered a serious form, one performed as an accompaniment to the greatest musical compositions. Tracing her inspiration to Greek myths and art, as well as to Sandro Botticelli's *Primavera*, Isadora danced as a free spirit; through her being, she sought to convey nature, love, and beauty. Her barefoot style, with its trademark flowing veils, found an especially receptive home in Russia both before and after the Russian Revolution.

As an example of personal courage and aesthetic adventurousness, Isadora deservedly had enormous influence. But her technical innovations were neither proficient nor consistent enough to spawn imitation; and most of her personal success came from her charismatic manner and her "bodily instincts," rather than by dint of any skills that she could pass on to her students or her "adopted daughters." For these reasons Isadora is generally considered an isolated pioneer rather than the progenitor of a new dance tradition. As the insightful dance commentator Agnes de Mille expressed it: "Isadora cleared away the rubbish. She was a gigantic broom. There has never been such a theater cleaning."

Whereas Isadora looked toward the Atlantic and Aegean waters, the other pioneer of dance, Ruth St. Denis, fixed her gaze primarily on the Orient. Almost an exact contemporary of Isadora, St. Denis was inspired by dance practices from Egypt, China, Japan, Java, Siam, and India. Like Duncan, St. Denis ardently wished that dance be taken seriously; she hoped to dislodge Americans from their puritanical aversion to the beauty and the functions of the body. Focusing on the specific details of the music, she strove to capture the sounds of instruments and the punctuation of rhythm in her dance movements. Rather than embodying the human passions or nature directly, as Isadora sought to do, the more mystically oriented St. Denis strove to capture the world of pure spirit.

St. Denis left her mark on the American dance scene. Remaining in America throughout her very long life, she exerted enormous influence as an inspirational figure and as a teacher, especially, through her collaboration with the dancer Ted Shawn. Indeed, most of the first generation of modern dancers trained with St. Denis and Shawn at their legendary school, Denishawn. Yet, St. Denis's actual legacies to the modern dance, like Isadora's, were modest, as much symbolic as substantive.

MARTHA GRAHAM'S AMERICA
AT THE TURN OF THE CENTURY

Born in 1894, Martha Graham spent her first fourteen years of life in Allegheny, Pennsylvania, a typical small American town, within the shadow of Pittsburgh. She had a comfortable home and a family reasonably secure and devoted to one another. Graham's father, George, the grandson of an immigrant from Ireland, seems to have had a wild streak; he liked to play music and to sing for Martha and her two sisters. As a physician, he took a special interest in matters of the mind; this "alienist" would today be called a psychiatrist. While he was very busy, and ultimately had, for professional reasons, to spend a few years separated from the family, Dr. Graham clearly exerted a major and largely benevolent influence on his eldest daughter.

On her mother's side, Martha Graham was vintage American. Her mother, Jane Beers, was a tenth-generation descendant of Miles Standish, the Puritan figure immortalized in Henry Wadsworth Longfellow's poem "The Courtship of Miles Standish." For Martha Graham, her maternal relatives, and particularly her forbidding grandmother, embodied the puritanical tradition. Martha was brought up very strictly, with daily prayers, compulsory church attendance, and Sunday school training. As with several of the other creators, the child received unconditional affection principally from a long-term nurse: "Maid Lizzie" seems to have served as a confidante and peer for the Graham children. Summarizing Graham's family background, the critic Walter Terry declared: "Martha turned out to be an even mix of the two parents, a stern indomitable Godfearing Puritan pioneer on one side, and on the other a wild tempestuous, moody, dream-obsessed and quick-to-anger creature of the Black Irish persuasion."

As befalls highly accomplished figures, and especially ones fated to live in the media-drenched twentieth century, the young Graham has come to be surrounded with a set of virtually mandatory anecdotes. Bored in church when but a toddler, Graham is supposed to have danced down the church aisle. Trapped on a train when it suddenly began to move, a young, but already decisive, Graham became upset and declared in no uncertain terms: "Man, I'm Doctor Graham's daughter and I want out of here." Since young Graham was inclined to lie and often got into trouble as a consequence, her mother encouraged her to pretend and even to set up a small theater at home. Most tellingly (and most often told!), Graham's father once discovered that she was not telling the truth: "Don't you know when you do something like this I always know? There is always some movement that tells me you are deceiving me. You see, no matter what you say, you reveal yourself—you make fists, you think I don't notice, your back gets very straight, maybe you shuffle your feet, your eye lids drop. Movement does not lie." This insightful parental response to a youthful peccadillo carried an important message for later life.

Because of young Mary Graham's asthmatic condition, the Graham family decided to move to Santa Barbara, California, in 1908, when Martha was fourteen. From all indications, the family found the change of climate and atmosphere congenial. After the strict Presbyterian regimen of a small eastern town, the casual pace and tropical flavor of southern California were a welcome relief. Graham did well in high school. A quick learner and well read, she was a competent editor of the literary magazine *Olive and Gold*. In addition, she acted as Dido in the school's production of *The Aeneid*, played basketball, wrote short stories and a two-scene play, and was an accomplished seamstress. Although a retiring person, she was esteemed by her classmates.

In April 1911, Graham noticed a poster that announced eight performances at the Los Angeles Opera House by an exotic dancer named Ruth St. Denis. The poster depicted St. Denis's famous characterization of Radha, the beloved of the Hindu god Krishna; Graham beheld a richly costumed St. Denis, wearing glistening bracelets and sitting cross-legged amid splendor on a small thronelike platform. Graham implored her father to escort her to this performance, and he agreed, perhaps to her surprise. He even gave his daughter a corsage of violets, which she treasured for many years thereafter. Graham was mesmerized by the spectacle of the solitary, attractive woman, who enacted various godlike figures and dominated the stage with her magnificent costumes, expressive eyes, and memorable gestures. "From that moment on," Graham recalled later, "my fate was sealed. I couldn't wait to learn to dance as the goddess did."

Being inspired by a dance recital was one thing; making a career choice, quite another. The thought of their eldest daughter becoming a dancer hardly pleased the straightlaced Graham family. Parents and daughter settled on a compromise in which Martha attended the Cumnock School, a junior college in which she could study liberal arts while pursuing her artistic interests. The relative freedom of the school pleased her, in much the way that Albert Einstein had appreciated the unpressured atmosphere of the Aarau school (see chapter 4). In 1914, when Graham was twenty, her father died of a heart attack. Graham now felt free to chart her own future course.

A NOVEL CAREER

In 1916, at age twenty-two, Martha Graham enrolled in the summer course at Denishawn, the unique dancing school in Los Angeles that had been founded a few years earlier by Ruth St. Denis and her husband, the dancer-entrepreneur Ted Shawn. The circumstances of her matriculation were hardly auspicious. Twenty is a very late age for an individual to become a dancer—most future performers begin rigorous practice before age ten. Graham was very small, not particularly attractive in a conventional sense, and seemingly not very malleable. As if to underscore this meager prospect, St. Denis decided she did not want this unimpressive person as a pupil. Accordingly, she assigned the new pupil to Shawn instead.

Despite her initial lack of promise, Graham soon shone in the dance troupe. Quiet and shy in class, she turned out to be a remarkably quick learner of difficult styles and techniques—a prodigy in the bodily-kinesthetic sphere. She enjoyed the regime of training and was pleased to see her body become strong and supple. She drove herself incredibly hard, working alone far into the night, making enormous demands on herself and equally sizeable claims on others. In one memorable defining event, Shawn expressed his regret that Graham did not know a certain dance: "She would look just right in it," he declared. "But I do know it," protested Martha, and soon proved that she had indeed mastered the dance merely by watching others perform it. In the short order that characterizes creators' mastery of their chosen domain, Graham became a lead dancer, an instructor in the group, and an emerging star in *Serenata Morisca*, a sensuous Moorish Gypsy dance. Having learned all that Shawn could teach, Graham continued to be inspired—probably for the rest of her life—by the figure, personality, and dances of St. Denis.

Graham's first starring vehicle was an Aztec dance called *Xochitl*. In this melodrama typical of the period, Graham assumed the role of a maiden who danced ferociously to protect her virtue, which was at risk in the presence of the demanding Tolchec Emperor. Graham proved to be a terrifying resister, tailor-made for the role, which she eventually danced opposite Robert Gorham, Charles Weidman, and Ted Shawn himself. Increasingly in command of this role, she performed it across America and even in London, in the process picking up considerable knowledge about how to direct a dance troupe. She received some encouraging early reviews; for example, the *Tacoma New Tribune* called her a "brilliant young dancer."

While being part of a dynamic troupe, with ample opportunities to teach, learn, and travel, proved congenial, Graham soon tired of the intrigues that pervaded the Denishawn "family." Her romantic involvement with one member of the troupe, who was married to another, no doubt aggravated an already tense situation. By 1923 she had broken off from the California-based dance company and moved to Greenwich Village, the magnetic "Left Bank" of New York City.

To make a living, Graham joined John Murray Anderson's Greenwich Village Follies, where she performed Denishawn-type exotic Oriental dances for Broadway audiences. Very quickly she became a Broadway star, earning as much as $350 a week, a remarkable sum for the time. And she had the opportunity to witness some of the great artists of the time, such as the Italian actress Eleonora Duse. Graham sampled the life of an individually sought-after performer working in a popular vein, and she determined that it was *not* what she wanted for herself. Nearly thirty by this point, she swallowed whatever doubts she may have felt and declared, in a manner worthy of Freud or Picasso: "I'm going to the top. Nothing is going to stop me. And I shall do it alone."

Having accepted a position at the Eastman School of Music, Graham moved to Rochester, New York. There, with Esther Gustafson, she directed a new dance department and found satisfaction in having her own students and her own position. Still, Rochester lay well outside of the center of things artistic, and the Eastman theater had its own artistic needs, which did not always conform with Graham's. Graham was struggling to forge an approach that epitomized her own desires and values; she had little tolerance for anything that detracted her from her long-term goals. And how to characterize these goals? She had formulated an answer upon seeing a canvas by the Russian painter Vasily Kandinsky that featured a splash of red against a field of blue: "I will dance like that."

A milestone occurred on April 18, 1926, at the Forty-Eighth Street Theater in New York City. Graham gave a first performance with her own tiny dance company. This concert had required much effort: Graham had worked for a year to amass the money, and even then she had needed a thousand-dollar contribution from Frances Steloff of the Gotham Book Mart. With her students Evelyn Sabin, Betty McDonald, and Thelma Biracree, Graham staged a full evening of performance. The dances themselves were similar to those of Denishawn, with names like *Three Gopi Maidens, Maid with the Flaxen Hair, Clair de lune*; they were decorative rather than deep. Martha wore a "white dress, a flaxen wig, and a demure expression" and was seen as "decorative, pretty and undisturbing." One critic said "she is in appearance a maiden after Rossetti, slender, unearthly, and exotically graceful." Graham herself was later to term the dances "childish things, dreadful"; a more sympathetic critic recalled that "the idiom was still prevalently romantic and eclectic but the spirit was new and as bracing as a salty sea wind." Clearly, this performance was closer to Stravinsky's first concerts of 1907 and 1908 than to the *Le sacre du printemps* of 1913; and yet within a few years, Martha Graham's dances were to prove as innovative as any artworks of the century.

THE NEW DANCE

The period from 1926 to 1935 represented a time of unique growth for dance in America and in the larger Western context. In the mid-1920s, dance was largely spectacle, with the repute of the dancers linked to their physical beauty and technical facility, and, to an almost equal degree, the exoticness and sentimentality of the works in which they performed. While this description could be applied to Martha Graham's first concert of 1926, it would become increasingly inappropriate for the performances over the next several years: Each concert—indeed, virtually each new dance—was to prove more daring than its predecessors. One by one, the romantic and nostalgic features of Denishawn were stripped away.

Distinctive Features of the Early Graham Dances

In 1927, Graham's *Revolt* jolted audiences with its direct depiction of human injustice. In 1928, Graham's close associate Louis Horst composed music for *Fragments*, a dance Graham had already created. (This was a reversal of the practice in which the dance was matched to already composed music.) *Dance*, of 1929, featured Graham in a narrow, tubular costume, standing atop a small two-step platform, which served as the sole performing area. Instead of concentrating on the movements of her feet over a wide stage, Graham made her torso the center of action, releasing forces of energy in a sequence of percussive gestures. In *Heretic*, of the same year, Graham dressed in white and with flowing hair, played herself off against a set of eleven women in black jersey tube dresses who tried in various ways to oppress the solitary rebel. With a repetitive Breton tune as background, the heretic sought several times, unsuccessfully, to penetrate a wall that the others had formed; at the end she fell to the floor, defeated by her oppressors. By this time Graham's dancers were devoid of glamour and makeup; the faces were masklike, with each mouth a Kandinskian gash of red, and the dancers' hair was drawn straight back and held firmly in a knot.

Inspired by statuary, such as the works of Ernst Barlach, *Lamentation*, in 1930, was the most haunting of the early dances (see figure 8.1). A solitary, grieving woman was encased in a tube of stretch jersey, with only hands, feet, and face visible. Seated on a low platform throughout, the mummylike figure

FIGURE 8.1.
Graham's writhing, mummylike figure in *Lamentation*.
© Photo Barbara Morgan: Willard and Barbara Morgan Archives, Dobbs Ferry, N.Y.

rocked with anguish from side to side, plunging her hand deep into the dark fabric. Barely perceptible, the body writhed as if attempting to break out of its habit. As the body moved, the tube formed diagonals across the center of the body. The movements, created through the changing forms of the costume, were prayerful and beseeching, not so much a re-creation of grief as its embodiment.

Providing a description of any of Graham's dances is not easy. To begin with, by her own wishes, very few of Graham's early dances have been filmed, and even fewer are available for public viewing. (Graham was not alone in this policy. Many other lead dancers scorned the celluloid image that could be viewed and reviewed, preferring to be remembered by the impression of the single performance.) Among the art forms, dance is especially difficult to capture in words, and few writers have succeeded in providing satisfying descriptions of important dancers and dances. I am too young to have seen Martha Graham in her prime (the years before 1950), and although I have watched—and in many cases, watched again—the films of her later dances, I am limited by my own lack of expertise in the dance and by the knowledge that the Graham I have seen is but a pale shadow of the great dancer from the 1920s through the 1940s. To convey the vibrancy of her work, I rely heavily not only on accounts given by critics, dance specialists, Martha Graham, and the members of her circle but also on the photographs of Graham during her early years, most especially the magnificent dance portraits by the photographer Barbara Morgan in the 1930s. Some of these portraits have been reproduced here.

An eclectic atmosphere surrounded dance in America—and especially dance in New York—during the late 1920s and early 1930s. A whole ensemble of talented young dancers (many, like Graham, trained by Denishawn) were breaking new ground at the time: Doris Humphrey, perhaps the most talented choreographer of the time and her long-time partner, Charles Weidman; the sexy jazz-and-spirituals dancer Helen Tamiris; and Agnes de Mille, a talented young dancer and ballet choreographer from a distinguished artistic family. De Mille has described the scene with characteristic vividness:

> This was a stirring period in American dance history—a period of revolution and adventure. There were at least ten soloists working in New York, each making experiments. There were a score more imitating, expanding, and developing our dances. . . . We all turned out en bloc for every occasion, wrangling and fighting in the lobbies, as though at a political meeting. . . . We risked everything; every one of us had thrown overboard all of our tradition. . . . I am glad I participated in the period of the origination. There are a force and a wonder in first revelation that have no duplication. Greater dancers may be coming, greater and more subtle choreography; but we worked where there was no pattern and no precedent. . . . It was kind of a gigantic jam session and it lasted nine years.

THE AMBIENCE OF THE MODERN DANCE DOMAIN

Martha Graham and her associates were laying out the domain of modern dance. Its scope was to be quintessentially American and indubitably modern; its aim was to capture the energy, the dynamics, and the social spirit of the country and, especially, the cities. As Graham herself put it: "Life today is nervous, sharp, and zig-zag. It often stops in mid air. . . . It is what I want for my dances." In lieu of fancy costumes and scenery, traditional stories, and lush vocal or orchestral accompaniment, modern dancers instead introduced themes dealing with modern life, social injustice, and the relationship between man and woman, or they spurned narrative altogether.

According to the new dispensation, dance movements ought to reflect these serious themes. They should be stark; the dancers sought to boil down moods and feelings, to lay bare their essentials. In contrast to the regal ballet, modern dance was earthbound and undecorative, and dealt with common people. In contrast to the romantic ornaments associated with Denishawn, modern dance embraced blunt gestures and prosaic expressions. Graham explained the transition:

> Once we strove to imitate gods—we did god dances. Then we strove to become part of nature by representing natural forces in dance forms—winds—flowers—trees. Dance was no longer performing its function of communication . . . [Modern dance] was not done perversely to dramatize ugliness or to strike at sacred tradition. . . . There was a revolt against the ornamented forms of impressionistic dancing. There came a period of great austerity.

With innovation came uncertainty. Few of the dancers had sufficient resources to underwrite their own concerts, let alone their own series or season. Hence, in 1930, the leading dancers banded together to form a Dance Repertory Theater. A week with different dancers performing each night proved an efficient, economical way to showcase new work. But it was not easy for the dancers to work together or to share resources and publicity, and so this "noble experiment" occurred for only two seasons.

Graham was generally considered the leader of the modern dancers' group, and symbolizing her special status, she was granted an evening for herself in this series. Graham was never certain what would be danced, how it would be danced, and when the scenery and costume would be completed; so the programs continued to change until the very day of the performance. Perfectionism existed alongside chaos. Often enough, the ultimate performance was on the mark, but this fact hid the enormous strain the entire troupe experienced up to the very moments when the curtain rose . . . and fell.

THE FIELD PROMOTES MODERN DANCE

While Graham and her associates were charting the domain, a small group of influential viewers comprised a self-styled field to judge the new forms. Rarely if ever has the action of a small group of critics proved as decisive in the course of a particular art form. (The closest analogy is perhaps the circle that had promoted cubism two decades before, as described in chapter 5.) In New York City throughout most of the 1920s, the major newspapers did not have regular dance coverage. If a well-known European ballet group, or an individual star like the ballerina Anna Pavlova or the theatrical Spanish dancer called La Argentina was to perform, the regular music critic would be assigned to cover the event. But in 1927, within a few weeks of one another, John Martin began to write regularly about dance for the *New York Times* and Mary Watkins assumed a similar function for the *New York Herald Tribune*. These enterprising journalists sensed that a momentous breakthrough was happening in American dance, and, often at the expense of Denishawn, they determined to aid this emergent movement in every way they could.

Martin took dance as his cause and Graham as his chosen vehicle. Articulate, intelligent, and energetic, Martin wrote regularly and eloquently about his discoveries and his discovery. Let me quote a few representative examples:

> The actual number of new creations produced in a single season by a dancer like Martha Graham, to take a striking example, is staggering even to the onlooker, who finds it difficult to digest at one seeing stuff of this richness of content; and if the onlooker is staggered by such a program, what of its creator and director.
>
> . . . . .
>
> Whatever the ultimate judgment may be as to the merits of the Dance Repertory Theater, the opening of its first annual subscription season tonight at Max Elliot's Theater constitutes a date of the utmost importance in the history of American dancing. Never before have dancers of the first rank been able to see the necessity for sacrificing their own sense of independence in the interest of a common cause—in this case, the integration of modern dance in America.
>
> . . . . .
>
> [On Martha Graham]: A distinguished American audience paid a memorable tribute to a native artist.
>
> . . . . .
>
> Audiences who come to be amused and entertained will go away disappointed, for Miss Graham's programs are alive with passion and protest. . . . She does the unforgiveable thing for a dancer to do . . . she makes you think.

257

. . . . .

When the definitive history of the dance comes to be written, it will become evident that no other dancer has yet touched the borders to which she has extended the compass of movement. Not only in a technical sense, though here too she has proved the body capable of a phenomenal range, but especially in the field of creative expressional movement has she made an incomparable contribution.

Martin was not alone in his enthusiasm. Watkins declared: "Dancing is no longer a step child of the arts" and, like Martin, singled out Graham for special praise. Others were generally supportive. But at times a note of caution was injected by other dance critics. Terry declared that "innocent people, who were introduced to modern dance by Martha in her sledgehammer, unsmiling, angular period sometimes said to themselves, if this is modern dance, no more"; and Watkins referred to "the cerebral, specialized, and laboratorial studies of Martha Graham." And some commentators were more pointed: Stark Young declared at one point: "If Martha Graham ever gave birth, it would be to a cube." Edwin Denby referred to her as "violent, distorted, oppressive, and obscure." And the noted essayist on dance, Lincoln Kirstein, reflected on his own changing attitudes:

> When I first saw Graham I mistook her attitude, confused her approach, and decided that it was all very old-fashioned, provincial, unresourceful and ultimately uninteresting. . . . This solitary dancer, not even a girl, with her Spartan band of girls seeming to me to press themselves into replicas of the steel woman she was, appeared either naive or pretentious. . . . I believe that in Graham's work of five and six years ago there were still elements of her own unachieved revolt, unassimilated and inorganic, which coincided with those insecure and immature philosophies of Spenglerian decay and European snobbery with which I was then equipped and which colored my opinion of her art.

Even less charitable were European commentators like André Levinson who saw American concert dancers as amateurish, eclectic, and excessively enamored of Oriental exoticism. The Russian choreographer Fokine, in one memorable exchange with Graham, criticized her ignorance of classical forms and called her use of the body "ugly in form and hateful in spirit." Graham responded simply: "We shall never understand each other."

In what I have written so far, it may appear that Martha Graham founded modern dance virtually alone and that the newly emerging field, led by Martin, had eyes only for her. Viewed at a distance, knowledgeable individuals might consider this summary at least superficially plausible. But, as with all of the other creators I am describing, the selection of a single individual is partially a convenience. A plausible story about the key figures in the development of modern dance undoubtedly could be written about the various predecessors of Graham, as well as about at least a few contemporaries such as Mary Wigman in Germany or Doris Humphrey in the United States. Each had her promoters as well as her followers; perhaps Graham had followers in part because of her tireless self-promotion. Moreover, because Graham's appeal lay so significantly in her person, she may have loomed larger as a figure during her life than she will in the next century; others were more determined choreographers, and only photographs can even begin to convey how Graham appeared in her prime.

ATTEMPTS AT COLLABORATION

If modern dance was to be put on the artistic map, the pioneers had to work together. Try they did, but with little success. It is difficult to know whether these difficulties reflect simply the tensions that characterize creative artists generally (such as those that colored the workings of Stravinsky and his collaborators [chapter 6] or those that eventually undermined the Picasso-Braque duo [chapter 5]) or whether particularly virulent personality conflicts were at work. In any event, a palpable tension surrounded the early modern dancers. Ruth St. Denis labeled Martha Graham's work as "the open crotch school of music," while Doris Humphrey wrote home, "I haven't much faith in Martha Graham; she is a snake if ever there was one." Students who worked with one of the dancers were forbidden to sit in on classes conducted by any of the other principal dance innovators. Whatever its financial merits, the kind of collaboration involved in Dance Repertory Theater performances proved unsustainable.

A Vital Relationship

But if Martha Graham was estranged to a greater or lesser extent from her chief dance rivals, one individual assumed incomparable importance in her life: Louis Horst, an American of German heritage, who had long been an accompanist, composer, and informal mentor at Denishawn. Graham and Horst had fallen in love during the Denishawn days, and while Horst remained married to the dancer Betty Horst, he and Graham had begun an affair that lasted some twenty years and an even lengthier professional relationship.

Unlike other composers of his era, Horst saw it as his mission to write for the dance; in fact, he composed pieces for most of the notable dancers of the time. Above all, he wanted to compose for Graham, and he was willing to submerge his identity as a composer in doing so. Serving as an all-purpose mentor, Horst introduced Graham to European music and to the pioneering dance practices

259

and notations of the Germans Rudolph van Laban and Mary Wigman; he also exposed her to the philosophy of Nietzsche and took her to baseball games and to prize fights. More generally, he served as her supporter and her sounding board, the individual with whom she could share her dreams and her doubts, the person present at the moments of creative tension.

Horst was Graham's alter ego: an individual on whom she could rely for affective support as well as a colleague on whom she could bounce off her most radical and still inchoate ideas. As her lover as well as her mentor, Horst was ideally suited in many ways to serve as a catalyst as Graham worked out her ideas. Graham was creating a new expressive language, one rooted in not only the use of the body but also musical understanding, sensitivity to contemporary events, and, eventually, mastery of pivotal American and classical texts. Horst had the task of keeping these diverse constituents of the emerging language in mind and helping make them accessible to an audience.

The Graham-Horst relationship was rocky. Graham had a volatile temper, and Horst did not always accept her abuse passively. They fought many vigorous battles. Graham declared: "You're breaking me. You're destroying me." Horst responded: "Every young artist needs a wall to grow against like a vine. I am that wall." On occasion, they would strike one another. Agnes de Mille recalls a time in the early 1930s when Martha Graham was very depressed. Graham declared: "The work is no good. I've destroyed my work. I've lost the year. I've thrown away the Guggenheim [fellowship]." Horst tried to console her, but to no avail, and finally walked out in disgust. Yet, in talking with de Mille afterwards, he relented, saying, "When you get down to it, there is no other dancer."

MARTHA GRAHAM'S DANCING IN THE EARLY 1930s

Working in tandem (but also in competition) with the other young American dancers, Graham had already established the legitimacy of modern dance by 1930—a dance stripped down to its essentials in movement and emotional expression, a dance that spoke to the issues of the time. The early works *Heretic* and *Lamentation* bore her unmistakable stamp, even as they spawned imitations by those who had worked closely with Graham and those who had seen her perform.

A Southwest Sojourn and Its Impact

But Graham, like other highly creative individuals, was never content to repeat herself: She had a virtual horror of any kind of self-imitation. In 1930 she visited native American lands in New Mexico. Just as Picasso was invigorated by his summer visits to sites like Gosol and Horta (see chapter 5), Graham was

FIGURE 8.2.
Graham and chorus re-create the force of primitive life
in *Primitive Mysteries*.
© Photo Barbara Morgan.

deeply affected by her time in this area. She was impressed by the closeness of the Indians to the soil, the importance accorded to spiritual life, and the ways in which the traditional Indian views meshed with Spanish and Christian cultures. On a more sensual level, she loved the vast sagebrush deserts, the wide spaces and open lights, and the dark skin of the people. The experience seemed to evoke aspects of her Californian adolescence that had perhaps been suppressed by the hectic pace of New York urban life.

Graham's sojourn in the Southwest influenced her dances in the early 1930s, especially *Primitive Mysteries* (1931), *Bacchanale* (1931), and *Primitive Canticles* (1931). *Primitive Mysteries* re-created an annual Indian-Hispanic ceremony in honor of the Virgin Mary (see figure 8.2). The movement of the dance itself was stark, simple, concentrated, and unadorned. Rather than seeking to re-create the ceremonial expression directly, however, Graham sought to capture and distill the force and power of the primitive life. She used rough textures and broken rhythms to convey the vitality of feelings; the repetition evinced unconsciously by the native Americans became a conscious tool of the modern dancers.

Simple in form, *Primitive Mysteries* consists of three parts, each opened and closed with a procession: "Hymn to the Virgin," an investiture of a woman marking the birth of Christ; "Crucifixion," a depiction of the spiritual crisis in His community upon the death of Christ; and "Hosanna," the delivery from crisis following the Resurrection. Dressed in a flowing white gown, Graham evokes an image of the Virgin in the adobe churches of the Southwest; the contrasting dance group of twelve, clad in tight dark-blue dresses and usually arrayed in groups of four, constitutes the background. This group acts as a chorus to the solo dancer, reflecting joy and glory in the opening part, horror and grief in the middle, and contained exaltation in the final part. The contrasting spirits of each part are conveyed by the different groupings of the dancers and the different movements: slow, emphatic steps in the opening hymn; tense, stretching, grief-filled movements and arched, suspended leaps in the middle section; and bodies tilting back exultantly in the last section. The dance is filled with isolated gestures—a lifted arm, flexed feet, clasped hands, pointing fingers—that are not coordinated with the rest of the body. But the body is also used as a metaphor, for instance, with the Virgin's symmetrical movements conveying her centrality in the rite.

The dance is rich in stirring moments. For example, to convey the sorrow and pain of the crucifixion, Graham stands transfixed for a time in the center of the stage, her hands pressed tightly into the sides of her face. At the same time, the remaining members of the chorus symbolize the crown of thorns, with palms pressing against foreheads and fingers distended like spikes. At the end of the composition all of the dancers stand in silence for a moment and then walk slowly in unison off stage to a heavily accented beat. As described by Barbara Morgan, "Each member seems to walk because of the beat whereas Graham steps because the beat itself has forced her to move."

Primitive Mysteries is one of Graham's most effective pieces. Horst's compelling orchestration involves dissonant chords on the piano, with flutes and oboe playing sustained high notes; it ranges from prayerful chords to sprightly staccato episodes, effectively punctuated by lengthy silences. The dance is notable for the tightly balanced construction across the trio of sections and for the reinforcing confluence of themes: the pureness of mystic rites juxtaposed with central points of the Christian doctrine; the sensitive interweaving of different roles in human life, including man, woman, and child, as well as mortal and divine; and the respect for stillness, silence, and formality. Critics agreed that this dance marked a new high for Graham and for the dance of the era. Watkins declared in the *Herald Tribune*: "The most significant choreography which has yet come out of

America . . . It is not only a masterpiece of construction but it achieves a mood which actually lifts both spectators and dancers to the rarefied heights of spiritual ecstasy." Martin called it "probably the finest single composition ever produced in America."

A Growing Versatility

Martha Graham was also revealing considerable versatility in her work. She displayed capacities not only for tragedy but for parody and comedy, as in *Four Insincerities* (Serge Prokofiev, composer) and *Moment Rustica* (Francis Poulenc, composer), both of 1929. *Ekstasis* (1933) revealed an exquisite awareness of the body in motion. Graham danced to new compositions by such respected composers as George Antheil, Henry Cowell, David Diamond, Hunter Johnson, and Wallingford Rieger. And she undertook a number of related activities that reflected her resourcefulness, including a performance in Stravinsky's *Le sacre du printemps* with Leopold Stokowski in New York and Philadelphia in April 1930. Dancing the role of the Chosen One under the guidance of the famed choreographer Leonide Massine, Graham showed that she could leap as high as a prima ballerina and realize the musical and dance ideals of an ensemble quite unlike her own.

She danced in Sophocles's *Electra* in Ann Arbor in 1931. The famed impresario Roxy (Samuel Rothafel) also arranged a landmark appearance in the winter of 1932 and 1933 at New York's new Radio City Music Hall. On this unusual occasion, Graham performed brilliantly. Martin reported that "she ran furiously, she leaped, she extended herself, she filled the space at her disposal with a composition worthy of its dimensions." Yet even Graham's bravura performance was not enough to wow the "pop" audiences at the Music Hall, and she was soon dropped from the bill.

At this time, Martha Graham's dancing was also at its most spectacular. One of her troupe members and her biographer, Ernestine Stodelle, writes:

> To remember Martha Graham the soloist is to relive visions of overwhelming but mystifying impact: the taut, strong torso with its deep contractions and spasmodic breath releases, the whipping arms and legs, the scampering sideways motions, the angular foot-flexed jump, the kick that stabbed the air, the monolithic fall and its quick recovery—mysteriously appealing movements that directly affected one's nervous system.

Martin recalled "her spacious extension and circling of legs, in turned twitchings, largely of the shoulders and upper torso; pelvic rotations, rollings on the ground, hopping with upturned toes on the extended free leg, semisquatting procession, rapid sequences of small walking steps, fluttering of hands, many cartwheels." In her combination of dramatic imagination, bodily inventiveness, memorable appearance, and sheer dancing skill, Martha Graham already stood apart from all other contemporary modern dancers.

AN AMERICAN PHASE

Martha Graham was quintessentially American. Her family genealogy dated back to the *Mayflower*. She had lived on both American coasts and had attended American schools. As a member of Denishawn, she had toured the country and had visited England, where she had been impressed but unhappy. Returning to the United States, she had slowly shed the layers of Denishawn's eclectic dancing style and, working in parallel with a small cohort of hardy pioneers, had sought to forge her own distinctive approach. While continuing to dance to music composed by Europeans (often introduced to her by Horst), she was increasingly estranged from this tradition; her tense encounters with the Ballets Russes choreographers Fokine and Massine exemplified her emerging independence. The visit to the Southwest had confirmed her love for a certain terrain in her country and for its native population.

Frontier *as a Breakthrough*

Graham's immersion in American culture gradually evolved into a recognition, celebration, and enrichment of that culture: She had danced for nearly a decade before explicitly tackling American themes. Among the early works in this tradition were *Act of Poetry* and *Act of Judgment* (1934), together called *American Provincials* (1934). Here the less palatable face of the Puritan tradition was directly exposed, as a crowd acted self-righteously and coldheartedly. The mixture of sex, pride, and demonism seemed out of the pages of Nathaniel Hawthorne's *The Scarlet Letter*.

America was embraced lavishly in *Frontier* (1935), which became Graham's signature work (see figure 8.3). With a stark, memorable set by Isamu Noguchi and music in loose rondo form by Horst, it was an arresting piece. Against a black curtain with no extraneous scenery, the stage revealed a vast vista that evoked the boundless plains of the American West. Center stage was a wooden fence, a little wider than it was high, which designated limited protection from unknown danger. The upper half of the stage featured a pair of ropes fastened to the floor and rising obliquely in opposite directions from directly behind the fence and stretching into the unknown reaches above; it anchored the fence yet suggested that the fence could travel toward infinity.

As the sole dancer, Graham was clad in a full-length dark jumper, with a light, full-sleeve blouse and a headband. She began the dance with one foot

FIGURE 8.3.
Martha Graham portrays "pure America" in
her creative breakthrough, *Frontier*.
© Photo Barbara Morgan.

planted firmly on the earth and the other high athwart the top bar of the fence. Her torso rotated, and as her head turned to scan the horizon, she broke out into a smile, indicating that the world of the frontier person, while perilous, also harbored pleasure. For a dancer known to be determinedly grim, this confirming sign conveyed acceptance of the possibilities of life. Graham's arms, legs, and torso extended in circular movements as far as possible to suggest endless space, only to curl up with small bursting movements at other points as if to signal a contrite reversion to womblike security. Comfortably anchored in a central space, she scanned widely with her eyes and ventured forth from the fence on three occasions. Graham flung bold, outgoing gestures toward the audience, at first slowly and then with greater force: She moved in simple, measured steps, as if pacing off the land; she leapt joyously through space with her head high and kicked sideways, making her apparently motionless body move rapidly across the stage up to a sudden pause and a final triumphant movement when she returned to rest contentedly against the fence. As befits a composition from the modern era, the parts were fragmentary, yet they added up to a singular encapsulation—a physical embodiment—of the American pioneer experience.

Frontier, a six-and-a-half-minute work, was a tour de force that redefined what the dance was and what a dancer could do. A dramatic Graham was conveying to her audience what it was like to be an American and what the world of America was like—it was "pure America: forthright, free—the very spirit of an indomitable westward moving people." In a brief and concentrated compass, but in a tone less austere than her earlier work, she conveyed the feelings of loneliness, distance, courage, and occasional moments of pleasure that defined the experience of American pioneer women.

At this time in her career, Graham became explicitly concerned with what it was like to be an American. As early as 1930 she pointed out: "The answer to the problem of the American dance on the part of the individuals who point the way is to know the land—its exciting strange contrasts of bareness and fertility." A few years later she declared:

> The American dancer owes a duty to the American audience. We must look to America to bring forth an art as powerful as the country itself. For Duncan or Denis a slow rising arm signified growing corn or flowers; a downward fluttering of the fingers perhaps suggested rain. Why should an arm try to be corn? Why should a hand try to be rain? Think of what a wonderful thing the hand is, and what vast potential personalities of movement it has as a hand and not as a poor imitation of something else. . . . Our dramatic force lies in energy and vitality.

And Lincoln Kirstein explained: "Martha Graham has a specifically American quality which cannot be ignored. . . . She has in *Frontier* . . . created a kind of candid, sweeping and wind-worn liberty for her individual expression at once beautiful and useful, like a piece of exquisitely realized Shaker furniture or homespun clothing."

A Series of Strong Pieces

Martha Graham's evolution between 1935 and 1945 was at least as astonishing as her progression during the first decade of her company. Forgiving the obvious pun, one can say that she truly hit her stride, broadening her repertoire and constantly expanding her range of expressed moods. Humor, satire, and brightness complemented the seriousness and barrenness of her earlier works. Graham recognized this shift. Looking back at her beginnings, she said, "I'm afraid I used to hit audiences over the head with a sledgehammer because I was so determined that they see and feel what I was trying to do." She went so far as to declare: "Now that we moderns have left our period of long woolens behind us, we must prove to our audiences that our theater pieces can have color, warmth, and entertainment value. . . . We must convince our audiences that we belong to the American theater."

Few other twentieth-century artists have been as prolific as Martha Graham. Works poured forth, one after another, at a feverish rate (sixty between 1926 and 1930 alone), and many represented distinctive advances in theme and techniques. Among them, *Steps in the Street* (1936), *Chronicle* (1936), and *Deep Song*

FIGURE 8.4.
Graham as Empress of the Arena in *Every Soul Is a Circus*.
© Photo Barbara Morgan.

FIGURE 8.5.
Graham dances in *El Penitente*, a narrative of penance
with Christian-Native American influences.
© Photo Barbara Morgan.

(1937) documented Graham's distress over the brutal Spanish Civil War. *American Document* (1938), a love dance in minstrel style, featured readings from important American texts, each with appropriate costuming, music, and dramatic dance action. *Every Soul Is a Circus* (1939) depicted a set of comedic situations, in which Graham played the role of a vain star, who was "tamed" by a stocky and sturdy ringmaster (see figure 8.4). *El Penitente*, an Indian-Christian dance in the format of a traveling show, depicts seduction, sin, flagellation, and contrition (see figure 8.5). *Letter to the World* (1940) depicted the life and world of Emily

FIGURE 8.6.
Graham portrays Emily Dickinson in *Letter to the World*.
© Photo Barbara Morgan.

Dickinson in three episodes, with three dancers depicting different parts of Dickinson's personality (see figure 8.6). *Deaths and Entrances* (1943), a wrenching portrait of Charlotte, Emily, and Anne Brontë, captured the dissipation of these exceptional sisters. Various meaning-laden objects—a goblet, a chess figure, a conch shell, a scarf, a fan, a vase—evoked charged scenes from different times in the life of the family. A final ambiguous gesture of triumph involved the placement on the chessboard of a mysterious glass goblet that had been held tentatively for close to an hour.

In these, and in other comparable works of the time, Graham was at the top of her form. She was in sufficient command of her medium that she could extend in new directions—comic as well as tragic, contemporary as well as classic, fictional as well as historical, explicitly sensual as well as sexually repressed—and consistently produce impressive works. Far from repeating herself, Graham was constantly surprising her audience with the fresh departures in each new work. She was always prepared to take a risk, and though she could be devastated by criticism, it never prevented her from taking comparable risks the next time around.

New Collaborative Ventures

By this time, Graham had already assembled around her a remarkably talented group of associates. She was now creating works with America's most outstanding composers, including her long-term composer-accompanist-lover Horst. She was collaborating with such gifted designers as Noguchi and Arch Lauterer. Her now expanded group of female dancers, featuring Jean Erdman, Sophie Maslow,

May O'Donnell, Anna Sokolow, and Ethel Winter, was probably the outstanding modern troupe; and she now added to her ranks such superlative male artists as Merce Cunningham and Erick Hawkins. Hawkins, a gifted young dancer trained in the Balanchine tradition, was to become Graham's lover, and, for a brief period, her only husband; and Cunningham, an even more gifted dancer and choreographer, often played the third member of a love triangle that featured Graham and Hawkins.

While initial efforts at collaboration in New York City at the start of the decade had not worked out, there was a coming together of modern American dancers at Bennington College in rural Vermont. Beginning in 1935, many of America's leading dancers, including Martha Graham, Doris Humphrey, Charles Weidman, and Hanya Holm, traveled to this campus each summer; they gave lessons to dancers from all over the country (including students like José Limón and Anna Sokolow) and developed new works for rehearsal and performance in the New England setting. Martha Hill was the impresario who made the dance school work, and she imported other artists such as the composers Otto Luening and Hunter Johnson, the poet Ben Belitt, and the photographer Barbara Morgan. Not only did modern dance move forward smartly as a result of the opportunities at Bennington, but the merger between modern dance and the women's liberal arts college energized both institutions.

Graham's world was expanding in other ways. Graham collaborated in dramatic productions with the actress Katherine Cornell and with the poet-playwright Archibald MacLeish. In 1937 Eleanor Roosevelt invited Graham to perform at the White House. An influential book of essays about Graham (on which I have relied) was issued by Merle Armitage in 1937, and Barbara Morgan's exquisite collection of photographs of major Graham performances appeared in 1940. The addition of male dancers to the company, and especially Graham's lover Hawkins, not only increased the range of works that could be performed but also gave Graham a chance to explore the world of sexual passions—a chance she relished. The addition of men was not without its costs: Four of her most important women dancers left the company shortly thereafter. It was hard to sever ties with Graham: While very demanding, she was also loyal to her dancers and made them feel that they were abandoning her personally if they left. Yet the biographer Donald McDonough comments that "whenever Graham took a new creative direction, some of her followers were unable to make the transition with her."

Appalachian Spring: *A Culminating Work*

If *Frontier* functioned as Graham's signature piece, *Appalachian Spring* was probably her most capacious work, and it may prove the most enduring. *Frontier* framed the beginning of the fertile decade in which Graham focused on the

American themes, while *Appalachian Spring* marked its end. Together the works exemplify a rough trend I have encountered in nearly all of the creative individuals being described: an initial decade in which a craft is first learned (in Graham's case, from 1916 to 1925); a second decade that extends until the moment of the most dramatic breakthrough (1926 to 1935); and a third decade until another major culminating work, which builds on the earlier breakthrough while linking it more definitively and comprehensively to the encompassing domain.

Appalachian Spring can be thought of as a kind of domesticated *Frontier*. It is set at a time when America was young, in the part of the world where Martha Graham spent her early years. In the compass of barely half an hour, against an arresting setting by Noguchi and with memorable music by Aaron Copland, life in the American past is evocatively conveyed.

As with other Graham works, the actual composition consists of a set of loosely coupled incidents depicting a marriage in frontier America, but the overall effect is integrated. At the beginning of the work, the main characters reveal their feelings for one another and their hopes for their future life. They explore the world alone and together, each revealing a strong sense of self-discipline in the process. With pride, they take possession of a newly built house. An authoritarian revivalist arrives on the scene, followed by a group of dutiful maidens. He promises blessings and hope to the faithful females but warns of damnation for the sinner.

With a wide spectrum of moods and configurations, the work features solitary dances as well as ensemble square dances, and the different dances bring out the traits and feelings of each character. The characters shift from playfulness to seriousness and back again. Graham appears in guises suggestive of a wife, a mother, a girl, a worshipper, and a neighbor; the husband and the revivalist are far more developed than earlier male roles. Complementing a feeling of loneliness and individuality is the suggestion of a rooted domestic life for the couple, with the prospects of a baby in the future. The same "husbandman" who peers out toward the distant unknown also touches the threshold of his home with quiet satisfaction. The air is mysterious, cool, and fresh. And, reliving the spirit of *Frontier*, the setting conveys a sense of both vast space and framed limitations—a sliced-off part of a simple country dwelling, with a small white cupboard wall and a carved rocking chair on a porch, plus a sloping tree stump on which the preacher can stand and address his followers. Martha Graham said that "behind the structure is the emotion that builds the house which is love."

Appalachian Spring confirms the central, celebratory role that America plays in Martha Graham's work. The ballet's setting, time, and background all relate to circumstances of her own life—her family's frontier background, the Puritan flavor of the revivalist, her growing love for Erick Hawkins, and even the season of spring, which reminded her of the end of the harsh New England winter and of her feelings when her two-year-old younger brother had died in the winter of

1906. But the power of the work derives equally from its most universal aspects: the feelings surrounding both religious and secular life, homestead and open space, isolation and intimacy, and the milestones of marriage, birth, and death, plus the possibility of an eternal existence thereafter. And in contrast to most of Graham's earlier and later works, each of which serves as a vehicle of one central character's feelings and projections, *Appalachian Spring* presents a panorama of life.

PEAKS AND VALLEYS OF CLASSICAL PROPORTIONS

If Martha Graham had neither danced nor choreographed after 1944, she would probably still be known as the most important figure in modern dance. She had led the revolt against the ancient traditions of Europe, as well as the reaction to the indigenous, but still romanticized, versions of Duncan and Denishawn; she had stripped dance to its physical and emotional essence; she had effectively revealed the potential of American dance; she had assembled the most impressive group of dancers; and she had exhibited a startling range of expressiveness, as well as a steady and daunting capacity for growth.

A woman's playing of a leading role in America—indeed, anywhere—was a notable achievement during the opening decades of the twentieth century. One need only reflect on the fact that American women achieved suffrage in 1920, when Graham was twenty-six, to realize how disenfranchised they had been. To be sure, dance was considered the realm of women, but those involved in the dance were thought of as "mere" entertainers rather than artists. Since Graham was universally considered to be the high priestess of modern dance, she became a particular target of parody and derision.

Graham did not waste time trying to defend herself as a woman or as an American. She simply accepted the marginal aspects of her persona and went on with her business, creating one significant work after another, withstanding the criticism and continuing to assume risks. She was a magnetic leader for her troupe, inspiring to those who knew her intimately, as well as those who knew her only from afar. Not particularly gifted in the realm of business, she gladly left the task of keeping the books and maintaining the financial well-being of her company to a string of managers, beginning with Horst, followed by Hawkins, and concluding with the manager-director who survived her, Ron Protas.

Emotional Turmoil

Artists often exist on a tightrope, and Martha Graham was no exception. She was a *performing* artist: Not only did she have to be "on" when performing, but she had to make sure that the rest of her group had been adequately rehearsed, that the musicians were there and paid for, and that the costumes were completed, the scenery set up, and countless other details executed properly. While an Einstein can work alone in his study, and even a Stravinsky can absent himself at the moment of performance, Graham had to be fully engaged, continually. In this sense, her creative domain most closely resembled that inhabited by Gandhi, who also had to "perform" in a certain way and who had even less control than Graham over what subsequently ensued (see chapter 9). To be sure, Graham craved to be there; to perform was to be alive, to realize her persona most fully. And yet, the tension of this form of life took its toll.

The depressions that had been manifest early in her career returned in the late 1940s, a period of emotional peaks and valleys unprecedented in Graham's life. After nearly thirty years of working closely together, Horst and Graham had a sharp parting of the ways. The precipitating event was a minor fracas at a rehearsal, but the relationship had been under increasing strains because of Graham's deep involvement with Hawkins. Barely a month after the breakup with Horst, Graham and Hawkins married. Graham's only marriage, which struck some as a replay of the warped relationship between St. Denis and Shawn, was destined not to last; Graham's insistence on domination of her surroundings soon alienated Hawkins.

In 1950 the Graham troupe traveled to Europe for its first international tour. When Graham injured herself in Paris at the start of her tour, and Hawkins suggested that the tour continue nonetheless, their marriage was effectively over. Such questioning of her centrality was unthinkable. "I was her equal," Hawkins later recalled, "and that created a lot of tension, which is why I finally left." Graham did not dance for two years, disbanded her troupe for a time, and fell into a lengthy and severe depression. After the death of her mother in 1958, and with Horst and Hawkins effectively gone, Graham seemed completely alone; from that time on, she would be surrounded chiefly by individuals who needed her—or at least her prestige—but on whom she could not depend and in whom she would not confide.

Works of the Greek Period

As was the case with other innovative artists, among them Eliot and Picasso, Graham's crises infused, rather than crippled her art. While not abandoning American themes or the lighter facets of her work, Graham in the later 1940s launched a series of works strongly influenced by the classical tradition, and particularly by the myths and gods of ancient Greece.

From childhood Graham had loved the myths and had sought to treat them in her early writings and dramatic play. Throughout her creative life she had also created powerful female roles, including the anonymous central figures of *Heretic*, *Lamentations*, and *Ekstasis*, and the identifiable historical figures of Emily Dickinson and Emily Brontë. She now combined these facets in a series of memorable roles, created for herself, where she captured the most tumultuous emotional and mental states and conflicts of the ruthless heroines of Greek tragedies.

Herodiae (1944), composed shortly after *Appalachian Spring*, can be thought of as the first work in this series. Attended by her maid, a woman, who resembles Salomé, sits in her boudoir and laments the problems of middle age. The neoarchaic *Dark Meadow* (1946) features a nameless heroine, who explores the mysteries of sex and life, of fertility and rebirth, devouring her own path in the course of her wanderings. *Cave of the Heart* (1946) relates the story of Medea, the personification of hate, driven to murderous revenge by her jealousy. Lying flat on her stomach, inside an elongated, sculpted form with copper wires raying outward, an enraged Medea plots her retaliation while her former lover, Jason, is making love to King Creon's daughter. In a chilling solo, danced partly on her knees and partly while squatting, Graham expresses her anger, jealousy, and frustration. In this morbid work, all the characters are doomed. The little snake of red cloth ultimately devoured by Medea dramatizes the tragic limitations of the human condition, duly commemorated by the impersonal and omniscient chorus.

Continuing in this vein, *Errand into the Maze* (1947) presents a loosely adapted story of the labyrinth. Ariadne, who seeks to help her lover, Theseus, is menaced by the Minotaur; the Picassoesque monster is enacted as a young man wearing a bull's head and winding his arms around a heavy yoke. Exploiting such extreme and bizarre images, *Errand* explores the emotional maze of the human experience and the monster of fear that must be confronted and defeated.

Finally, based on Sophocles's *Oedipus Rex*, *Night Journey* depicts the double tragedy of Oedipus and Jocasta. Noguchi's powerfully evocative setting features a twisted, uneven, sloping golden bed; it harbors aspects of a royal bed, a torture rack, a gridiron, and a promontory on which one can stand. Ropes represent the bonds of wedlock at one moment and the umbilical link between Oedipus and Jocasta at another. As the young and lithe Oedipus (played by Bertram Ross) cavorts and Jocasta seeks to embrace him, the blind seer Tiresias (Paul Taylor) slowly wanders around, tapping his lengthy staff.

Reliving her tortured past in events that unfold at an unrelievedly feverish pitch, Jocasta becomes aware of the enormity of her crime. As the program notes indicate, Jocasta "sees with double insight the triumphal entry of Oedipus, their meeting, courtship, marriage, their years of intimacy which were darkly crossed by the blind seer Tiresias." Graham explained the sexual content of her movements: "I felt that when Jocasta became aware of the enormity of her crime, a cry from the lips would not be enough. It had to be a cry from the loins themselves, the loins which had committed sin." Events unfold inexorably, culminating in Oedipus's self-blinding and Jocasta's suicide by hanging.

For many viewers these works in the Greek tradition remain Martha Graham's most memorable. Few have had the opportunity to watch the earliest Graham works; and the works of the American period have acquired a certain familiarity, if not tameness, over time. But the power and the anguish of the Greek works remains. The Greek works appeal particularly to those of a Dionysian temper, and those who are drawn to the dark and tragic corners of life, and particularly the life of the aging heroine; for them, the American works seem, in contrast, more Apollonian and optimistic.

Graham continued to dance these works until she was in her seventies, so many individuals have seen these works either in person or on film. Indeed, some younger dancers in the 1970s and 1980s equated the Greek works with the entire Graham corpus, failing to appreciate that the movement to plot and to graphically depicted (rather than formally conveyed) emotions actually represented a significant shift from the relatively plotless, more abstract works of the early 1930s.

In adapting mythic themes, Graham was realizing a neoclassical phase analogous to the neoclassical periods through which Picasso and Stravinsky had passed twenty years before. This lag between music and painting, on the one hand (where the decisive breakthroughs occurred around 1910), and dance, on the other (where the breakthrough occurred around 1930), makes sense; after twenty years of deliberately defying the past, an artist may find it natural, or at least tempting, to return to classical themes and traditional forms. However, differences also exist. The neoclassical periods for Picasso and Stravinsky represented a kind of extended pause, in which they dealt with lighter themes or with established themes in a measured way. In contrast, Graham's classical period was one of her lengthiest and most elaborated, one perhaps more all-encompassing and self-revelatory than any other in her career.

This was also the time in which Graham became most directly involved with written texts. Not only did she read copiously in Greek literature and in literature about the classical period, but she also immersed herself in writings about myth, ritual, and the unconscious, particularly as captured in the works of Freud and Jung and their followers. Much evidence of her reading and thinking can be found in her notebooks of the time, which were published in 1973 under the editorial direction of Nancy Wilson Ross.

The Enigmatic Notebooks

It is not easy to know what to make of Graham's notebooks. While some have seen them as a treasure trove of information, providing privileged access into Graham's polymathic mind, others have considered them pompous, long-winded, and more misleading than revealing about Graham's animating genius. *The Notebooks of Martha Graham* are a pastiche of castings, delineations of dance step sequences, and an occasional diagram or drawing, embroidered by snippets of quotations from a wide range of literary and philosophical sources. They do capture something of Graham's usual speech, characterized by de Mille as "elliptical and in no way logical or ordered."

As an example, the notes accompanying *Night Journey* begin by describing some of the opening movements, both in language and in terms of steps. The instructions are quite literal: "Runs with tip 3X 1—r—1; two darts and turn to stage r . . bourrée turn to stage r. left hand holding right elbow." The notes for *Cave* describe a "snake solo" complete with "rise into Bali turns." Notes for *The Eye of Anguish* are punctuated with quotations from *Puritan Oligarchy*; Plato's *Republic,* Book Ten; works of the English poet and novelist Walter de la Mare, the English essayist and critic Thomas De Quincey, T. S. Eliot (*Burnt Norton*), and the American poet Hart Crane; and text materials from a work called *Mona Lisa.*

For reasons about which one can only speculate, Graham destroyed most of her early notebooks but sanctioned the publication of notebooks compiled in conjunction with many of her later works. (This was essentially her policy as well with respect to the filming of her dances.) What she was apparently ashamed of in her early period gave way to at least a certain pride of accomplishment in the later period. I do not find that the notebooks provide much additional insight into the development of Graham's dances as a visual-gestural-bodily expression of her own person. As Hawkins points out: "What was most important for Martha was her own emotional life. She could tap her own emotional roots for her solos. But it was very hard for her to convey those feelings to another dancer." But, whatever their limitations as a primer for dance, Graham's notebooks do provide considerable insight into the more general preoccupations of her mind.

Specifically, I see the notebooks as a place in which Graham developed the "space" of her works—that space between the literal, step-by-step description of what each dancer should do (worded in plain English rather than in any kind of personal or established dance notation) and the evocative quotation of literary texts whose animating ideas and emotions she was presumably trying to embody in her works. On their own, neither the quotations nor the mechanistic instructions are particularly revealing. But taken together, one senses the kind of work that Graham was trying to create between these "poles"—a work where the bodily movements, facial expressions, sets, props, and musical accompaniment captured the ideas embodied in the text.

Perhaps Graham herself concluded that providing these two sets of clues to the puzzle would be as revealing to the reader as a film of the dance; especially when united with the still photographs, they do convey a sense of what a dance would be like, while allowing readers to fill out the dynamic forms in their minds. And since Graham was already in late middle age (quite elderly for a dancer) when her notebooks were written and published, they may have helped to convey what she wanted to look like, rather than how she actually appeared.

Still, what is left out of these notebooks is the central part of her creative breakthrough: the evolving conception of how to convey, in dance, powerful emotions like pride or jealousy or fear; strong personalities like Emily Brontë or Medea; compelling physical settings like the plains of the American Southwest or the hills of New England; and above all the dynamic quality of bodily movement. My own speculation is that Graham worked these out through experimentation with her own body and with the bodies of her dancers—alone, in front of mirrors, in the presence of a few trusted souls like Horst or Hawkins, and, eventually, in the presence of various kinds of audiences.

The crucial bodily-kinesthetic intelligence was represented in the course of its own experimentation, its transformations and retransformations, rather than thought through or encoded in a self-standing symbol system. As the dance historian Lynn Garafola puts it: "Graham *was* her body; she became who she was because of it and through the discipline that made it strong, eloquent and beautiful. What it could and could not do defined the limits of her invention and prompted the exercises that became the basis of her technique." Regrettably, no written records of Graham's experiments-in-movement exist; one must rely on the recollections of individuals who danced with Martha Graham and on the descriptions of successive dances as they were created in the middle decades of the twentieth century. Those dances that were less successful, in particular, may be seen as drafts of later, more comprehensive and more effective works.

By the time of her Greek works, Martha Graham had gone beyond the status of high priestess and had become legendary. She reveled in this role. But her status was strange because, even more so than was the case with other modern masters like Eliot, Stravinsky, Joyce, Picasso, or Schönberg, many individuals did not understand Graham's work at all or actively disdained what she was trying to do. Some were brutal in their comments. A writer for the *Nation* termed *Deaths and Entrances*

a deadend road which like the earlier fad of "socially significant" is being followed by many practitioners of the arts. Its disastrous results in the field of the dance were tellingly demonstrated by Martha Graham's latest program. . . . It is Miss Graham's indulgence in the exhibition of neurotic conflicts, for the purposes of display rather than of communication, that marks her basic failure as an artist.

Of the same work, Henry Simon of *PM* opined that "*Deaths and Entrances* is a long work . . . full of sound and fury, signifying something, but just what I do not feel able to say," and he referred disparagingly to "the somber long-skirted ladies who stalk and leap meaningfully across a barren stage to the accompaniment of equally sterile music." A *Detroit News* writer recorded frustration:

Miss Graham is the most perplexing of American dancers and her programs stir controversy. A reporter who expresses bewilderment over Miss Graham receives letters comparing him, unfavorably, with Ivan the Terrible. It is perilous work, reporting on Miss Graham, for no one is dispassionate about her. . . . She was just as puzzling Monday evening . . . moving from downright ugliness to an occasional cold beauty.

A LIFE IN DANCE

Despite (or perhaps in part because of) the controversy she generated, Martha Graham maintained her customary practices for seventy years. She danced, choreographed, and developed techniques, working styles, theories, and philosophies, while adhering to a yearly performance schedule with her company. Many who now follow similar practices do so in the belief that these are intrinsic to all modern dance, rather than inventions of one luminous figure.

Techniques and Work Habits

Inspired by the choreographer Wigman and a few other figures, Graham developed her characteristic dancing technique. Her approach, which grew out of her choreography, was designed in part in contrast to traditional ballet. Ballet has a highly linear design, whereas Graham's dances emphasize dynamic, irregular forms. Ballet highlights legs and arms used as separate revolving members, and it is taught through a number of fixed positions. Graham's idiom strives to keep the body in constant flux, with movement flowing from the pelvis to the head. Rather than movement appearing effortless, as in balletic arabesques or entrechats, the strain is supposed to be manifest to the viewer, revealing the self to be a motivated, disciplined, striving agent. Despite these and other contrasts, Graham commented: "It would be a criminal waste not to take advantage of three hundred years of ballet development. My quarrel was never with ballet itself but that, as used in classical ballets, it did not say enough, especially when it came to intense drama, to passion, it was that very lack that sent me into the kind of work I do."

Graham's technique is based on a central contrast: When you release, you inhale; when you contract, you exhale. Contraction originates in the pelvic region, whence comes its explicit sensual tension as well as the percussive angular movements that flow through the body and the lengths of the arms and legs. Both contraction and release are taken with a sudden impulse that can send the body falling to the floor, jerk it into a turn, or have it thrust in a leap. Tension occurs between the back and waist, head and chest, or shoulder and body against the floor.

Graham devised exercises to convey her principles. She invented a series of falls to the floor, where the dancer sinks backward, with nothing to break the fall. Students were made aware of the power of the back, the originating role of the pelvis, and the press of the floor; they were shown how feelings can emerge through contractions, releases, stretches, and pulls and spasms of the muscles of the torso, rather than primarily from gestures of the hands or the arms. An important part of training was the snakelike spiral movement, in which one coiled and uncoiled on the floor, giving strength and flexibility to the torso. Graham used vivid images to convey the desired moves: Contraction was like seeing the heavens; release, like viewing the earth from a cliff.

Becoming a Graham dancer was hard work. Graham believed that it took ten years to build a dancer (which fits with the ten-year rule for creative breakthroughs that I have described): "The body must be tempered by hard, definite technique—the science of dance movement—and the mind enriched by experience." Students worked every day on "the torture," becoming muscular and hardened in the process. After ten years a student could leave the ensemble and join a group of four. Graham commented that "it took years to become spontaneous and simple. Nijinsky took thousands of leaps before the memorable one." She once added: "The difference between the artist and the non-artist is not a greater capacity for feeling. The secret is that the artist can objectify, can make apparent the feelings we all have." (I am reminded of W. H. Auden's advice to an aspiring poet: "Poems are made not of strong feelings but of words.")

A modern dance company is an extension of its leader, and Graham cast a very long shadow on her company. She made enormous demands on her students, terrifying many of them, alienating some, insulting and cajoling as necessary, but also building enormously strong ties to them. The dancer Elisa Monte recalls: "Nothing was ever 'just fine.' There was never the easy triumph. If you could stand up under it, you could triumph. I've seen Martha cut so many heads off." Many of her female students identified with her to such an extent that they even started to look like her—a primitive mask, sucked-in cheeks, hair pulled back, expression frozen into mystery. Life as an ascetic was a virtual necessity, because dancing paid next to nothing, the performance season might be as short as a week, and Graham herself believed in and exemplified a life of spareness and sacrifice. Always a perfectionist, she would often change major roles, movements, or costumes at the last moment (expecting her troupe to take these sometimes major changes in stride), thus engendering a constant crisis atmosphere around her—a Picasso-like atmosphere that she probably prized, at least unconsciously.

Graham's dance exercises and techniques were far more than purely technical. Much of what Graham created and passed on to her dancers was a particular way of dancing. Graham danced better than anyone else of her generation, and a significant part of her legacy consists in her hundreds of performances. But particularly because her dances are not available for examination, her legacy rests primarily in the techniques that she passed on to her students.

Ideas and Philosophy

While Graham's belief in the expressive power of movement was deeply held, she never embraced movement for its own sake. She always related motion to feelings experienced, to feelings she wished to express. "I have to have a dramatic line even in the most abstract things I've done," she declared. "It has to come from one person's experience. I have never been able to divorce the dancing from life." In this sense, Graham, like the other modern masters, avoided the attraction of pure abstraction. "I don't want to be understandable," she once declared. "I just want to be felt."

Graham worked intensively, late into the night, thinking through her dances, paring them down unceasingly: "I would put a typewriter on a little table on my bed, bolster myself with pillows, and write all night." Her heroine St. Denis had always scribbled down words, essays, and poems, from which her dances had somehow emerged, and Graham engendered the same process in her own creative life. Everything became grist for the mill of her poetic and bodily imaginations:

> I get the ideas going. Then I write down, I copy out of any books that stimulate me at the time many quotations and I keep it. And I put down the source. Then when it comes to the actual work I keep a complete record of the steps. I keep note of every dance I have. I don't have notations. I just put it down and know what the words mean, or what the movements mean and where you go and what you do and maybe an explanation here and there.

Like the other modern masters I am considering, Graham spoke openly about her incorporation of the ideas and images of others: "I am a thief—and I am not ashamed. I steal from the best where it happens to be—Plato, Picasso, Bertram Ross. I am a thief—and I glory in it. . . . I think I know the value of what I steal and I treasure it for all time—not as a possession but as a heritage and a legacy."

Graham found nothing miraculous in her style of working. On the processes involved in her choreography, she declared: "You draw from memory, from your understanding of life, and others' understanding. All you've read and absorbed falls like a jewel into your being." The beginning of a dance came in various ways; Graham found it terrifying, "a time of great misery." She was fond of repeating some words that the composer Edgar Varèse had spoken in her presence: "Everyone is born with genius but most people only keep it a few minutes."

For most of her life, Graham saw herself as a dancer and an actress. She felt that she had been born as a dancer: "I didn't choose to become a dancer. I was chosen to be a dancer." She discouraged young people from becoming dancers, insisting, "Only if there is just one way to make life vivid for yourself and for others should you embark on such a career." She added: "I live and work out of necessity . . . as deeply and committedly as an animal. There is no choice. As an animal lives, eating, drinking, begetting one's young without pretense or ambition."

Most of Graham's dances were constructed around heroic characters. Often they centered on a kind of priestess, a woman renowned for her power and her passion who orchestrates the fates of a community. As an actress, Graham sought to capture every strength and blemish of her characters—indeed, to become each character. Referring to Jocasta in *Night Journey*, she quipped, "I want to know what this woman ate for breakfast." As with other versatile actresses, Graham was able to find in herself the traits of varying characters, from the humorous, scatterbrained woman in *Every Soul Is a Circus* to the evil, malicious Medea in *Cave of the Heart*. Characters like Jocasta and Clytemnestra were queens of antiquity who emerged with a shard of triumph from their tragic situation by dint of their painful efforts at self-understanding. In creating these characters, Graham returned time and again to the writings of Nietzsche and Schopenhauer on the power of human will, as well as the writings of Jung and Freud on the unconscious. But one can see that Graham was attracted to these writings because their ideas were already embodied in her tempestuous personality and in the roles she chose to invent or to re-create.

Graham devoted her life to her work. As her friend Agnes de Mille remembered her: "Martha felt that she must cut from her life all deep emotional involvements, all attachments, all comforts, even moments of leisure, and beyond that, love involving family and children. She gave everything to her work, withheld nothing, kept nothing apart. She was obsessed." Like Picasso, she could be cruel, vicious, and vindictive if she felt that such behavior was in the service of her work. "I know I am arrogant and vain and must be adored," she once confessed. "An element of destructive savagery is very, very close to Martha," declared her student Pearl Lang. The dancer Robert Cohan observed that "she had to live out her drama with the people who were offering themselves as parts of her drama." The narcissistic and egotistical Graham seems to have forged the same kind of Faustian bargain that I have described in other creators: If she was to retain the privilege of continuing her work, she would sacrifice on the altar of her dancing all worldly pleasure and all hope for intimate relations.

Involvement with the Graham company was more like a crusade than a profession. Graham had little understanding of individuals who lacked dedication and audacity: "Everyone has the right to fail. You fail and from your failure you go up one more step—if you've got the courage to get yourself up. . . . I happen to believe that there is one cardinal sin and that is mediocrity." She related to other individuals chiefly—perhaps even exclusively—in terms of whether they could help realize her vision of the dance. Cohan commented: "While you were in the company, Martha would die for you. The moment you left, you were gone." De Mille adds more brutally that "whenever she was done with an idea or a person, she cut that idea or person off." And she did not mind controversy: "I want people to think. I welcome the arguments that follow my concerns, for if people don't discuss my dance work, why then I shall have failed in what I set out to do."

A Peripatetic Pattern of Life

Armed with her techniques, her students, and her philosophy, Martha Graham established a pattern of life. It involved regular teaching, sometimes at New York's Neighborhood Playhouse and ultimately at her own dance studio, which was located in a town house purchased for her on East Sixty-Third Street in Manhattan. Graham directed her company, with little help from other individuals and with very few vacations or holidays. There would be one or possibly two brief seasons in New York City each year (perhaps eight performances over the course of a week), a national or international tour, and a summer at one of the dance schools around the country. Starting in the 1950s, Graham began regular tours abroad, and these earned her enormous acclaim not only as a dancer and choreographer but also as a goodwill ambassador. Indeed, her final illness began after her return from a fifty-five-day tour of the Far East in the fall of 1990.

Graham developed her new dances chiefly in the course of her daily work with students and associates. Originally, both the dancing and the choreography were built entirely around her person, but by the 1940s, she allowed others to dance important roles and also allotted a certain amount of the choreography to her most gifted students. This latitude was not without strings: Graham could become furious if the dance was not created to her liking.

Graham was able to keep up her life as a dancer for an amazingly long time. One of her greatest triumphs occurred in 1958, when she was sixty-four. Following three years of relatively sparse dance creation, she composed *Clytemnestra*, a full-length dance drama that occupied an entire evening's program. This portrait of the doomed House of Atreus featured Graham as the aging queen, "an angry wild wicked woman" with the mind of a man and the beguilement of a woman. Consigned to the Underground, the queen looked back on the events of her life, not in a logical order, but rather through a free association based on the queen's own consciousness of her heinous misdeeds. By observing how she "progressed" from the unfaithful wife of a king to the roles of plotter, usurper, and murderess, she discovered why she had been damned in hell. In the end she came to the realization that she could break this chain of revenge only by acknowledging her pivotal guilty role in the chain of events.

Deliberately choreographed so that it could be acted as much as danced, *Clytemnestra* stood as a remarkable aesthetic achievement. Graham's portrayal was widely acknowledged as masterful, combining her highly developed acting skills with her still formidable physical mastery. At the same time, it was a stark reminder to all—and perhaps most especially to Martha Graham—that she could not go on indefinitely as a dancer.

Panicked by the painful thought that she might no longer be able to dance, Graham sought to prolong her life in dance as long as possible. In the years following her most intensive Greek period, Graham had continued to choreograph; indeed, she had created thirteen new roles for herself from 1947 to 1969. Graham's later works fell roughly into two categories: heavier theatrical pieces about a heroic figure baring his or her soul, and lighter pieces of poetic fancy, including ones that poked fun at Graham's own earlier work. Overall, there was a neoclassical flavor; there was renewed attention to formal matters, a greater symmetry and distance, and less sheer exoticism, gore, and violence, though often a suggestion of despair and disintegration. The works created just before her retirement exhibited a kind of nostalgia for Graham's lost powers: Typically, Graham would place herself in the center of the action, but almost immobile, in a way reminiscent of the spectator role assumed by Picasso in his late works and the commentator stances associated with other aging creators, like Eliot and Einstein.

DECLINE AND RENEWAL

By the 1960s, Martha Graham's personal and professional life had reached a crisis point. Several of her best dancers and a number of her most gifted collaborators had left, charging that she used them and did not give them sufficient credit. Following the loss of the individuals to whom she had been closest, she felt truly alone. Her dance troupe continued to perform, but critics and audience members began to whisper that Graham could no longer dance. True, she was an inspired choreographer; she was also a formidable and influential actress, perhaps even, as Katherine Cornell had maintained, the greatest actress in America. No fool, Graham had continued to create roles that highlighted her acting abilities and her charismatic persona, while minimizing the demands on her body. Still, there was quite a difference between holding one's body *as if* one were old and holding one's body in a certain way because one *was* old. Martha Graham, variously believed at the time to be sixty, sixty-five, or even over seventy years of age, was now too old to dance publicly.

Graham took this realization very hard. She drank regularly, sometimes to excess, sometimes so much that she could barely make her way onto the stage. Other forms of destructiveness appeared. Graham threw away records, letters, and notes from early life. While she gave many interviews, she made it difficult to write about her in her absence. For twenty years, virtually the only biographical information available was contained in a portrait by Angelica Gibbs in the *New Yorker*. She discouraged filming of her dancing and did not wish her early roles to be danced by anyone. She objected to the camera because, in her view, it just followed the dancer passively: in contrast, in a live performance, "when you get Nureyev in the air, there he is. In the air like an insect." Like Isadora Duncan, Martha Graham preferred to be remembered as a legend, rather than as a filmed icon, let alone as an aging dancer. As she had once declared: "A dancer's instrument is his body bounded by birth and death. When he perishes, his art perishes also."

Vivid in Graham's mind was the model of St. Denis, who had danced (and apparently danced well) until her eighties. Graham conveniently passed over the examples of Wigman, who had stopped dancing at fifty-six, and of Humphrey, who had been forced by illness to stop at an even earlier age. But finally matters came to a head. Members of her company—the very individuals who had been at her beck and call for decades—came to her and demanded the right to perform without her. Angered and hurt, Graham did all that she could to change this startling sentiment, but to no avail. And then one day in 1970, an announcement appeared in the *Berkshire Eagle* and the *New York Times*, without her permission, that Martha Graham had retired.

The announcement infuriated Graham. Indeed, the day afterward, she rejected the report defiantly and, to the acclaim of her fans, announced that she would continue to dance. In actuality, however, the announcement only confirmed what had already come to pass. Eventually Graham was able to be somewhat philosophical about this painful turn of events:

> The decision made me physically ill. . . . I had to retreat to the country until I made certain adjustments within myself. Someone told me, "Martha, you are not a goddess. You must admit your mortality." That's difficult when you see yourself as a goddess and behave like one. . . . In the end I didn't want people to feel sorry for me. If I can't dance any more, then I don't want to. Or at least I won't.

A New Life in Old Age

Following her retirement, Graham gave no indication that she would continue her life in dance. And, indeed, for the succeeding two years, when she was quite ill, Graham remained absent from the public arena. But in 1973 she announced that she was reemerging as the director of her troupe. The Martha Graham Dance Company was now the single oldest existing theatrical institution in the country except for the Metropolitan Opera. To the amazement of many, Graham took charge, eliminating many of the personnel and vestiges of her former troupe; and she began to choreograph, for the first time in her life, without seeing herself as the major embodiment, the raison d'être, of her choreography. The later dances included *The Owl and the Pussycat*, a parody of Graham classics, with animals portraying various stock Graham characters; and her last piece, *Maple Leaf Rag* (1990), another parody of human foibles set to the music of Scott Joplin, a work that recalled her warm relationship with Horst during a seemingly less complex era. Perhaps not surprisingly, the new dances following her formal retirement did not center on a particular protagonist; Graham may have found it too difficult to place in the body of another the movements and roles she would previously have reserved for herself. "I would much rather be dancing. I will always miss it," she declared.

In addition to designing dances for others to perform, Graham remained active in other ways. One of the most theatrical guises was her perfection of the art of the lecture-demonstration, where she, still an astonishingly striking figure, would talk to the public about art in general and then reflect on her own experiences and accomplishments. These spellbinding performances were elaborate spectacles, and Graham exploited them to the hilt. She could tell jokes, quote lengthy passages of literature from memory, and speak extemporaneously about her own life. She now gave many interviews, wrote occasional articles, and even permitted some filming of her dance company.

Of greater importance, Graham finally allowed the re-creation of some of her most notable works. Where, before, she had preferred to limit their existence to the memories of those who had been fortunate enough to dance or to be an audience member, Graham now worked diligently with her dancers to bring them to the attention of a new public. More often than not, these dancers were very well received by the public and by critics; they gave a second, if not a permanent, life to creations that would otherwise have disappeared from human consciousness. Finally, the oldest art form was yielding its secrets to an era devoted to techniques of work preservation.

Financing her enterprise had never been easy. Even when her classes, recitals, and lecture-demonstrations were sold out, as they frequently were, there were still sizeable deficits to be made up. The impresario Sol Hurok had taken great pride in organizing Graham's tours, but he had stopped doing so when he was constantly unable to turn a profit. Throughout her career, Graham had depended on individual philanthropists to aid her efforts, dating from the time of Frances Steloff's support of her initial concert, to Martha Hill's presiding over the summer dance workshops, to generous sustenance from patrons like Bethsabee de Rothschild and Lila Acheson Wallace. In late life, the designer Halston, who outfitted Graham in dashing style, had been particularly generous. But as deficits grew from hundreds to thousands of dollars, and then, ultimately, to millions, the shaky status of the Martha Graham Dance Company became a widespread concern.

Despite her avowed lack of interest in money—for herself, her dancers, or her company—Graham had herself been a magnet for financial support. Like many of the greatest artists of the past, she had managed to intersect with patrons who allowed her to follow her muse, if sometimes at a price. When government funding for the arts began after the Second World War, Graham was one of the major beneficiaries. But in 1983, the Martha Graham Dance Company received a widely publicized rejection of its application for matching funds from the National Endowment for the Arts (NEA). It is difficult to determine how appropriate this rejection was. Sitting on the panel were a number of individuals who could have been expected to be sympathetic to her application: It is reasonable to assume that they would not have rejected the application without cause. (Weak financial management and an uncertain future for the company were the reasons reported in the press at the time.) But ever the performer, Graham seized the occasion to launch a public protest, accusing the panel of ageism, and taking her case "to the people." The charge of having turned down an American institution was too painful for the NEA, and in subsequent years the Martha Graham Dance Company always received at least some money from this flagship governmental agency. Yet upon Graham's death in 1991, there was still a huge deficit, and the ultimate fate of the Graham company remained as uncertain as ever.

GRAHAM'S ACHIEVEMENTS

Martha Graham's long life effectively spanned the twentieth century, a period marked by an incredible expansion of America's political role in the world from a relatively simple, basically rural land to the most highly industrialized world power. There was an equivalent shift in the arts, with America gradually assuming an influential role in forms ranging from the traditional (painting and orchestral performances) to the contemporary (movies and the broadcast media).

Nowhere was the trend more dramatic than in dance. To the extent that dance was recognized in the past century as an art form, that recognition was restricted to the increasingly old-fashioned imperial ballet. Thanks to the efforts of Diaghilev's Ballets Russes and, more recently, to Balanchine's stunning achievements at the New York City Ballet, ballet has assumed new vitality in recent years, with its influence extending even across the Broadway stage. And of course, movies at mid-century featured many varieties of dance.

But the singular dance story of the twentieth century lies in the rise of the modern dance. Beginning with the inspirations of Duncan and the spectacles of Denishawn during the opening decades of the century, modern dance took off in the late 1920s in New York City. Within a decade, Denishawn seemed as outmoded as the traditional ballet, and audiences instead flocked to the contemporary works created by a set of serious young dancer-choreographers. By the second half of the century, many dance companies had emerged: Dance was now a fixture not only on the campuses of women's colleges but also in many major universities and virtually every metropolitan area of the country, as well as in many distant lands.

No movement of this magnitude is the product of a single person, but while it is possible to envisage the enterprise of modern dance in its current dimensions without Humphrey or Holm or Wigman, it is inconceivable to do so without Graham. For well over a half century, she was its leading figure, its inspiration, its conscience. Her accomplishments—about two hundred separate dances—merit the descriptor "legendary." As the biographer Stodelle put it, "Here is the largest, most awe-inspiring unique and diverse repertoire by one choreographer ever to have been produced, comparable only to Shakespeare in dramaturgy and Picasso in painting." The dance critic Clive Barnes declared: "It is Miss Graham's fortune to become a legend in her own lifetime and indisputably the greatest figure ever to have developed within the American dance movement." The composer Virgil Thomson called her the major American actress of the era. Of all contemporary commentators, fellow dancer-choreographer Agnes de Mille has been most eloquent on the subject of Graham's achievements:

> In this hundred years, the five shaping artists are, I believe, Igor Stravinsky, Béla Bartók, Frank Lloyd Wright, Pablo Picasso, and Martha Graham. As far as dancing and theater are concerned, this is her century. And of all five giants, Graham made the greatest change in her art—in the idiom, in the technique, in the content, and in the point of view—greater, finally, than any other single artist who comes to mind. . . . It is probably the greatest addition to dance vocabulary made in this century, comparable to the rules of perspective in painting or the use of the thumb in keyboard playing. . . . Martha is the supreme force. Her compositions are masterful. Martha is the one.

Graham's achievements are particularly remarkable given that she lacked the assets generally expected in an individual who rises to the top of a field and then refashions it in her own image. While her childhood was comfortable and her family supportive, Graham had had no financial resources to bring to her profession; to make matters worse, the social group from which she came had looked down on dance. Because Graham was short and hardly beautiful, St. Denis had not even taken her seriously as a new student. The avant-garde arts were widely considered a European phenomenon, and Americans who wanted to participate typically moved to Europe, sometimes remaining there permanently. Finally, while the dance itself was seen as a realm suitable for women, the domain of creating in general was seen overwhelmingly as one reserved for men, rather than one open to "Shakespeare's sister."

But Martha Graham was able to turn these disadvantages—in my terms, these asynchronies—to her advantage. She made her exotic appearance a central part of her appeal. She drew shrewdly on the various strands of her stern Puritan heritage in developing the charged themes of her dancing—at times miming it, at other times undermining it. While cognizant of the classical past and of more recent European developments, she fixed on the special flavors of the American landscape, democratic ideology, and contemporary rhythms of living in our time. She built her company around strong and talented women and then opened it up at an early point to strong and talented men. While not especially occupied with politics, she led by example in the search for equality of opportunity, having dancers of racial and ethnic diversity and casting them where they best fit, independent of what the roles canonically stipulated. Above all, she was unfailingly adventurous, never resting on her laurels, always ready to take risks, willing to fail, and then prepared to try again with renewed energy and dedication. As with the other creators I am considering, she was able to make the most of her marginality, and to continue to raise the ante when it appeared that she might be accepted uncritically.

Because her body remained young, and her mind even younger, Graham was able to continue creating, and creating well, for as long as Picasso—a period far longer than for the other masters under consideration. Because she had virtually created a new domain and had stimulated the development of a complementary field, Graham had available a capacious expanse on which to express herself, and she took full advantage of its breadth. Her originality continued unabated for many decades. In some ways her task was more difficult than Picasso's, because as a performer—and, later, as a leader of her performers—she had to be able to mobilize herself and her group at a specific historical moment, and to be in full readiness in a way that creators in other domains need not be. Her creative essence could not be dissociated from her physical presence at a given instant. In this respect, as I have noted, an artistic performer like Graham resembles more closely a political personality like Gandhi than a "distanced" artist like Picasso, Stravinsky, or Eliot or a scientist-in-isolation, like Einstein or Freud.

Graham reveled in this centrality of her physical being and sought to prolong it as long as possible. From all indications, however, the emotional and personal costs were severe, and perhaps particularly so for a pioneering woman. While she had the support at crucial times of Horst, the Faustian bargain she ultimately struck entailed the sacrifice of personal happiness and intimacy. When the critic Joseph Campbell once spoke about devoting oneself fully to life, Graham retorted, "If I were to take that step, I would lose my art." But as one with a phalanx of pioneering women in her own background, Martha Graham was carrying on a family (and a national) tradition. Perhaps it was this sense of rootedness in a tradition that emboldened her to blaze new frontiers and to continue to do so until the end of her life.

9

MAHATMA GANDHI:
A HOLD UPON OTHERS

Gandhi, ca. 1914

In 1600 THE BRITISH East India Trade Company was formed, and a decade later the Crown gave the company unlimited authority to trade throughout Asia. Over the next two centuries, the company's power and influence grew steadily. Indian agricultural products and textiles were exported, while English manufactured goods were imported, duty free, to the Indian subcontinent. In the wake of the decline of the Moghul empire, the company brought a firm ruling hand, a measure of stability, and some industrialization to India, even as it provided huge profits to its foreign owners.

INDIA UNDER BRITISH RULE

The British enterprise succeeded largely by exploiting the lingering tensions among the various warring political and religious factions on the Indian subcontinent. A population of a quarter of a billion individuals, living in a patchwork of states, was subjected to the domination of a few thousand British merchants, civil servants, and military personnel. In the historian William Shirer's words, "It was the only instance in history, I believe, of a private commercial enterprise taking over a vast, heavily populated subcontinent, ruling it with an iron hand and exploiting it for private profit."

While the British East India Company continued to handle the commercial arrangements, the British government had begun to assume a major administrative hand by the late eighteenth century. The tension between the ruling Britons (the "raj") and the Indian populations came to a head in 1857. Indian soldiers in the Bengal army of the British East India Company led a bloody revolt, which eventually involved a large section of India and resulted in the takeover of Delhi and the temporary installation of an Indian emperor. At great financial and human cost, the British eventually put down the so-called Sepoy rebellion. Thereafter, the British East India Company was replaced by the Queen of England—now the Empress of India—and by a succession of viceroys who administered the territory in her name. Following these wrenching hostilities, stirrings of Indian nationalism emerged, talk of an independent India increased, and some efforts toward home rule began. Still, throughout the nineteenth and well into the twentieth century, it was assumed that Britain would continue to rule India indefinitely; that a single determined Indian could lead his fellow citizens to independence was unthinkable.

GANDHI, A MORALISTIC CHILD

Two hundred fifty years after the British began to control India and a mere decade after the Sepoy rebellion, Mohandas K. Gandhi was born in 1869 at Porbandar on the Arabian Sea. His family came from the Vaisya caste, a middling stratum of Indian society engaged in trade and agriculture. Though not highly educated or wealthy, Gandhi's family had been prominent at the provincial levels. Six generations had served as home ministers or prime ministers on the Kathiawar peninsula, and a number had stood out by virtue of their high ethical standards. The society within which Gandhi grew up was quite conservative in religious and political matters; Gandhi himself felt it most improbable that he would ever assume a national leadership position, insofar as his family came from a small town and had available limited material and social resources. (Most of the other creators I am treating would have described their backgrounds in similar terms.) However, Gandhi's parents appear to have been unusual Hindus, open to a range of religious behaviors and beliefs.

The senior Gandhis set a high moral tone. Mohandas's mother, dutiful and saintly, fasted regularly and selflessly and displayed much courage—"to keep two or three consecutive fasts was nothing to her." Mohandas's father was stern and sometimes ill-tempered, but he adjudicated legal matters admirably and helped out with domestic responsibilities willingly. Once when young Mohandas confessed that he had stolen a golden chip from his brother's amulet, Gandhi's father took the guilt upon himself: Rather than punishing his wayward child, the father cried. This example of an injured individual who refrained from lashing out at others made a deep impression on the younger Gandhi.

As a child, Gandhi was puny, inclined to solitude, and reluctant to engage in sports. Not a particularly good student, he found school unappealing. He once described himself as follows: "I am an average man with less than an average ability. I admit that I am not sharp intellectually. But I don't mind. There is a limit to the development of the intellect but none to that of the heart." From an early age Mohandas displayed a notable interest in issues related to right and wrong. When playing games, he naturally gravitated to the role of peacemaker. At school, when asked to fake an answer to protect his teacher from a public embarrassment, the young boy refused to participate in this scheme. Instead, he sought a position of moral authority vis-à-vis his parents and other elders. Perhaps impressed by his evident talents in this area, his parents allowed him to serve as a moral arbiter even in realms where they might have been expected to assert their authority.

Precocity or prodigiousness can be readily recognized in certain domains. If the society provides opportunities, one is not surprised to learn of a child of five or ten who is outstanding in mathematical powers, musical performance, chess-playing skills, or mechanical or spatial abilities. When it comes to an understanding of other people and an ability to deal effectively with them, however, the markers are elusive. Childhood limitations in size, power, emotional breadth and subtlety, worldly experience, and knowledge of motivation all minimize the likelihood that the young child may appear precocious in the social, political, religious, or ethical realms.

That said, certain children no doubt are particularly attracted to issues that involve relations among other individuals, including questions of morality. Still, there is something a bit ludicrous (from a Western secular perspective) in the prospect of an eight- or twelve-year-old positioning himself to advise or handle squabbles among his elders. For whatever reasons (including a belief in the possibility of reincarnation), Gandhi seems to have been such a child; fortunately for him, his family gave him latitude in probing family relations and in developing his own responses to the social and ethical problems that arose each day. Thus, Gandhi had repeated opportunities to test himself as a moral agent.

Individuals who eventually become religious, social, or political leaders often seem to have been highly self-censorious; in Freudian terms, they have powerful superegos. Experiences that others might have forgotten or considered trivial take on a weighty significance. Such disparate figures as Saint Augustine, Martin Luther, Jean-Jacques Rousseau, and Abraham Lincoln dwelled on their youthful foibles and sought to atone for them even decades later. Gandhi was strongly affected by a youthful friendship with a Muslim youth named Sheikh Mehtab, who convinced him to violate Hindu principles and eat meat. The two boys also stole money to buy cigarettes and felt such guilt over this act that they contemplated suicide. Gandhi was also mortified that he had gone to a brothel, even though he had been rendered speechless and immobile and had made no advances to the woman provided for him. From his more worldly stance, the British essayist George Orwell attempts to put these youthful peccadillos in perspective: "a few cigarettes, a few mouthfuls of meat, a few annas pilfered in childhood from the maid-servant, two visits to a brothel (on each case, he got away 'without doing anything'), one narrowly escaped lapse with his landlady in Plymouth, one outburst of temper—that is about the whole collection."

Adult Responsibilities

Growing up in a society in which marriages were arranged during childhood, Gandhi wed Kasturbai when they were both thirteen. In some ways the marriage was appropriate; Gandhi's wife proved almost as tough and stubborn as her husband, and their marriage lasted for over a half a century. But still quite immature, Gandhi resented many aspects of this forced alliance in what he later called "the cruel custom of child marriage." Gandhi ardently desired his young wife and yet felt guilty about his lustful thoughts and deeds. These feelings were fanned enormously when Gandhi, who had been at the bedside of his dying father, retired to have sexual relations with his pregnant bride, only to learn shortly thereafter that his father had died. Gandhi never forgave himself for this act of filial disloyalty and considered the subsequent death of his first child shortly after birth as a suitable punishment for his misdeed.

An indifferent student, already married and with increasing family responsibilities, Gandhi might well have sunk into oblivion. Barely successful in passing an entrance examination, Gandhi was admitted to a small, inexpensive college in Bhavnagar in 1888. A family friend, Mavji Dave, convinced Gandhi that it made much more sense for him to travel to England to become a barrister at law. Accepting this advice and leaving behind his wife and first surviving son, Harilal, Gandhi embarked on a journey that was to alter his life.

A CASCADE OF LIFE OPTIONS

By electing to study in England, Gandhi had chosen to follow a forbidden course. The headman of his community admonished: "In the opinion of the caste, your proposal to go to England is not proper. Our religion forbids voyages abroad. We have also heard that it is not possible to live there without compromising our religion. One is obliged to eat and drink with Europeans." When Gandhi vowed to maintain Hindu practices and sought to defend his decision, the headman responded decisively: "This boy shall be treated as an outcast[e] from today. Whoever helps him or goes to see him off at the dock shall be punishable with a fine of one rupee four annas."

Whatever the psychic costs of traveling to England, of deliberately choosing a marginal path, the experience opened Gandhi up to a far greater set of options than would have been conceivable in provincial India. Like Eliot visiting Europe or Picasso traveling to Paris, Gandhi soon attained a more distanced view of his home base. Initially he was attracted to some of the surface features of his new world. He dressed like a dandy and installed a large mirror, in front of which he could spend minutes each day arranging his tie and parting his hair in the approved fashion of the day. To gain entry to desirable social circles, Gandhi also studied several languages, including French, and took dancing, locution, and violin lessons.

The attraction to a dandy's life proved relatively short-lived; fascination with a whole gaggle of new ideas proved more enduring. Gandhi read widely in Christianity, as well as in Hinduism. He learned about emerging ideological movements, such as Theosophy and pacifism. He joined organizations like the Vegetarian Society of England; indeed, the writer Ved Mehta notes that Gandhi seems to have developed an early taste for eccentricities of all kinds. He carried out experiments in eating, dieting, exercising, and otherwise maintaining his health. He kept careful records of everything that he did and all the money that he spent, in the process gaining and applying organizational talents. Though he was already quite abstemious (partly out of financial necessity), he permitted himself a trip to Paris to see the Great Exposition of 1889.

Paradoxically, the three years that Gandhi spent abroad in Europe both underscored the parochialness of his background and confirmed his Indian identity. He was learning about European practices, behaviors, and legal systems, not so that he could "pass for" an Englishman, as he might initially have wished, but rather so that he could one day stand on equal terms with individuals from all over the world. In this respect he was like Martha Graham, who also defined much of her subsequent creative activity in opposition to the formidable foreign models she had encountered as a young woman traveling through Europe.

The biographer Louis Fischer saw little resemblance between the "mediocre, unimpressive, handicapped, floundering M. K. Gandhi, attorney at law, who left England in 1891 and the Mahatma leader of millions." And, indeed, except for an incident in which Gandhi defended one Dr. Allinson of the Vegetarian Society against attempts to expel him for his espousal of certain contraceptive methods, there are few overt signs of political courage and leadership.

However, I think that this characterization misses an important dimension of Gandhi's British experience. In his sojourn abroad, Gandhi took advantage of the situation to absorb enormous amounts of written materials as well as generous dollops of lived experience. Upon his arrival in England, he had been familiar with only a small part of one country and a tiny sample of the world's religious ideas; he departed as someone relatively at home in a center of European civilization, who had read widely, had been exposed to a broad, if motley, array of current views, and had managed to break bread and hold his own in conversation with individuals from diverse lands and backgrounds. Such experiences prepared Gandhi to deal with the spectrum of individuals whom he would subsequently encounter, and to understand in an intuitive way the British and British-educated leaders with whom he eventually had to negotiate. In this relative cosmopolitanism, he resembled Chinese leaders like Zhou Enlai and Deng Xiaoping, and Soviet leaders like V. I. Lenin who had spent time in Western Europe; and he differed in key ways from leaders like Mao Zedong and Josef Stalin, who had never ventured beyond their homelands and lacked intuitions about other societies.

In 1891, Gandhi returned home to India, only to be informed upon his arrival that his mother had died almost immediately after learning that her son had passed the bar. Shortly thereafter, he met a young Indian named Raychandbai (later called Rajchandra by Gandhi) whose worldly success, spotless character, religious knowledge, philosophical thirsting, and capacity for self-reflection impressed Gandhi greatly. "No one else has ever made on me the impression that Rajchandra did," Gandhi was later to comment. The time spent with the mentoring figure Rajchandra helped convince Gandhi, at a vulnerable moment in his life, to remain a Hindu and to pursue a life of good works in his profession.

Nonetheless, unsuccessful efforts at litigation, an unfortunate effort at influence peddling on the part of a family member, and some advice from a British official convinced Mohandas that he had no future in India and that he ought to seek his fortunes elsewhere in the British empire—specifically, in South Africa. When an opportunity arose to travel to Durban to provide advice about a lawsuit, Gandhi hesitated little before accepting it and once again abandoning his growing family. One can see at work here an important facet of Gandhi's personality: When opportunity knocked, no matter at what distance and cost to self and family, he seized it.

MATURING IN SOUTH AFRICA

Gandhi arrived in South Africa as a young, inexperienced lawyer, hoping to establish his competence and some measure of worldly achievement before returning to his native land. He was in fact successful in his first legal encounters, thereby reconfirming his impression that compromise and reconciliation could be more effective than exploitation of an adversary's weakness. A series of unexpected events soon intervened, and Gandhi was to spend most of the next two decades engaged in political struggles in that distant land.

The Train Experience in Natal

A defining event occurred when, after a week in Durban, Natal, Gandhi decided to take a train to Pretoria in the Transvaal. In Maritzburg, Natal, a white man entered the compartment in which he was seated and refused to spend the night in the same space as the dark-skinned Gandhi. The conductor ordered Gandhi to a third-class compartment, but Gandhi refused to comply. Gandhi was removed from the train and forced to spend the night freezing in a railway station. During the remainder of the trip to the Transvaal, Gandhi continued to insist that he be allowed to travel first-class and to stay in a first-class hotel. Unsuccessful in his protestations, he became increasingly angered by this mistreatment. Gandhi decided then and there that the position of Indians as second-class citizens within South Africa was unacceptable. Within a short time he had organized a meeting of all the Indians in Pretoria to discuss a position that he considered to be untenable.

In later years Gandhi traced the origins of his political mission in life to the night that he spent shivering in the railway station at Maritzburg. As he recalled: "I thus made an intimate study of the hard conditions of the Indian settlers, not only by reading and hearing about it, but by personal experience. I saw that South Africa was no country for a self-respecting Indian and my mind became more and more occupied with the question as to how this state of thing[s] might be improved."

Working for the Indians of South Africa

The position of Indians in South Africa was nebulous at this time. Slavery had been abolished in the British Empire in 1833, and thereafter a supply of cheap labor was not available in South Africa. Indians, originally recruited to fill this void, had exceeded their charge. The Indians often competed successfully with European businessmen, a fact that alienated the British and Dutch power brokers, who worried about becoming a minority in "their" own country. Chiefly intent on making money, so that they could support their families and ultimately return home, the Indian residents concentrated scant energies on achieving political rights. Though technically considered citizens equal in status to those of European descent, in effect they came to be treated as "coolies," a form of labor to be disdained and exploited. Described in statute books as "semi-barbarous Asiatics," they were not allowed to walk on footpaths or to remain out at night without a permit.

The Indians living in South Africa represented a microcosm of the range of political, religious, and social groups that inhabited the Indian subcontinent. Most of them would have been inclined to ignore or forget mistreatment of the sort encountered by Gandhi on his trip from Natal to the Transvaal. Gandhi could not. He felt personally humiliated; the moralistically inclined young child, now abetted by firsthand knowledge of the rights and "civilities" of English life and by newly acquired skills of the courtroom, sprung into action. At the meeting that he organized, he impressed on Pretoria's Indians that they were a mistreated minority; their only hope for improving their political lot was to band together and thereafter to adhere to the highest possible standards of conduct. The first meeting with Pretoria's Indians was followed by a swarm of organizing activities, which led to certain modest victories. For example, the Indians were granted the right to travel first- or second-class, if they were "properly dressed."

Though Gandhi had expected to return to India within the year, he soon became convinced that he must remain in his new residence and he ultimately brought his family to South Africa. This change of plans was occasioned by his discovery that the South African government planned, in effect, to disenfranchise all Indian citizens. For the next twenty years Gandhi plunged into an almost ceaseless round of activities, all designed to improve the condition of Indians in South Africa. The issues that prompted struggle included the government's efforts to deprive Indians of the right to vote, to restrict emigration from the subcontinent, and to make Indians register, be fingerprinted, and pay taxes on indentured labor. The lines of authority were widely dispersed in South Africa. This state of affairs meant that Gandhi had options on how to proceed, but it also implied that a victory at one level might soon be undercut by a contrary decision at another level or in another arena. For instance, Gandhi attained some concessions from the British Secretary of State for colonies, Lord Elgin, only to have them abrogated in the wake of a governmental reorganization that permitted the state of Transvaal to enact whichever procedures it liked.

As much as possible, Gandhi proceeded through peaceful and legal means—writing petitions, holding meetings, launching organizations, arguing cases, and looking for legal loopholes and means. He traveled back and forth to England and to India to secure support. He was initially a novice in each of these activities; where he could not find a master on whom to model himself, he sought to improve his performance through self-study and self-observation. In this process Gandhi discovered a remarkable capacity to rely on himself when necessary. While he was not always loved, he was widely respected for the dogged yet calm way in which he pursued his ends.

Gandhi often placed himself at risk. In 1897 he was beaten almost into unconsciousness and virtually lynched by a white mob on the streets of Durban. Characteristically, he felt sorry for the ignorant individuals who had attacked him and did not press charges. In 1908 he was sent to jail for the first time. The conditions there were quite primitive, and Gandhi felt desolate because the supports on which he had come to rely had been removed; he was not able to attend meetings, negotiate with supporters, or plan for future confrontations.

Far from intimidating Gandhi, these encounters with harsh and sometimes unyielding reality strengthened the resolve of the maturing barrister. Despite temptations, he refused to sacrifice his principles. Once extraordinarily shy, he became an increasingly accomplished public speaker. He formed a number of organizations that allowed him to practice his emerging leadership skills. He became a persuasive writer and contributed to many periodicals both in South Africa and abroad, most notably *Indian Opinion*, which was published in English and in the Indian language of Gujarati. He learned how to raise funds for his political efforts. In ever-widening circles he was respected as a man of honor and as a person who could get things done in a way that minimized ill will.

A Change of Course in Midlife

According to Hindu teaching, the period of early adulthood is a time when a man is very much in the center of activity. Gandhi was certainly a prototypical Indian householder in this sense. He was sufficiently successful as a lawyer that he could live well, raise his children comfortably, and assemble a large and skilled staff. (Significantly, he did not take fees for his public-interest work.) He lived a highly organized life, timed almost to the minute, and scrupulously kept track of all the activities that took place under his aegis. He belonged to many organizations, ranging from the Natal Indian Congress and the Indian Educational Association to the Esoteric Christian Union and the London Vegetarian Society.

It might seem that by 1905 or 1906 everything had come together for Gandhi, the once lackluster young attorney with few prospects. Indeed, the now successful, but still young, Indian might have either rested on his laurels in

South Africa or returned home triumphantly to India, to build up his law practice further or, following in the steps of his forefathers, to assume a supporting
role in politics.

Gandhi did not feel a sense of accomplishment, however. If anything, he
felt frustrated and in the deepest sense, unfulfilled. He lamented the fact that
others did not follow his personal example and blamed this state of affairs on
his failure to exemplify his own life principles. Hindu thought features a role
contrasted to the engaged householder—that of the withdrawn religious ascetic,
or Vanaprastha. Gandhi concluded that this portion of his persona was underdeveloped. Around 1905 he began to read a different vein of literature, especially
the more spiritually oriented writings of the English social theorist John Ruskin
and the Russian novelist Leo Tolstoy, plus the writings on civil disobedience by
the American social thinker Henry David Thoreau. And soon enough, this
activist was seeking opportunities to translate his evolving philosophical and religious views into practice.

Abandoning a busy home and professional life in Johannesburg, Gandhi
moved with his wife and four sons to a farm he called Phoenix House on the
outskirts of Durban. Gandhi sought consciously to simplify his life. He performed daily exercises and began to prepare his own food. He immersed himself
in issues of health and medical care, even delivering his last children. Then in
1910 he founded Tolstoy Farm, an eleven-hundred-acre development twenty
miles from Johannesburg.

At Tolstoy Farm all seventy-odd inhabitants, representing different religions
and drawn from many regions of India, were expected to live as members of a
joint family, in an ascetic, cooperative, and morally exemplary fashion. Placing
himself in charge of spiritual and intellectual education of the youngsters,
Gandhi promoted a Gujarati, rather than a European, model of education.
When two boys engaged in sodomy, Gandhi hit on a solution that felt right to
him: He himself undertook a fast. When a young boy and girl slept together, he
cut off the girl's hair and again fasted. He formed alliances with Sonya Schleslin
and Henry L. S. Polak, two gifted, idealistic individuals who sacrificed an independent existence to help Gandhi realize his ideals in living. And he took vows
of self-control, or *brahmacharya*, according to which he had to relieve himself of
all possessions, remain poor, and refrain from sexual involvement. "It became
my conviction," he later recalled "that procreation and the consequent care of
children were inconsistent with public service."

This shift to the life of a spiritual leader is perhaps less strange for an Indian
than for a Western man of affairs, since numerous models of such a retreat from
the external world can be found within Hindu culture. Nonetheless, it calls for
some explanation. My own view is that Gandhi did not feel that he could proceed as an ethical agent, attempting to achieve a better position for his people,
unless he himself had attained and come to embody some kind of moral author-

ity. Only by leading an exemplary life, and by attempting to influence those around him to do the same, did Gandhi feel like he had attained the necessary degree of spiritual purity; and only when such purity had been achieved did he believe that he had the moral authority to make demands on others within the public arena.

As he made these crucial decisions about how to lead his life, Gandhi was forging a number of bargains—with his people, his God, and himself. In effect, he was publicly renouncing many of the pleasures of life and assuming the existence of an ascetic in order to model for others the highest standards of conduct. While creators engaged in isolated work can forge such a covenant privately, those whose work directly affects the behavior of others may have to realize their Faustian bargains in a distinctly public way, embodying what they preach.

Part of Gandhi's personality led him to an ascetic life, removed from the centers of power; yet he also felt strongly attracted to political involvement and protest. In 1909 he wrote "Indian Home Rule" ("Hind Swaraj") a political tract in which he propounded an aggressive defense of nonviolent protest; argued against a machine-centered civilization; criticized the advent of a European secular society, whether in Britain or India; and called for a life of simplicity, traditional values, and abstinence, all to be pursued in small villages. As he put it: "The true remedy lies, in my humble opinion, in England's discarding modern civilization which is ensouled by this spirit of selfishness and materialism which is purposeless, vain, and . . . a negation of the spirit of Christianity."

Refining the Protest Process

During the final years of his South African stint, Gandhi became significantly more confrontational. In the first of a series of potent symbolic actions, he burned his registration certificate publicly. He led a series of protest marches in which five thousand individuals functioned as an army of peace. Gandhi sought to fill the jails with his compatriots in order to expose the anti-Indian nature of the laws, including one especially nefarious statute that decreed Indian marriages illegitimate. For the first time this protest came to include women. Gandhi was arrested several times and sent to jail. When Indians heard about these indignities toward the leading figure of their community, many of them went on strike.

Perhaps most important, the method on which Gandhi was working at this time ultimately came together in his practices of *satyagraha*, a genuinely new form of protest that he was to perfect during his first years back in India. As early as 1906, Gandhi asked his followers to resist laws that they believed to be unjust, by picketing and disobeying ordinances, but also by refraining from violence and not resisting arrest. The followers stayed the course, even if it cost them their lives. Gandhi began to sense that this resistance should not be merely a passive stance; it should be an active, life-affirming, positive force.

Gandhi's activities in South Africa made headlines in England and in India. Through regular statements to the press and through the judicious timing of cables, Gandhi insured that all who should know were kept informed about the mistreatment of Indians in South Africa and about their peaceful resistance. Pressures mounted for some solution to a situation that many saw as unfair and that nearly all saw as acutely embarrassing for the government.

Though holding no official portfolio, Gandhi was able to work toward a settlement with General Jan Christian Smuts, the head of the South African government. All Indian marriages were recognized as legitimate, and various taxes and indignities were canceled. On the surface the relations between Smuts and Gandhi were cordial, and on more than one occasion, Smuts indicated his respect for the Indian political leader. But when Gandhi returned home to India for the last time in 1914, Smuts declared: "The saint has left our shores, I hope forever."

Gandhi had achieved some success in aiding the Indians in South Africa, but probably his greatest accomplishments were personal. During his two decades away from home, he had evolved from a shy and ineffectual barrister to a political force to be reckoned with. Alternating his focus between actions in the public arena and private spiritual growth, he was always fully engaged. He developed philosophical positions and then lived them. His experiments were carried out partly inside his own head, but partly too in consort with other individuals, whom he attempted to engage in his several missions. Gandhi studied and learned from each of these organizing activities, whether successes or failures, and made use of lessons during the succeeding encounters. Scientific creators like Einstein work primarily with the organization of concepts; artists like Stravinsky or Picasso are engaged with conception, execution, and revision of works in an already evolved symbol system; for people like Gandhi, political creators, the core of their creative work lies in the capacity to mobilize other human beings in the service of a wider goal, often undertaken at great personal risk. Their personal actions are their medium of expression, central to their mission.

LEARNING THE LAY OF THE LAND IN INDIA

Although he had spent his childhood in India and had returned home regularly, Gandhi was a virtual stranger to the land that he reached in 1914, on the eve of the First World War. Now forty-five, he had spent most of the last twenty-six years away. Gandhi was known in political circles because of his well-publicized protests in South Africa, and the dominant Congress party had actually offered him a leadership position upon his return. Wherever he went, people flocked to see him and to receive his *darshan*, or blessing.

At the time, his closest political ally and his mentor in matters Indian was a leading Congress party official named Gopal Krishna Gokhale. Gandhi had known Gokhale since 1896; they had remained in close contact, and Gokhale had often represented Gandhi in India during the South African years. In a well-motivated but ambivalent gesture, Gokhale had extracted from the recently returned Gandhi a promise that he would not speak publicly for a year. This political respite gave Gandhi an opportunity to travel widely throughout India (third-class railway coach—by choice, of course), to familiarize himself with the conditions in the country, and to evaluate the courses of action he might follow in coming years and in light of whatever opportunities were to present themselves. Unfortunately, Gokhale died within the year, so Gandhi lacked a trusted older person on whom he could try out his still-untested ideas.

A peculiar coincidence complicated Gandhi's first years back in India. On the one hand, he knew that he wanted somehow to use the methods of protest developed in South Africa in order to better the conditions in India. As Gandhi put it: "I wanted to acquaint India with the method I had tried in South Africa and I desired to test in India the extent to which its application might be possible. So my companions and I selected the name 'Satyragraha Ashram' as conveying both our goal and our method of service." At the same time, however, Britain was at war against the Axis powers, and Gandhi felt as a matter of principle that Indians should be loyal to their government of record. Thus, Gandhi desisted from engaging in any kind of civil disobedience against Britain during the period of the First World War. Some saw this decision to suspend a struggle as a sign of weakness within the Indian ranks. As Lord Lamington declared: "The real guarantees of our stay in India remain as strong as ever, viz., the caste system, the diversity of nationalities and creed, and the lack of confidence and trust of one native for another." Still, the war forced Britain to give India some political concessions and to delegate some powers, steps both necessary and appropriate at a time when over one million Indians were fighting under the Union Jack and over a hundred thousand were to lose their lives.

In the case of political and ideological innovators like Gandhi, especially ones who lead by personal example, it is possible to point to dozens, if not hundreds, of points when their creed has been articulated and their principal practices have solidified. We have already encountered a number of such defining moments, beginning with Gandhi's moralistic encounters during childhood, his defense of the hapless Dr. Allinson at the Vegetarian Society, the night at the train station in Maritzburg, the initial protest meetings in Durban, the kernels of the idea of satyagraha, the first fasts at Tolstoy Farm, and the jailings and public marches during the last years in South Africa. We can also point to early events upon Gandhi's return to India, such as civil disobedience conducted with peasants in Champaran, nestled at the foot of the Himalayas. In this instance, Gandhi called attention to an unfair practice whereby peasants had been legally compelled to dedicate 15 percent of their land to producing indigo for their landlord. During this proceeding, Gandhi came to a realization: "I had to disobey the British law because I was acting in obedience with a higher law, with the voice of my conscience. This was my first act of civil disobedience against the British."

Gandhi himself contributed to this impression of a long series of "experiments with truth," each of which might be deemed an additional stone placed on the emerging edifice of satyagraha. And indeed, in the case of individuals who are conducting their experiments in the public arena, where events necessarily consume much time and have an uncertain course, the ultimate shape of practices will emerge only from many tentative and partial rehearsals and attempts over the years. In this sense, their creative growth appears more gradual, less epochal, than that of creative individuals who traffic exclusively in the world of concepts or in the production of discrete artistic works, an occasional one of which seems pivotal.

A Pivotal Event

All this said, I agree with Erik Erikson that events in Ahmedabad in West Central India in 1918 were central to the formation of the Mahatma (or "great soul"), as he came increasingly to be called by his admiring countrymen, including the great poet Rabindranath Tagore. It is worth taking a close look at the particulars of what Erikson dubbed "The Event"; we can then step back to consider more broadly the nature of Gandhi's considerable achievement and his deepening understanding.

On the surface, The Event was a labor dispute in an area often dubbed "The Manchester of India." During a time of inflation, high profits, and high taxes, textile workers at the mills of the Sarabhai family and of other nearby mill owners felt that they were not receiving adequate compensation. Anger, unrest, and an ardent desire to correct this sensed inequity emerged. At the urging of local authorities (but probably, as well, with a strongly held intuitive sense that he could make a difference here), Gandhi intervened; he forged an agreement that both the laborers and the management considered satisfactory.

This event was notable in a number of respects. The first unusual feature was the identity of the mill owners. The Sarabhais were a distinguished Indian family who had carried out many good works and were in fact serving as hosts to Gandhi during his residence at an ashram, a kind of reborn Phoenix farm, in this locale. The landowner, Ambalal Sarabhai, led the mill owners, while his sister Anasuya was sympathetic to the strikers and a strong supporter of Gandhi and his methods. As the biographer Robert Payne describes the confrontation, "For Anasuyabehn Sarabhai it was a question of justice; for the mill owners it was a question of profits; for Gandhi it was a question of testing the resources of Satyagraha."

A second distinguishing feature was the nature of a strike that was ultimately undertaken when early attempts at arbitration had failed. Gandhi asked the workers to take a pledge to undertake no violence, no molesting of strike breakers, no begging; they consented to remain firm and to earn bread by other means during the strike. For two weeks the strikers showed exemplary courage and self-restraint: They attended daily meetings where Gandhi, seated under a spreading banyan tree, exhorted them to remain faithful to their pledge. But then, amidst signs of weakening, Gandhi began to fear that the laborers' resolve would not hold.

The third unexpected feature was the turn taken by the protest. Instead of continuing the strike, with its risks of escalating demands or of unacceptable violence, Gandhi decided to fast. Having analyzed the situation carefully, Gandhi thought that he had arrived at a solution that was truly fair to both sides, as well as the patterns of behavior that each antagonist should follow. To underscore the strength of his conviction, he elected to put his own being, his own life on the line:

> In my opinion I would have been untrue to my maker and to the cause I was espousing if I had acted otherwise. . . . I felt that it was a sacred moment for me, my faith was on the anvil, and I had no hesitation to rising and declaring to the men that a breach of their vow so solemnly taken was unendurable by me and that I would not take any food until they had the 35 per cent increase given or until they had fallen. A meeting that was up to now unlike the former meetings, totally unresponsive, woke up as if by magic.

The actual decision to fast had come unannounced: "The words came to my lips," Gandhi reported. Gandhi rejected the strikers' interest in fasting alongside him, preferring to have them adhere to the pledge that they had already taken. It was his first fast for a public, as opposed to a private, cause. And it revealed to Gandhi the drama, potency, and simplicity of a new weapon.

The final remarkable feature was the outcome. At first the mill owners were infuriated by the strike: Gandhi said, "[They] received my words coldly and even flung keen, delicate bits of sarcasm at me, as indeed they had a perfect right to do." After three days of the fast, the mill owners set about to discover some means of settlement. The mill owners had insisted that they would give no more than a 20 percent increase in pay, while the mill hands were holding out for a 35 percent increase. Gandhi craved a solution in which each party felt that it had achieved moral legitimacy, and so he hammered out an agreement in which the mill hands received an increase of 35 percent one day (thus satisfying their own analysis of the situation), an increase of 20 percent the next day (thus satisfying the mill-owners' analysis), and then a perpetual increase of 27.5 percent (the arithmetical compromise). With the settlement reached, the mill hands returned to work. And more satisfyingly, a method of arbitration that lasted for decades was put into place shortly thereafter.

From this seemingly accidental concatenation of events, both Gandhi and India achieved much. As the biographer Judith Brown puts it:

[Gandhi's] Ahmedabad campaign demonstrated not only the viability of satyagraha in a further type of conflict but also many of the characteristics of his campaigns which were to recur wherever he had some real control—the search for a peaceful solution at the outset, the sacred pledge as the heart of the struggle, strict discipline and self-improvement among the participants, effective publicity, generation of an ambience of moral authority and pressure, and finally a compromise solution to save the face of all concerned.

The Significance of the Ahmedabad Event

If we identify the Ahmedabad event as a pivotal one in the development of the Mahatma, the question arises about the kind of support that Gandhi needed to carry it out successfully. My earlier examinations of creative breakthroughs would suggest the presence of a single individual (like Georges Braque) or a small group (like Einstein's Olympiad or Diaghilev's circle), who provide the affective support as well as a cognitive sensitivity to what is being undertaken. To the extent that the creators are working out a new language, the sustainers must share in the decoding process and help the creators see that what they are attempting to express can make sense to others, as well as to the creators themselves.

From one perspective, Gandhi's experiences at Ahmedabad may appear remote from those of Eliot in writing *The Waste Land*, or of Freud in working out the details of "The Project" or interpreting his own dreams. I do not wish to diminish the differences. What, for Freud or Eliot, is a largely private activity, one occurring over a relatively short time, is a distinctly public activity for Gandhi, one with stronger links to earlier (and to later) protests.

Yet, at a deeper level, there may well be strong affinities between the two kinds of experiences. By his own testimony (which I examine in a moment), Gandhi felt that he was working out the principles of a new language, or even a new axiomatic system, during the period when satyagraha coalesced. Having pondered issues for years, the decision to fast came to him just as the "sacred chord" came to Stravinsky's prepared ear. Under such stressful conditions, it may well have been necessary for him to have a supportive, familylike environment in which to accomplish this feat. Here is where the Sarabhai family becomes crucial. Gandhi was not dealing with a remote or impersonal combatant; rather, he was in the daily company of members of a family, with one of whose members he was closely allied, while he found himself temporarily at loggerheads with another, respected member of that same family. These individuals could read one another in the same way that mother and child can readily interpret one another's sounds and gestures: Gandhi could be confident that his "language" was understood by friend and adversary. From one perspective, Gandhi may look like the creator who invented himself; but from another perspective, he benefited from a familylike support system during the moments of his most daring and decisive breakthrough.

The Ahmedabad situation had been remarkably successful, though Gandhi came to acknowledge later both the unique circumstances there (his personal friendship with the leading protagonists), and the coercive nature of the fast, which he had undertaken to capture the attention and conscience of his friends. In the years immediately following the Ahmedabad strike and fast, Gandhi encountered—and perhaps encouraged—a number of situations that called on the full arsenal of his powers.

The Expansion of Militant Efforts

The record of this period was mixed. Gandhi proved only partially successful in dealing with the Rowlatt bill, a hostile continuation of wartime restraints on Indian political life. In what he termed "the greatest battle of my life," the man who was emerging as India's central protestor led a nationwide *hartal,* or strike. The first national organizational effort directed explicitly against the British, the Rowlatt protest ultimately ended in violence: Indians killing one another, Indians killing British, and the notorious Amritsar massacre, in which Brigadier General R. E. H. Dyer's troops opened fire against the assembled crowds, injuring or killing almost sixteen hundred individuals.

In the course of an amazingly full three years after Ahmedabad, Gandhi made several additional attempts to unite Indians, for their own sake, and to confront British hegemony. Gandhi sought to unite the Muslim and Hindu forces in the Khilafat agitation of 1919 and 1920, and he defended two Muslim journalists who had been interned during the war; but he never succeeded in reconciling these implacably opposed religious parties. He was particularly distraught by an Indian mob that committed murder in Bardoli early in 1922. Gandhi conceded his errors in some of these organizing activities, including a confession of "a Himalayan blunder": He had assumed that Indians were capable of nonviolence, but he had instead discovered that many were incapable of maintaining this demanding form of self-restraint.

In March 1922, Gandhi paid a heavy price for his increasingly militant organizing activities. He was placed on trial for sedition—for inciting Indians to riotous and murderous acts through his written words, his confrontational stance vis-à-vis Britain and its representatives in India, and, implicitly, for his increasing authority over the Congress party. Judge Robert Broomfield knew and liked Gandhi and paid tribute to a man of "high ideals and of noble and even saintly life." At the same time, the judge signaled that his personal feelings could not weigh in the trial or the sentencing, and he forecast the ultimate outcome of his distasteful task. Gandhi cooperated by pleading guilty and waiving the right to counsel.

Gandhi made one request: that he be allowed to speak on his own behalf. Granted this wish, Gandhi spoke with unprecedented eloquence about his position. While affirming that nonviolence was his creed, he took responsibility for the violent crimes that had taken place. He announced his willingness to accept the harshest penalty: "The only course open to you, the Judge, is, as I am just going to say in my statement, either to resign your post, or inflict on me the heaviest penalty." He then laid out step-by-step the entire history of his time in South Africa and India, indicating his loyalty to Britain throughout the period and his shock at the Rowlatt Act—"a law designed to rob the people of all real freedom." He chronicled the massacres that had taken place in India and reported his conclusion that the British connection had made India helpless. Documenting the exploitation of the masses, Gandhi concluded, "I have no doubt whatsoever that both England and the town dwellers of India will have to answer, if there is a God above, for this crime against humanity." Claiming to have no personal animus against any particular administrator or the king, Gandhi declared: "I hold it to be a virtue to be disaffected towards a Government which in its totality has done more harm to India than any previous system."

The diplomatic, but deeply felt, exchange between defendant and judge ended in the foreordained way. Noting that Gandhi had made his job easy by pleading guilty, Judge Broomfield said: "It will be impossible to ignore the fact that you are in a different category from any person I have ever tried or am likely to have to try. It would be impossible to ignore the fact that in the eyes of millions of your countrymen, you are a great patriot and a great leader." Broomfield went on to pronounce the lightest sentence that he thought permissible—a sentence of six years. Gandhi responded: "So far as the sentence is concerned, I certainly consider that it is as light as any judge would inflict on me, and so far as the whole proceedings are concerned, I must say that I could not have expected greater courtesy."

While Gandhi's followers embraced him and viewed him with great admiration, he was about to suffer the most awful penalty—being removed from his political movement at the time when his skills of leadership were most needed. Such a decapitation of the Indian protest movement had been precisely the aim of the entire judicial proceeding. Never one to fail to seize the opportunity, however, Gandhi treated his enforced imprisonment as a respite: He could reflect on what had happened, read and think widely, and prepare for the next steps in a campaign that would not cease until India had achieved complete separation from England. For, in the years since the war, Gandhi had reluctantly but firmly come to the conclusion that no lesser goal would suffice.

THE PRINCIPLES OF SATYAGRAHA

Gandhi once testified that action was his domain; in so doing, he was designating the core of his being. His most enduring legacies to posterity are his specific achievements in helping liberate India from British rule and his universal demonstration that vexed problems can, at least at times, be addressed and alleviated through methods of nonviolent resistance. Yet Gandhi could appropriately have enlarged his domain to include reflections and writings: An equally important sphere of operation was his deep thinking about human nature. As his biographer Brown puts it: "Like many great visionaries he combined the contemplative calling with a capacity and need to be involved in active, even frenetically active work." Any effort to understand Gandhi must encompass the constant and productive dialectic between reading, writing, and reflection, on the one hand, and leadership by courageous personal example, on the other.

The Philosophical Base

While hardly a systematic or exhaustive scholar, Gandhi did read widely, especially at times of imprisonment or of voluntary withdrawal from the active life. Early in his life he read religious writings—the Bible, the Koran, Theosophy, and especially the Bhagavad-Gita—as well as secular texts of the period, such as the writings of Thomas Carlyle, Ralph Waldo Emerson, and Thomas Huxley. By his own testimony, as already noted, he was most influenced by three writers encountered during his South African years. From Thoreau he received insights into the philosophy of civil disobedience and the ways that the individual can withstand pressures from the state. From Ruskin, still alive when Gandhi studied in England, he gained an understanding of the social dimensions of human action: Ruskin alerted him to the centrality of labor (from the most exalted to the most humble forms), relations among the privileged and the poor, and the tension between economic and human factors.

But Gandhi was most profoundly affected by the writings of Tolstoy. His reading of Tolstoy not only weaned him forever from the recourse to violence as a means of achieving ends but also impressed on him the need to focus on one's duties rather than on one's rights, the importance of one's relation to one's God, and the centrality of love in all human affairs. Gandhi was proud that he had been able to establish a correspondence with the great Russian thinker in the period immediately preceding the latter's death. Tolstoy was also impressed: Praising Gandhi personally and to others, Tolstoy said that Gandhi's approach to nonviolent protest was "a question of the greatest importance not only for India but for the whole of humanity."

I do not believe that Gandhi needed to read any—let alone all—of these texts to arrive at his principles and his principal modes of operation. Over the decades they sprang organically from his being. Nonetheless, such writings inspired Gandhi and helped him place himself within a field of roughly contemporaneous figures who had wrestled in different corners of the world with many of the same issues that occupied him. At greater remove, Gandhi also identified with the most important religious and spiritual leaders in human history, though he scrupulously denied the existence of any godlike characteristics in himself. Paradoxically, while relatively hermetic artistic revolutionaries were actively engaged with their immediate contemporaries, the creator whose contribution was most intimately associated with other human beings received the greatest inspiration from individuals whom he had never met.

Asked at various times about his own creed, Gandhi stressed a few basic dimensions: a quest for truth, morality, and spiritual renewal—stressed in the Bhagavad-Gita—were fundamental to his being. Gandhi could not treat as separate the search for a meritorious life in the personal sphere from the search for an exemplary life in service of one's community. Personal freedom would coincide with freedom in the service of society; personal nonviolence would be equated with nonviolence in the larger arena; a personal quest for truth, knowledge, and understanding would take its place within a community dedicated toward those same ends.

As Erikson stresses in his study of Gandhi, the religious innovator is an individual whose solution to personal dilemmas ultimately makes sense to a much wider community. Certainly, the struggles within Gandhi were eventually to resonate with struggles preoccupying India and, indeed, reverberating throughout much of the developing world. The burdens he took on himself ultimately became ones that others could assume as well. Erikson ponders rhetorically about "why certain men of genius can do no less than take upon themselves an evolutionary and existential curse shared by all . . . and why other men will be only too eager to ascribe to such a man a God-given greatness surpassing that of all others."

Most individuals think of Gandhi primarily as a religious figure, but Gandhi's view of religion was very different from the kind of sectarian territoriality that label often evokes today. He believed that having a religious vision was a mark of being human. Far from having a theological axe to grind, he endorsed a study of all the major religions, discerning good as well as limits in each, and ultimately favoring a generic creed that recognized a fundamental divinity pervading all things. Consistent with the Hindu upper-caste background, he stressed a life in which the more privileged would aid the less fortunate; but he became increasingly critical of the notion of castes and devoted much energy to the restoration of dignity to the so-called Untouchable population. "I regard untouchability as the greatest blot of Hinduism," he once asserted.

Perhaps Gandhi's most provocative ideas concern the relation between religion and politics. We in the West honor—at least in theory—a longstanding dissociation between church and state. Not so for Gandhi. Throughout his writings, Gandhi underscored the indissolubility of these two realms—in our terms, the two usually estranged domains. On the one hand, he declared: "For me politics bereft of religion are absolute dirt, ever to be shunned." But he was equally dismissive of religion that takes no account of practical problems and does not help to solve them: "I take part in politics because I feel that there is no department of life which can be divorced from religion and because politics touch the vital being of India almost at every point." He once remarked pointedly: "I can say without the slightest hesitation, and yet in all humility, that those who say that religion has nothing to do with politics do not know what religion means."

An individual of Gandhi's complexity and subtlety cannot be reduced to a unique philosophical message or a single political or religious practice. Even his own attempt falls short: "What I want to achieve—what I have been striving and pining to achieve these thirty years—is self-realization, to see God face to face, to attain Moksha [roughly, "oneness with God"]. . . . All I do by way of speaking and writing and all my ventures in the political field, are directed to this same end." (This expression resonates astonishingly with Einstein's hubristic desire to know God's thoughts.) Yet, situated at the center of Gandhi's being is satyagraha, the practice he developed and refined over several decades. As much as Einstein is forever yoked with relativity, Freud with the unconscious, or Picasso with cubism, to that extent is Gandhi the master of satyagraha.

The Nature and Practice of Satyagraha

Gandhi first spoke explicitly of satyagraha in South Africa. He employed the phrase to express the force, born of truth and love, that Indians had mobilized over the years to call attention to the injustices inflicted on them and to bring about a more humane and equitable set of relations among the players in the region. Satyagraha assumes the existence of a community in which two or more parties find themselves in disagreement or opposition. Instead of confronting one another through violence, pain, or the threat of harmful encounters, proponents of satyagraha mobilize the reasons and conscience of their opponents by inviting suffering on themselves. The satyagrahis hope thereby to convert the opponents and make them willing allies.

As Gandhi saw it, satyagraha is a form of purification, an assertion of the spirit. Self-inflicted suffering dramatizes the sincerity and plight of the satyagrahi and seeks to convince the opponent of the legitimacy of the satyagrahi's case: "No matter how badly they suffered, the Satyagrahis never used physical force." Gandhi expresses succinctly the difference between traditional confrontations and those governed by satyagraha: "If by using violence I force the government to repeal the law, I am employing what may be termed body-force. If I do not obey the law and accept the penalty for its breach, I use soul force. It involves sacrifice of self."

In general, satyagrahis have a particular goal; they may compromise in the achievement of that goal so long as fundamental principles have not been undermined. But the struggle also has wider political significance: It is used to educate all members of the community and to encourage decision making and government by consensual means. Ideally, satyagrahis display a belief in themselves rather than a coercion of the opponent, but Gandhi acknowledged the coercive element latent in a nonviolent protest.

The fundamental aspects of satyagraha were not original with Gandhi; indeed, they could be traced not only to his own mentors, Tolstoy and Thoreau, but to religious leaders like Christ and to philosophers like Socrates. What Gandhi contributed was a detailed working out of the scenarios by which satyagraha could work, at least in the Indian-British contexts with which he was familiar. Indeed, he actually specified codes that followers were expected to adhere to—for example:

- Harbor no anger but suffer the anger of the opponent. Refuse to return the assaults of the opponent.

- Do not resist arrest nor the attachment of property, unless holding property as a trustee.

- As a member of a satyagraha unit, obey the orders of satyagraha leaders and resign from the unit in the event of serious disagreement.

Gandhi identified the conditions under which satyagraha made sense: (1) a situation of complexity where return to an elementary "first principle" can be clarifying and helpful, and (2) an issue where one's opponent is vulnerable and where there is widespread moral agreement that something is amiss. He specified the crucial ingredients of self-discipline, self-control, and self-purification, as well as loyalty and obedience to the leader. Individuals must understand their own motives for being involved in a satyagraha and their obligations to others involved in the encounter. Gandhi explained: "A satyagrahi differs from the generality of men in . . . that, if he submits to a restriction, he submits voluntarily, not because he is afraid of punishment, but because he thinks such submission is essential to the common weal." Gandhi underscored the need to be sensitive to one's opponents' needs and even to help them when they lose their freedom to be autonomous counterplayers. Above all, the objective evil must never be equated with the persona of the individual in the group associated with the injurious practice.

The satyagraha process begins with efforts to persuade through discussion and reason. Early stages should allow accommodation and compromise. Should such efforts at conciliation break down (as they did in Ahmedabad, for example), one evolves to persuasion through suffering (for example, through fasting). At this time one tries to dramatize the issues at stake and to gain the attention of the wider community; the hope is to be able to return once more to a process of discussion. If persuasion by reasoning or suffering still fails, the satyagrahi may resort to nonviolent coercion, characterized by techniques such as noncooperation or civil disobedience. The risk here, of course, is that nonviolence may give way to violence, or that the two parties may become so estranged that any nascent agreement will prove difficult to maintain in the long run. In a compelling figure of speech, Gandhi described the delicacy of this process: "The rope dancer, balancing himself upon a rope suspended at a height of twenty feet, must concentrate his attention upon the rope and the least little error . . . means death for him. . . . [A] satyagrahi has to be, if possible, even more single-minded."

Finally, Gandhi noted the limitations of satyagraha. It cannot be adopted in every situation, because some situations lack the moral clarity, some lack the disciplined actors or followers, and all too many lack an opponent with a sense of fair play: Satyagraha, Gandhi noted wistfully, does not work with tyrants. The coercive elements latent in satyagraha must be used sensitively, or else one risks destroying the opponent or rendering subsequent encounters dysfunctional.

Despite these limitations, Gandhi had perfected a method of enormous precision and power, one that had incalculable consequences in India during the first half of the twentieth century and one that may yet exert even greater influence in the world community, given people who can adopt it and use it with conviction. Gandhi declared: "I have the unquenchable faith that, of all the countries in the world, India is the one country which can learn the art of nonviolence . . . that . . . if the test were applied even now, there would be found, perhaps, thousands of men and women who would be willing to die without harboring malice against their persecutors."

An Unpredictable Human Drama

How, in a study of human creativeness, should we think of satyagraha? It is misleading to view it simply as a set of interlocking concepts, and equally misleading to think of it solely as a body of established practices to be enacted by a master performer with supporting cast. What distinguishes satyagraha and makes it an impressive human accomplishment is precisely the fact that it represents a philosophy in practice. Through his analysis of human interests and his experiments with forms of interaction, both conducted over a period of several decades, Gandhi arrived at a process that was as precise in its way as a ballet that has been choreographed and performed, or a mathematical expression that has been formulated and tested. Each of the features at a given historical moment, and each of the features as it evolves over the course of the interaction, must be scrupulously enacted and carefully monitored; but in contrast to a stylized ballet or a worked-out equation, there can be no algorithm for a successful outcome. Rather, on the basis of the underlying principles *and* the spontaneous actions and reactions of the participants, the next step of the course must be planned and executed. Thus, the performance is truly a high-stakes one.

We may think of satyagraha, then, as a process that is in some respects ritualized but that also must in ways be improvised as well. It is perhaps best conceptualized as a human drama—a new form of ritualization in which, guided by strict principles of conduct but cognizant of the unpredictability of events, individuals can relate to one another confidently. Gandhi's understanding of the personal realm—that of one's self and the others around one—was crucial to this process, as were his abilities to reason logically about options, to put his ideas into words, and to alter course when indicated. The symbol system with which he worked out the dynamic principles of satyagraha necessarily entailed these human intellectual competencies.

THE PERSONAL SIDE OF GANDHI

Few other leaders in any era have laid out their innermost thoughts more candidly than Gandhi. Throughout his writings, and particularly in his autobiographical *Experiments with Truth*, Gandhi reviewed his own behavior, thoughts, and motivations with precision and sincerity. Such confessional writing had a double effect. First, it allowed Gandhi to reconcile himself with his own history, his current being, and his aspirations for himself, for his people, and for the rest of humankind. In addition, it served as a kind of sharing of his life with his personal companions and with others anywhere who had an interest in his methods.

A comparison with Freud comes to mind. Both men spent much time analyzing the events of their lives and writing about their conclusions. They often seized upon apparently small matters—slights, anxieties, dreams—and strove to make sense of them. And as intuitive psychologists, they were curious about the mechanisms that motivated human behaviors. His biographer Bal Ram Nanda has characterized Gandhi's method of reflection: "In every case he posed to himself a problem for which he sought a solution by framing a proposition in moral algebra. 'Never again' was his promise to himself after each escapade. And he kept his promise." Gandhi himself commented: "Such experiments are an integral part of my life; they are essential for my mental peace and self-realization."

Except perhaps in certain specific respects, Freud did not view himself as a model for other human beings. Gandhi did set himself up as a model, though *not* in a self-serving way. He constantly experimented and wrote about himself as a means of inspiring others to behave uprightly and of warning them away from suspect or destructive practices. In everything that he did—traveling third class, spinning his own cloth, speaking and writing in the vernacular, surveilling his diet—Gandhi was testing his own limits, reflecting on them, and investigating by trying out alternative life patterns, thereby exploring the possibilities of all human endeavors.

Simplicity and Self-control

Part of Gandhi's experimentation involved the pursuit of a life of saintlike simplicity. He shrugged off worldly possessions, ate and drank abstemiously, wore few pieces of clothing, and lived with as few creature comforts as possible. His decision to stop sexual relations, one consonant with Indian mythology about abstinence and self-control, represented another effort to strip down his life, though this particular decision received relatively little sympathy from most outsiders. (As mentioned in chapter 3, Freud had also decided to cease all sexual relations.) Gaining sustenance from his self-restraint, Gandhi declared: "I own no property and yet I feel that I am perhaps the richest man in the world. The life I am living is certainly very easy and very comfortable. . . . I am a poor mendicant . . . prayer has saved my life." Some observers on the scene adopted a more worldly view of this life-style. One declared that "it cost a great deal of money to keep Gandhi living in poverty."

Gandhi believed that individuals should take control of their own lives, as much as possible. His dedication to spinning—it amounted to an obsession— was not only a form of daily self-discipline but also a statement that all Indians could support themselves if they were to master such a craft. His enduring interests in medicine, health, food, and diet represent efforts to figure out how one might best survive in the world, especially in the absence of significant material resources. Gandhi not only handled his own medical care but insisted on supervising the treatment of others for whom he felt responsible. He felt confirmed in his practices when his young son was cured of typhoid; but his obstinacy about medical care resulted in unnecessary suffering for other family members, whom he kept from receiving the medical care they needed.

The most extreme form of self-restraint involved fasting. In refraining from eating, Gandhi was revisiting a practice of purification with a long tradition in India, one in which his mother had in fact indulged on numerous occasions. For many years Gandhi activated the technique of the fast chiefly as a means of testing himself or those in his immediate environment. But with the events of Indian liberation accelerating, Gandhi came to invoke fasting as a powerful weapon of political persuasion: "Fasting cannot be undertaken against an opponent. . . . Fasting can be resorted to only against one's nearest and dearest and that solely for his or her good . . . it has its own science. No one so far as I am aware has a perfect knowledge of it." Given this view, it became necessary for the mature Gandhi to think of the whole of India as his family—a mental leap he undertook with ever-greater readiness.

Most people came to know Gandhi through participation in his ever-expanding social circle and his elaborate network of enterprises. Those who encountered him in the context of a protest were impressed by his meticulous concern for every step: the initial meeting, the search for common ground, a careful marshaling of properly documented evidence, the work with trusted associates, and mobilization of the media. During extended periods of reflection, he anticipated problems that might arise each day and ways he would deal with them. If he expected to be arrested, he would lay out his plans for the period of incarceration. He shared his evolving thinking with both friends and opponents. Even in the midst of chaos he dealt assiduously with daily matters—writing letters and articles, attending meetings, keeping journals, maintaining contact with his secretaries and his lieutenants "in the field"; in short, he held all details about persons and events in his mind. Most impressively, he balanced this local attention with a global awareness, taking care to ensure that journalists around the world were constantly being fed the information they needed for their stories. To the marvel of all, he kept going on three to four hours of sleep for days at a time.

Membership in Gandhi's Circle

To those who became part of his circle, Gandhi was a daunting and powerful figure. Though unattractive in terms of facial features, he was well built and had magnetic eyes that often twinkled. His soft and teasing mien was endearing. In a manner reminiscent of his own parents, he maintained tremendously high expectations of his associates, and in return, he contributed significantly to their growth and showed compassion for their welfare. He spoke sincerely and directly to each of them, always in a conversational manner, and he maintained this tack even when confronting large crowds. He augmented his inner circle by forming judicious alliances with the local leaders at each protest site. Gandhi exerted a charismatic hold on people through both his manner and his reputation. "In his presence," declared one follower after another, "one could not tell a lie." His political protégé Jawaharlal Nehru recalled:

> It is nearly three and a half decades now that I first came in contact with this strange personality and his stranger ideas. The effect was almost instantaneous, as if an electric shock went through the system. And yet the shock was a soothing and, at the same time, an enlivening one. The mind struggles with these new ideas often put out without much method or logic. But the whole system reacted to them and grew under their impress.

Being part of Gandhi's circle had its costs. In effect, Gandhi made all of the major decisions and, so far as I can determine, rarely allowed competing viewpoints to alter his course of thought or action. It was as if he expected others to allow the difficult pondering and thinking to take place chiefly within his head. Gandhi got away with this dictatorial arrogance because he at least listened carefully to what others said, he found a useful niche for each person, he had a sense of humor about himself and others, and he could admit mistakes and change his mind, though not without stubbornness. Gandhi explicitly denied that he had any special powers or divine authority, but in practice he had a very high opinion of his own judgments and was reluctant to jeopardize them. He may well have been more open to the input of others as a young man in South Africa or during the time when his ideas of satyagraha were first coalescing in India; but by the time he had become the Mahatma he behaved confidently and decisively. He once declared to those who sought to imitate satyagraha: "They should leave it to me alone."

The implicit requirement to bracket one's own point of view and suppress one's own resistances was not for everyone; those who remained with Gandhi clearly had strong needs for guidance from a charismatic figure. As one follower said: "Of course when we were living with Bapu [Gandhi's familiar name] we felt we could do anything and yet we also felt we needed him to solve all our problems. We became so dependent on him that we couldn't make any decision without consulting him." Only when dealing with external political leaders, who had their own constituencies, did Gandhi feel any pressure to compromise; and on many occasions he refused to be flexible.

Virtually unique was Gandhi's hold on the Indian masses. Something, perhaps everything, about the way in which Gandhi held himself and conducted his life resonated with *his* people. His simple dress, spinning, drinking of goat's milk, vegetarianism, and asceticism appeared absolutely genuine to these impoverished individuals; they felt that he understood them as no one else did. They eagerly sought his blessing. As the journalist William Shirer put it, "They felt in the presence of the great man that something immense was suddenly happening in their drab lives, that this saintly man in his loincloth cared about them, understood their wretched life and somehow had the power, even in the face of the great white sahibs in Delhi and the provincial capital, to do something about it." Gandhi's genius was to link the traditional appeal of the man of spirit to a dramatically new course of action in which individuals might be called on to sacrifice their lives while helping achieve a national ideal, and to do so selflessly rather than vengefully.

The most problematic aspect of Gandhi's character is the contrast between his relations with his most trusted intimates in the political struggle, on the one hand, and his relations with family members, on the other. To his innermost circle, Gandhi was the all-providing parent, concerned with their every need, always solicitous about their well-being. His power over them was so great that they merged their identities with his, sacrificing everything to help realize their mutual ideals. Gandhi; his intimate secretary, Mahadev Desai; and his coolly efficient organizer, Vallabhbai Patel, made a tremendously effective synergistic team. Undoubtedly, Gandhi's intense presence solidified these relationships. Anasuya Sarabhai declared: "There was something in Bapuji that was irresistible. . . . Most of us think in one way and speak in another way and act in yet another way. Not so with Bapuji. . . . He said what he believed and put into practice what he said, so his mind, spirit, and body were in harmony."

Strained Family Relations

Gandhi's relations with his family members, and particularly with his children, were an entirely different matter. Gandhi struggled against his wife for years, trying to get her to educate herself (perhaps in retaliation, she remained illiterate) and to do his bidding (including cleaning chamber pots and mixing with members of the Untouchable caste). When she became ill, he was matter-of-fact, almost cruel, allowing his idiosyncratic medical theories to take precedence over what common sense and prudent medical practice indicated. He paid grudging respect to Kasturbai, and, in his own way, may have loved her, but her life with him was long-suffering.

Gandhi had very poor relations with his children and particularly with his eldest son, Harilal. Gandhi held very high expectations for his children, and when they failed to meet these expectations, he turned against them, at least unconsciously and sometimes quite directly. Harilal led a dissolute life, challenging his father's beliefs and practices in every conceivable way, exemplifying what Erikson calls a "negative identity" with reference to family expectations. While Kasturbai came to Harilal's defense, Gandhi himself could not—or would not—do so. In effect, he disinherited him on any number of occasions, accused him of not being a son, and pointedly referred to others, such as Desai, as his ersatz children.

How does one piece together this disturbingly mixed picture of relations to other people in an individual whose very métier is the human sphere? Dating back to his connection to his own parents, Gandhi seems to have had unusual difficulties in sustaining long-term intimate relationships with those who should have been closest to him. He respected his parents, but by no indication (except, perhaps, his sense of guilt) was he particularly attached to them or to other members of his immediate family. He lacked any direct mentor, though he felt close in different ways to the poet-philosopher Rajchandra (who, as mentioned, died when both were still young) and to the politician Gokhale (who introduced him to Indian political life and, as noted, died shortly after Gandhi's return to India); and he never personally met Thoreau, Tolstoy, or Ruskin.

Gandhi's difficulty with intimacy seems to have been revisited with a vengeance in respect to his own family. From the very start, his actions toward them seem to have been motivated either by professional goals or by philosophical principles, rather than by a sustained capacity to love them unconditionally and to empathize with them. Asked once whether a genius might leave a legacy through his family, Gandhi answered, with perhaps unintended candor: "Certainly not. He will have more disciples than he can ever have children." Gandhi was able to feel close to those who worked with him on projects, and those relations in some cases continued for decades and involved great personal fondness. But Gandhi's most crucial relations were to individuals whom he would never meet: Like the greatest political leaders of any era, he maintained a mysterious and enduring link to the masses of his nation, to all those who functioned as part of Gandhi's mission rather than as people with whom he interacted.

In the interpersonal realm Gandhi's relational style may represent an idiosyncrasy rather than an incipient principle of creativity. Or, as I have just suggested, Gandhi's relationship with the Sarabhai family or with other members of his ashram may have proved an adequate substitute for a more intimate kind of cognitive and affective support system of the sort that surrounds our other creators. Possibly, the scheme that worked with the other masters proves less appropriate for an Indian than for a Westerner. But my own speculation is that a difficulty with the most intimate personal relations may be a concomitant of great skill at dealing with large masses; and that, to the extent that Gandhi had mentors, they were individuals at a remove, like Tolstoy, or the founders of the great religions, like Christ, Buddha, and Mohammed. Gandhi's most important conversation took place with the God that he had internalized.

Criticisms and Shortcomings

A man so complex in both philosophical and personal terms was bound to have his critics. British politicians were predictably condemnatory. Lord Irwin spoke of him as "rather remote and moving in a rarefied atmosphere divorced from the practical facts of the situation." Lord Frederick Birkenhead said: "Poor Gandhi has indeed perished! As pathetic a figure with his spinning wheel as the last minstrel with his harp, and not able to secure so charming an audience." The classicist Gilbert Murray spoke of "a dangerous and uncomfortable enemy because his body which you can always conquer gives you so little purchase over his soul." And in perhaps the most famous (or notorious) dismissal of Gandhi, Winston Churchill referred disparagingly to the "seditious fakir, striding half naked up the steps of the Viceroy's palace, to negotiate with the representative of the King-Emperor." Even those close to him sometimes despaired. In the early 1930s Nehru declared: "I am afraid that I am drifting further and further away from him mentally, in spite of my strong emotional attachment to him. His continual references to God irritate me exceedingly. His political actions are often enough guided by an unerring instinct but he does not encourage others to think." This air of lingering doubt or incomprehensibility surrounded Gandhi, whether one admired or feared him.

Although farseeing, Gandhi was by no means infallible in his political judgments. The most notorious area of his blindness was in dealing with tyrants. While conceding that satyagraha was not suited to dealing with totalitarian figures, he continued to hope that he could adapt his methods to hostile contexts: "I seek entirely to blunt the edge of the tyrant's sword, not by putting up against it a sharper-edged weapon, but by disappointing his expectation that I would be offering physical resistance. . . . The resistance of the soul that I should offer instead would elude him. It would at first dazzle him and at last compel recognition . . . which . . . would not humiliate him but uplift him." And he added elsewhere: "Hitler . . . Mussolini . . . Stalin are able to show the immediate effectiveness of violence . . . but the effects of Buddha's nonviolent action persist and are likely to grow with age."

Gandhi seemed unable to grasp that some individuals are totally immoral or amoral. He encouraged the Jews in Europe to go quietly to slaughter in the belief that this reaction would unleash sympathy in their tormentors. He wrote a direct appeal to Hitler, addressed "Dear friend," calling on him to change his tactics and promising him forgiveness. Hitler's response is nowhere recorded.

A NATIONAL AND INTERNATIONAL LEADER

The Salt March

If Ahmedabad was the crucible in which Gandhi first blended the ingredients of satyagraha, the so-called salt march of 1930 represented the apotheosis of the Indian independence movement. Thanks to the rendering of his life in the movie *Gandhi*, the facts of the salt march have once again become well known today, even as they were worldwide news at the time, over sixty years ago.

In a manner reminiscent of the Boston Tea Party, the march was triggered by the imposition of a tax on all salt, a measure widely perceived as unfair and regressive in a society where that condiment was literally a matter of life and death. The government had a monopoly on salt production, and no Indians were allowed to make their own salt, though they could easily have done so and thereby secured one of life's necessities at little or no cost to themselves.

Gandhi had been searching for a dramatic act of protest to indicate once and for all the necessity of an end to British domination in India. On March 12, 1930, he and a small group of followers set out from Ahmedabad, heading toward Dandi on the coast near Jalapur. At sixty-one, Gandhi, the oldest marcher, led the procession. The crowd grew in size at each stop on the march until it extended for nearly two miles. Gandhi trudged all the way, from ten to fifteen miles a day, resting frequently but never giving way to the temptation to sit in the ever-present cart that followed the queue. At each stop he patiently explained to crowds the rationale behind the march, asked the local leaders to abandon their loyalty to the government, and watched as people of different ages, backgrounds, and belief systems joined the throng or at least gave it their blessing.

From one vantage point, the march to the sea seemed illogical. Why lead a huge number of Indians merely to pick up a few grains of salt? The British leaders were mystified by the process and, perhaps proving Gandhi's point, exhibited no real understanding of the power of the demonstration. (Some Indian leaders were equally uncomprehending.) Convinced that the march would peter out by itself, the government made a tactical error; it elected not to interfere in any way with Gandhi and his forces.

From the point of view of Gandhi and his followers, however, the march was both logical and powerful. Its symbolism was bound to have a great effect on Indians and on sympathizers all over the world. Gandhi reported his intentions to the viceroy directly: "It is, I know, open to you, to frustrate my design by arresting me. I hope that there will be tens of thousands ready, in a disciplined manner, to take up the work after me, and, in the act of disobeying the Salt Act, to lay themselves open to the penalties of a Law that should never have disfigured the Statute Book." To his followers, Gandhi was resolute: "For me there is no turning back whether I am alone or joined by thousands. I would rather die a dog's death and have my bones licked by dogs than that I should return to the ashram a broken man." Fischer commented that "to pick up a pinch of salt in defiance of the mighty government and thus become a criminal . . . required imagination, dignity, and the sense of showmanship of a great artist. It appealed to the illiterate peasant and it appealed to a sophisticated critic."

After a march of 241 miles in twenty-four days, the Gandhi procession reached Dandi. At night Gandhi led his followers in prayer. On the morning of April 6, Gandhi walked into the sea, bent down, and picked up a small lump of natural salt. By this act he had technically become a criminal, inasmuch as he had been making salt from seawater without carrying out this action in the service of the government. At first nothing happened and there was a slight feeling of anticlimax. But on the succeeding days, protests mounted all over India. As Nehru later recalled:

> It seemed as though a spring had been suddenly released. . . . As we saw the abounding enthusiasm of the people and the way salt-making was spreading like a prairie fire, we felt a little abashed and ashamed for having questioned the efficacy of this method when it was first proposed by Gandhiji. And we marvelled at the amazing knack of the man to impress the multitude and make it act in an organized way.

Many individuals gathered saltwater in pans, allowed the water to evaporate, and retained the precious condiment. In sympathy, others burned foreign cloth, picketed liquor shops, or engaged in other acts of civil disobedience. Gandhi had succeeded in his stated goal; he had written on April 5, 1930, at a time when he believed that he was fated to die on this mission: "I want world sympathy in this battle of right against might."

But then events began to spin out of control. Hindustani terrorists conducted a raid in which they murdered six people. Military battalions were called in to quell angry demonstrations, but the troops broke rank and refused to fire on the crowds. British and Hindu soldiers clashed, and courts-martial resulted in heavy sentences for the protestors. The viceroy clamped down on the press, making it a felony to report on acts of civil disobedience. Gandhi protested loudly, seemingly to provoke the government to employ even more violent measures. He declared his intention to "raid" a government-owned saltworks. Gandhi was then arrested, and the government began to gather up the others involved in the campaign, one designed for "nothing less than to cause a complete paralysis of the administrative machinery."

The climax of the salt march soon occurred. Leadership of the ranks fell to a poet named Sarojini Naidu. She led a group of about twenty-five hundred members of the Congress party. The expected confrontation came on May 5, 1930. In one of the most famous journalistic dispatches of the century, a United Press reporter, Webb Miller, described what happened when the police confronted the lines of disciplined satyagrahis:

> Suddenly at a word of command, scores of native policemen rushed upon the advancing marchers and rained blows on their heads with their steel-shod latha. Not one of the marchers even raised an arm to fend off the blows. They went down like ten pins. From where I stood I heard the sickening whack of the clubs on unprotected skulls. The waiting crowd of marchers groaned and sucked in their breath in sympathetic pain at every blow. . . . They marched steadily, with heads up, without the encouragement of music or cheering or any possibility that they might escape serious injury or death. The police rushed out and methodically and mechanically beat down the second column. There was no fight, no struggle; the marchers simply walked forward till struck down. The police commenced to savagely kick the seated men in the abdomen and testicles and then dragged them by their arms and feet and threw them into the ditches. . . . Hour after hour stretcher-bearers carried back a stream of inert bleeding bodies. . . . By 11 A.M. the heat had reached 116 degrees and the assault subsided.

Within a few months, most of the political leaders and close to a hundred thousand of their followers had been gathered up and placed in Indian jails. The British had broken the rebellion, but at a severe cost—the moral hold Britain had exerted over India had been shattered. Gandhi did not miss the opportunity to draw an analogy with an earlier confrontation between colony and empire. Invited for tea at the viceroy's palace, he dropped a pinch of illegal salt that he carried with him into the tea "to remind us of the famous Boston Tea Party."

In a number of ways the salt march represents a model of an effective satyagraha. First, it was organized around a simple and immediately understood cause—a protest against an unjust tax. Every step in the campaign was carefully laid out, orchestrated in conjunction with appropriate local and national Indian leaders, and given the widest publicity. At each point in the ritualistic script, the British had opportunities to change their tactics or position, but in the absence of such a shift, the size and stake of the protest gradually increased. The satyagrahis maintained their nonviolent stance with admirable consistency. There was a dramatic climax in the sifting of salt off the coast of Dandi. The pace of protest then continued with increasing intensity until the final, bloody, and perhaps inevitable denouement, where the police troops turned ruthlessly on the unarmed, nonviolent resisters. Finally, while the salt tax was not repealed, the viceroy addressed most of the grievances against it in the succeeding months.

The London Conference

The moral algebra worked out over the past decades by Gandhi was applied scrupulously. Partly in response to the roller-coaster events of the salt march, Viceroy Irwin announced that a meeting would take place in London to consider constitutional reforms for India. In fact, a series of meetings eventually were held. At a first meeting, undertaken while Gandhi and most Congress party leaders were still incarcerated, some steps were taken toward the creation of a constitution. Heightened expectations surrounded the second meeting, which the Indians saw as an opportunity to set into motion the decisive steps toward independence.

Once in Britain, Gandhi was treated as a tremendous hero. He met other celebrities, toured the Lancashire Mills, and had tea with the king and queen. The reaction to Gandhi was virtually unprecedented: "The Mahatma kept breaking into the headlines as crowds turned out by the thousands to greet him wherever he went and hordes of photographers and reporters kept at his heels." He kept an unbelievably burdensome schedule, even for him, having a dozen daily meetings, yet finding time to spin, read, write, pray, and meet with his associates.

To reporters, Gandhi spoke powerfully but unyieldingly: "I have come here to win freedom. . . . I claim to represent the nation of India. I claim to represent the dumb, semi-starved paupers who made up India. I have not come to London to bargain." He rejected the option of dominion status (like Canada or South Africa) and instead endorsed complete independence. This tactic was bound to antagonize the British, who had little intention of letting go of the most important possession—the jewel—of their empire.

Gandhi rose to the occasion at the conference, delivering perhaps the most brilliant political speech of his life. He laid out the history of Indian-British relations and explained why any solution less than complete independence could not work:

India has been held by the sword. I do not for one moment minimize the ability of Great Britain to hold India in subjection under the sword. But which would be better—an enslaved but rebellious India or an India as an esteemed partner with Great Britain to share her sorrows and take part side by side with her in her misfortunes? An India that, if need be, of her own free will, can fight side by side with Britain, not for the exploitation of a single race or a single human being, but, it might be, for the good of the whole world.

Gandhi espoused nonviolence, but he indicated that his patience was wearing thin: "Freedom is our birthright as it is yours. It is true we do not wish to spill blood in winning it. But I must tell you frankly that if any sacrifice can win our freedom, we will not hesitate to let the Ganges run red with blood to obtain it."

And yet, by general consensus, the so-called Round Table Conference itself turned out to be a failure—even a disaster—from the Indian point of view. Though clearly the central figure, Gandhi was still seen as an outsider both by the British and by many members of the Indian delegation, and discussions were inconclusive. Gandhi's message fell on deaf ears, for despite his eloquent arguments, his British hosts were simply unprepared to relinquish control over India, especially not to the likes of this improbable spiritual figure. Magnifying the emerging debacle, the Indians began to quarrel with one another. The continued factionalism—and, most especially, the smoldering tension between Hindus and Muslims—gave the British the perfect pretext for maintaining their domination over the Indians. Gandhi noted sadly: "By our internal squabbles we are playing right into the hands of the British. It is a humiliation for all of us Indians." The British press, which had initially built up Gandhi's reputation, now turned on the Indian leader and blamed him for the failure of the conference. Gandhi condemned himself, declaring, "This has been the most humiliating day of my life." While expressing the hope that Indian-British differences could one day be ironed out, he was deeply shaken by this failed opportunity.

Returning to India, Gandhi was newly aware of two facts. First, Britain would never voluntarily give up India out of any feeling of morality or charity; India would have to seize its independence. Second, he could not expect the different factions, and especially the Muslim and Hindu groups, to bury their differences easily. As William Shirer expresses it:

I think it was this mistrust of him by so many of his fellow countrymen, which had become much more outspoken in London than it had at home, especially on the part of the Moslems and the untouchables, that hurt his feelings and discouraged him more than anything else that had happened in England. He was used to the taunts of the British, who were his adversaries. But not of the Indians, who, despite his differences with them, he regarded as his brothers.

Returning to Village Life

In the broadest sense, Gandhi's program, his mission, and the tempo of his life remained the same. He continued to lead protests; he undertook fasts, coming close to death on more than one occasion; and he kept abreast of events at home and abroad, attempting to influence them as only he could.

But now, as he reached his mid-sixties, following two decades at the center of Indian political life, Gandhi revealed a shift in interest and focus. He was becoming disillusioned with what could be achieved by constitutional politics and felt that the Congress party was out of touch with rural India, where 90 percent of the population lived. His health was also more fragile, and he was despondent about the failure of the London Conference. Gandhi decided to withdraw from the day-to-day activities of the Congress party so that he could pursue his vision in an untrammeled way. He undertook special efforts on behalf of the Untouchables. He spelled out his increasing alienation from the Western world and his feeling of remoteness from the mounting tensions in Europe. And he laid out in greater detail his plans for a new kind of India—what he called "spiritual socialism" or "the constructive program."

Spurning industrialism and Western society, which he saw as synonymous with materialism and violence, Gandhi discerned a way for the India of the future in the revitalization of her seven hundred thousand villages. Settling himself in a village in 1936, Gandhi sought to introduce a balance between cultivator and soil, herdsman and animals, craftsman and craft. Gandhi's new ideal featured communal unity, removal of untouchability, production of sufficient khadi (homespun cloth) for everyone, setting up of sanitation, improvement in the condition of women, propagation of a national language, establishment of universal basic education for children, and encouragement of adult education. As biographer Nanda indicates, "There was hardly an aspect of village life—whether it was housing, sanitation, medical aid, fertilizers, cattle care or marketing—which did not engage Gandhi's attention." Once again, Gandhi was assuming responsibility for the care of an entire population, even as he remained estranged from members of his own family.

At the very time that Gandhi was returning to village life, epochal events were occurring in other parts of the world. In Russia, Joseph Stalin had become a ruthless dictator, wiping out every shred of opposition. In China, civil war had already broken out between Communists and Nationalists. In Western Europe, the Fascists of Germany, Italy, and Spain were gradually revealing their aims of wide conquest, while the democratic countries, obsessed by their own financial and social problems, did little to underscore their opposition to totalitarian ambitions. These contrasting agendas have come to be symbolized by Munich, where, in order to stave off the prospects of immediate war, Prime Minister Neville Chamberlain of Britain appeased Adolf Hitler's growing territorial appetite. Gandhi knew of these events, but at this time his concerns lay as much with the purification and idealization of a single village as with the liberation of his country or the reformation of the rest of the world.

LAST YEARS: THE MAN AND THE LEGEND

While Gandhi and his followers had clearly set the stage for Indian independence, the pressures toward it, as well as the obstacles in the way of achieving it, came from unexpected sources. Once the Second World War had begun and British resources were being strained to the limit, there was simply insufficient resolve and energy within Britain to clamp the lid on pervasive Indian strivings for independence. It was as if England had attempted to maintain the increasingly obstreperous American colonies while waging the Napoleonic Wars. By the conclusion of the Second World War in 1945, it was not a question of *whether* India would achieve independence but *when* such independence would be achieved. When Lord Louis Mountbatten was selected as the last viceroy, the timetable sped up and the actual transition occurred with relative smoothness: By this time, India and Britain held virtually the same goals.

The chief obstacles toward the achievement of independence came not from England, surprisingly, but from within India. Long-festering tensions between the Hindus and Muslims came to a head during the last years of the independence struggle. The formidable leader of the Muslims, Mohammed Ali Jinnah, made it clear that he and his people were unwilling to live in a single India under Hindu domination. Displaying brilliant tactics and sticking resolutely to his position, Jinnah succeeded in having India partitioned into two nations, with Pakistan becoming the homeland for the majority of Muslim Indians.

Gandhi found little cause for celebration; he quipped, "Would it not be more appropriate to send condolences?" Instead of a triumphant ceding of leadership to a single India, united under his vision (though not under his leadership), Gandhi instead had to contemplate the grim scene of his compatriots fighting bitterly with one another. His last fasts were directed not against the imperial British but against his own people, Hindus and Muslims, who proved incapable of sustaining nonviolent techniques within their own country and who were brutally killing one another. In the wake of independence, Gandhi wandered from one province to another like a prophet of yore, attempting to stem the tide of violence through exhortation and example. As Brown describes it: "He went into the eye of the storm, into places where violence was most terrifying: He became the sole satyagrahi, a man who pitted himself totally against violence by personal demonstrations as well as preaching and fasting."

At times Gandhi was surprisingly successful: His moral authority remained immense, and even his enemies preferred that he not starve himself to death as a direct consequence of their own waywardness. In the words of a *Times* correspondent, he achieved what several divisions of troops could not have done. But in the end Gandhi was not capable of stopping the violence. Indeed, one of his own coreligionists, an overzealous Hindu named Nathuram Vinayak Godse, shot the great leader, then seventy-nine, to death as he was walking to a prayerground.

Gandhi left an enormous and, for the most part, inspiring legacy. Through the power of his writings and the strength of his personal example, he had affected thousands of individuals personally, as well as millions of others in India and elsewhere. As happens after the death of a great spiritual leader, many have sought to maintain his influence by writing about him, building memorials to him, and creating mechanisms that would preserve his memory and his program. Reviewing these efforts to sustain the Gandhian contribution, Ved Mehta, an Indian-American writer, commented in 1971 that "Gandhi was Christ and they themselves are now his apostles, trying to spread the word through an indifferent world." Mehta noted further that Gandhi's own writings already amounted to ninety volumes, that over four hundred biographies had been published in a single year, and that nearly everything he had said or done was recorded and preserved for endless review and unrestrained hagiography.

Of course, a man of Gandhi's stature does not escape condemnation—nor should he. To the criticisms of his imperious ways with his own followers, his strange and often cruel treatment of his family, and the naïveté of some of his political views was added a new charge: that Gandhi in the twilight of his life insisted on having naked young women sleep next to him. Gandhi never denied the facts, though he heatedly denied the implication that he was sexually abusing them or that they had been forced against their will to sleep next to an old man. But to many observers, this Gandhian practice not only confirmed the bizarreness of this man but also underscored the way in which his personal pathologies and preoccupations were often acted out at the expense of others.

For my consideration of creative masters it remains striking that an individual so brilliant at reaching millions of other persons could exhibit such difficulties with those closest to him and, in this last instance, could show so little sensitivity to his own possibly coercive practices and selfish motives. Even if this nighttime practice was an old man's way of seeking to rejuvenate or to test himself, Gandhi carried it out with singular clumsiness. At least in this instance, Freud seems to have been more unsparing of himself than was Gandhi. As I suggested earlier, the capacity termed interpersonal intelligence may well harbor a number of relatively separate components, with the gift of reaching the masses being dissociated from the abilities to relate to those closest to an individual or to empathize with others. This dissociation of components of interpersonal intelligence can also be discerned in Albert Einstein, a creator at work in a remote domain.

It would be gratifying to report that Gandhi's satyagraha methods exerted long-lasting effects on India. India has remained a democracy of sorts, and the descendants of Gandhi's close associate Nehru were able to maintain for some time a dynasty that is at least politically related to Gandhi's person. But few would claim that legacies of satyagraha endure in the India of today. India has displayed its share of belligerence toward Pakistan and other neighbors; the strife among the various ethnic and religious groups remains as constant and as violent now as at any time in the twentieth century; two members of the Nehru-Gandhi family have been assassinated; neither the village life nor the urban sprawl of India can serve as a model to other developing countries. Many Indians continue to admire Gandhi, but it could well be argued that his principles have had more application in the United States, during the civil rights movement, and in occasional other circumstances, such as the Greenham Common Action in Great Britain or the student protests in China, than in his homeland.

Still, Gandhi's achievement is widely seen as having been optimal for his time and circumstances. Drawing on Western as well as Indian ideas, he created an original philosophy that was simultaneously an effective program of action. Nehru said that Gandhi had accomplished "a psychological change, almost as if some expert in psychoanalytical methods had probed deep into the patient's past, found out the origins of his complexes, exposed them to his view, and thus rid him of that burden." Martin Luther King Jr., perhaps the most direct inheritor of Gandhi's mantle, said that Gandhi had provided "the only morally and practically sound methods open to oppressed people in their struggle for freedom."

In some ways, Gandhi seems anomalous in the company of great creators. Some students of creativity would question whether political innovation ought to be considered in the same terms as the creation of a scientific theory or a piece of music; and those who might accept Gandhi conditionally could legitimately point out the many disparities between the manipulations of symbols involved in creating new physics or new poems, and the relations to other human beings that lie at the heart of Gandhi's contribution. Moreover, even in the terms of my own study, Gandhi emerges as quite different from the mixture of scientists and artists examined here: For instance, the identification of a single pivotal creative moment in his career proves difficult, the distinction between domain and field is not as immediately helpful, and the kinds of mentors and supporters who suggest themselves turn out to be uncharacteristically remote from the individual creator. Even the fact that Gandhi is the only creator who was not raised in the West complicates my analysis and comparisons.

Yet, on balance, this study is enriched, rather than compromised, by the inclusion of Gandhi in this sample. On so many dimensions, many to be reviewed in the concluding chapters, Gandhi is a prototypically creative master. He is prototypical in his precocity in his chosen domain (the moral domain); his obsessive search for opportunities and his capacity to exploit them; his studied marginality; his oscillation between attachments to the many and the need for isolation; his essential selfishness; his asceticism; his Faustian sacrifice of personal intimacy in favor of political effectiveness; and the persistence of childlike features in his philosophy and in his person. In many ways childlike, he shared fundamental insights that were both simple and revolutionary; no racial or ethnic group is inherently superior to any other, conflicts need not be settled with violence, compromises can strengthen both parties. Perhaps most revealingly, the experiences surrounding the strike and fast at Ahmedabad have the familiar ring of core components of a creative breakthrough: One encounters the cognitive and affective support provided by the Sarabhais, the tentative working out of a new language, and the forging of a Faustian bargain.

In 1949, the year following Gandhi's assassination, Prime Minister Jawaharlal Nehru visited Albert Einstein at the Institute of Advanced Study in Princeton. Einstein took out a pad of paper and wrote down a number of dates on one side and a number of events on the other. Decade by decade he showed a parallel evolution of the nuclear bomb, on the one side, and of Gandhi's satyagraha methods and accomplishments on the other. The quite amazing parallels served as a list of human options available in the nuclear age. One might say Gandhi had achieved through his manipulation of relevant variables the kind of insight into human beings that Einstein had achieved about the natural order through his abstract conceptual experiments. Einstein was struck by the analogy: "Gandhi had demonstrated that a powerful human following can be assembled not only through the cunning game of the usual political maneuvers and trickeries but through the cogent example of a morally superior conduct of life. In our time of utter moral decadence, he was the only true statesman to stand for a higher human relationship in the political sphere." Einstein added more poetically: "Generations to come, it may be, will scarcely believe that such a one as this ever in flesh and blood walked upon this earth."

INTERLUDE THREE

"Action is my domain."
—M. Gandhi

In many ways Graham fits comfortably into the company of the other artists described in this book. Like Picasso, Stravinsky, and Eliot, she forged a new artistic language that exerted enormous impact on a domain throughout the twentieth century. Along with a few talented colleagues, she also stimulated the creation of a new field—the set of writers and judges who came to preside over modern dance. Having created epoch-making works in the 1930s and 1940s, Graham continued to grow as an artist, real-izing a set of brilliant neoclassical works in later middle age. And though she had to abandon dancing at the age of seventy, she continued to choreograph and to perform in other ways until virtually the time of her death. Certainly, her case could as readily have been discussed in the second interlude as in the final one.

Yet, in at least one crucial respect, the kind of creation exhibited by Graham merits consideration in tandem with that embodied by Gandhi. While the other artists could readily retain a distance from their work, Graham was inherently a performer. Her contribution was inextricably linked with her actions at a specific historical moment and with the immediate reactions of the audience to those behav-iors. In this crucial respect, she resembles Gandhi as a creator: The Indian leader also succeeded or failed depending on his performances at particular historical mo-ments and the reactions to those performances, both by those in his presence and by those who learned about what had happened.

In a very concrete sense, then, Gandhi and Graham created with their bodies. Their actual physical appearance and their uses of that guise stand at the center of their creation. For Graham and Gandhi, neither of whom was especially attractive in a conventional sense, it was essential that their presence be arresting and that oth-ers be caught up in their performance and affected by it.

A life centered around the behavior of one's body is a very different kind of life than the studio- or laboratory-based existence of other creators. One must care about one's health. One must have—and indeed revel in—a certain degree of exhi-bitionism. One must be able to sample the reactions of the audience and of co-performers and respond virtually instantly to them. One stands in constant fear that one's body will fail—that the fast will cause death, that a step will be mis-taken, that one will appear ludicrous rather than convincing. A tremendous weight hangs on the existential moment.

Of course, there is a crucial difference in the stakes confronted by these two performers. Though Graham changed her performances until the last moment, essentially they were ritualized; the differences between performances of the same work were usually modest, and, ultimately, others were able to follow at least partially in her footsteps. For Gandhi, on the other hand, the stakes could not have been higher; not only his own life but that of his beloved people hung in the balance. To put it graphically, Graham could afford to make mistakes, while mistakes by Gandhi would be much more costly. This characterization is not to ignore the ritualistic aspects of a Gandhian protest or the improvisational potential of the Graham troupe, but one should not confuse a difference in degree with a difference in kind.

I have deliberately stressed body and the existential moment in order to bring out the intriguing link between the otherwise disparate domains of modern dance and modern politics. However, as should be clear, the breakthroughs associated with Gandhi and Graham go much further than the motions of the body. They are not merely athletic specimens. Graham was an artist, and much of her work operated through the symbol systems of her art form—through language, music, concepts, design, and the ways in which her own body interacted with and communicated these elements. As she was conceptualizing her works, she explored each of these symbol systems alone and together, testing out her ideas with Horst and with her troupe. For his part, Gandhi was a thinker of the highest order. The conception of satyagraha was worked out as carefully as a philosophical system, with every step and its possible consequences carefully calibrated.

Both Graham and Gandhi operated in the space framed by their ideas and their bodies, that is, in a dialectic that spanned their conceptions and the way these could be realized physically at a particular historical moment. Given the draining nature of their respective activities, both creators found it necessary to withdraw periodically from the fray, from live performances, so that they could reflect on the course of their work. Existing to an extent as a separate entity, "the work" can be captured apart from the performance. Others are able to dance Martha Graham dances, just as others can participate in or lead a satyagraha.

Yet at the same time, it would be unwise to underestimate the importance of the person himself or herself and of the specific performance. Not only did Graham live to perform, but those privileged enough to have seen her do so believe that no one could ever duplicate her performances. By the same token, Gandhi's person and his personal history were integral to his protests, strikes, and fasts; once he died, the movement to which he had given his life quickly collapsed. It would need to be re-created by an agent of the quality of Martin Luther King Jr., not simply copied by a disciple.

Of course, other resemblances and differences between these two individuals suggest themselves, and some of them may have been important for the life of the performer. For Graham, a performance probably was an end in itself, while for Gandhi it probably was the means to political, social, or religious change. One might also ponder the significance of the strict and moralistic households in which both individuals were raised, or the effects on them of the early deaths of their fathers.

However, another point deserves mention in conclusion. In one way, the first six creators I have described can all be thought of as individuals who rose to the head of a culturally defined domain *and helped direct it along fresh lines. With a stretch, one could also nominate a domain for Gandhi—say, religious leadership, political revolution, or even philosophical writing. But to do so, I believe, would distort an important difference between Gandhi and the others. Alone among these creators, Gandhi sought to speak directly to other human beings, not as members of a group or domain but rather by dint of their humanness. He sought to create a story, a conception, a way of being that could make sense to every other individual irrespective of his or her particular history or craft. Difficult as it is to change a domain, it is far more challenging to create a new human narrative and to render it convincingly to other individuals. For these reasons, Gandhi's achievement is especially notable, though it may take centuries—it did for Christ and for Buddha—to determine whether his religious and political breakthrough can take hold in a world so different from that inhabited by his predecessors.*

PART III

CONCLUSION

10

CREATIVITY ACROSS THE DOMAINS

IN THE PRECEDING CHAPTERS, I have related the stories of seven remarkable human beings, each of whom made an indelible mark in one or more domains while also contributing uniquely to the shape of the modern era. Their stories are, I trust, of interest in their own right. Yet, given my focus on the conceptualization of creativity, I need to step back and discuss which lessons hold for the study of creativity in general.

REVISITING THE ORGANIZING FRAMEWORK

In chapter 2, I introduced a framework for treating the complex issues of creativity. Explicitly developmental, that framework features a concern with the creators' childhoods, as related to their adult creativity; an interest in phases of development across the life span; and a focus on the finer-grained steps that characterize moments of breakthrough. I posited a dynamic that appears to characterize all creative activity: an ongoing dialectic among talented *individuals, domains* of expertise, and *fields* charged with judging the quality of creations. According to my formulation, this dynamic is often characterized by various kinds of tensions and asynchronies: Provided that the asynchronies are not overwhelming, they should prove conducive to the fostering of creative individuals, processes, and products. Finally, I suggested a set of guiding themes, most of which provided background for the study, but two of which emerged, unexpectedly, from the study itself.

That framework has now been put to work, implicitly in the case studies and more explicitly in the three interludes. In this concluding chapter, I examine explicitly a number, but certainly not all, of the issues raised thus far. I touch on the major questions that motivated the study, providing, when possible, a rough quantitative survey of "data" relevant to the issue in question.

Earlier I indicated that current work on creativity ought to be framed by two stances: the detailed views of individual creators, of the sort undertaken by Howard Gruber and his associates; and the large-scale quantitative studies undertaken by Dean Keith Simonton and his colleagues. In my work I have sought to integrate these stances, which have traditionally been termed *idiographic* and *nomothetic*, respectively; I have approached this integration by looking at seven individuals deliberately chosen from diverse domains and yet searching for generalizations applicable to all or at least most of their cases.

One person's generalization is another person's exception. Depending on how one defines a term or carves out a category, one can either collapse individuals together or cleave them apart. In what follows I offer my current impressions about which findings are likely to qualify ultimately as reliable generalizations, and which are better described as either domain-specific or unique to particular individuals. Those who have detailed knowledge about specific individuals, who have available arrays of data and powerful statistical techniques, or who are wedded to different conceptual frameworks may well carve the pattern somewhat differently. I hope to have at least set up a structure that merits debate.

A PORTRAIT OF THE EXEMPLARY CREATOR

I need not focus here on the many important ways these seven creators differ. Even to place Gandhi and Stravinsky, or Graham and Einstein, in the same comparative study involves a suspension of customary categorical schemes. Moreover, my own theoretical bias has predisposed me to look for differences across domains of accomplishment; I believe that this study confirms the distinctive character of the activities typical of each of the creators.

That said, I have been struck by the extent to which common themes nonetheless emerge in the lives of these creators. While no theme emerges with equal force for all the creators, and an exception can be found to each of the emerging generalizations, I feel comfortable in putting forth a portrait of the Exemplary Creator, whom I shall nickname E.C. and speak of as female.

E.C. comes from a locale somewhat removed from the actual centers of power and influence of her society, but not so far away that she and her family are entirely ignorant of what is going on elsewhere. The family is neither wealthy nor in dire financial straits, and life for the young creator is reasonably comfortable, in a material sense. The atmosphere at home is more correct than it is warm, and the young creator often feels a bit estranged from her biological family; even though E.C. has close ties to one of her parents, she feels ambivalence, too. Intimate ties are more likely to exist between E.C. and a nanny, a nursemaid, or a more distant member of her family.

E.C.'s family is not highly educated, but they value learning and achievement, about which they hold high expectations. In a word, they are prototypically bourgeois, holding dear the ambitions, respectability, and valuing of hard work that have come to be associated with that class, particularly in the late nineteenth century. E.C.'s areas of strength emerged at a relatively young age, and her family encouraged these interests, though they are ambivalent about a career that falls outside of the established professions. There is a moral, if not a religious, atmosphere around the home, and E.C. develops a strict conscience, which can be turned against herself but also against others who do not adhere to behavioral patterns she expects. The creator often passes through a period of religiosity that is rejected and that may, but need not, be revisited in later life.

There comes a time when the growing child, now an adolescent, seems to have outgrown her home environment. E.C. has already invested a decade of work in the mastery of a domain and is near the forefront; she has little in addition to learn from her family and from local experts, and she feels a quickened impulse to test herself against the other leading young people in the domain. And so, as an adolescent or young adult, E.C. ventures toward the city that is seen as a center of vital activities for her domain. With surprising speed, E.C. discovers in the metropolis a set of peers who share the same interests; together, they explore the terrain of the domain, often organizing institutions, issuing manifestos, and stimulating one another to new heights. Sometimes E.C. proceeds directly to work in a chosen domain although she might just as well have flirted with a number of different career lines until a crystallizing moment occurred.

Experiences within domains differ from one another, and there is no point in trying to gloss over these here. Still, with greater or lesser speed, E.C. discovers a problem area or realm of special interest, one that promises to take the domain into uncharted waters. This is a highly charged moment. At this point E.C. becomes isolated from her peers and must work mostly on her own. She senses that she is on the verge of a breakthrough that is as yet little understood, even by her. Surprisingly, at this crucial moment, E.C. craves both cognitive and affective support, so that she can retain her bearings. Without such support, she might well experience some kind of breakdown.

Of course, in the happy circumstances that we have examined, E.C. succeeds in effecting at least one major breakthrough. And, the field rather rapidly acknowledges the power of the breakthrough. So special does E.C. feel that she appears willing to enter into special arrangements—a Faustian bargain—to maintain the flow that comes from effective, innovative work. For E.C., this bargain involves masochism and unbecoming behavior toward others, and, on occasion, the feeling of a direct pact with God. E.C. works nearly all the time, making tremendous demands on herself and on others, constantly raising the ante. In William Butler Yeats's formation, she chooses perfection of the work over perfection of the life. She is self-confident, able to deal with false starts, proud and stubborn, and reluctant to admit mistakes.

Given E.C.'s enormous energy and commitment, she has an opportunity for a second breakthrough, which occurs about a decade after the first one. The succeeding breakthrough is less radical, but it is more comprehensive and more intimately integrated with E.C.'s previous work in the domain. The nature of

E.C.'s domain determines whether an opportunity for further breakthroughs arises. (Remaining highly creative is easier in the arts than in the sciences.) E.C. attempts to retain her creativity; she will seek marginal status or heighten the ante of asynchrony to maintain freshness and to secure the flow that accompanies great challenges and exciting discoveries. When E.C. produces an outpouring of works, a few of them stand out as *defining*, both for E.C. herself and for members of the surrounding field.

Inevitably with aging, limits on E.C.'s creative powers emerge. She sometimes exploits young persons as a means of rejuvenation. Finding it increasingly difficult to achieve original new works, E.C. becomes a valued critic or commentator. Some creators die young, of course, but in the case of our E.C. she lives on until old age, gains many followers, and continues to make significant contribution until her death.

I am well aware of the limitations of this hypothetical portrait. Behind each sentence are arrayed not only the seven individuals in the study but also many others, at least some of whom appear directly to contradict this composite portrait. If most creators come from an intact and reasonably supportive family, certainly the Brontë sisters did not; if many live to an old age, Keats and Mozart certainly did not; if a majority come from somewhat marginal backgrounds, most members of the Bloomsbury set certainly did not. Thus, when it comes to offering generalizations about creativity, one must assess how essential each generalization is. In all probability, no single one of the factors just highlighted is critical for a creative life; but it may be that one needs at least a certain proportion of them, if the chances for a creative breakthrough are to be heightened. To evaluate the importance of different factors, I move now to a more explicit consideration of the central issues that guided my research. It should be stressed that the patterns proposed here are illustrative rather than definitive; one would need larger samples, and more precise measures, to establish the validity of any proposed pattern.

MAJOR ISSUES: A REPRISE

Individual Level

Cognitive. My slant in this study has been determinedly cognitive. A major assumption has been that creators differ from one another in the kinds of intelligences that they exhibit; and indeed, each of the creators was selected because he or she was thought to exemplify one of the seven intelligences that I detailed in *Frames of Mind.*

I conclude that creators differ from one another not only in terms of their dominant intelligence but also in terms of breadth and the combination of intelligences. Freud and Eliot had strong scholastic abilities (which reflect linguistic and logical intelligences), and they presumably could have made contributions in many academic areas. Picasso, on the other hand, was weak in the scholastic area, while exhibiting quite strongly targeted strengths in spatial, bodily, and personal spheres. Stravinsky and Gandhi were indifferent students, but one senses that their lackluster performances arose more out of lack of interest in school than out of any fundamental intellectual flaw. Graham had broad intellectual strengths but was never fully engaged until she encountered the world of dance. A rough summary of their intellectual profiles follows:

	Strength	*Weakness*
Freud	linguistic, personal	spatial, musical
Einstein	logical-spatial	personal
Picasso	spatial, personal, bodily	scholastic
Stravinsky	musical, other artistic	
Eliot	linguistic, scholastic	musical, bodily
Graham	bodily, linguistic	logical-mathematical
Gandhi	personal, linguistic	artistic

Just as the creators exhibited distinctive intellectual strengths, so, too, their relation to prodigiousness was also quite different. Freud was precocious in scholastic matters but did not discover his true vocation until in his late thirties; Graham did not begin to dance until she was over twenty; Gandhi ambled from one role to another until he discovered his political-religious calling; Stravinsky did not compose seriously until he was well into his twenties. Einstein and Eliot can be seen as having done important work when still quite young, but neither was seen as a prodigy in his chosen area. Indeed, of the seven creators, only Picasso comes close to the classic view of the prodigy—an individual performing at a master level while still a child. The other creators were distinguished chiefly by rapid growth, once they had committed themselves to a domain.

Personality and Motivation. In many respects, the picture of creators that emerged from the study closely parallels that reported in the classic empirical studies emanating from the Institute of Personality Assessment at the University of California at Berkeley and from other research centers. Individuals of the E. C. type are indeed self-confident, alert, unconventional, hardworking, and committed obsessively to their work. Social life or hobbies are almost immaterial, representing at most a fringe on the creators' worktime.

I have reluctantly concluded that these characterizations may traditionally have been taken in too positive a way. That is, the self-confidence merges with egotism, egocentrism, and narcissism: Each of the creators seems highly self-absorbed, not only wholly involved in his or her own projects, but likely to pursue them at the cost of other individuals. The British psychologist Hans Eysenck has suggested that there may even be a genetic basis to this amalgam of creativity and hard-headedness.

Nuances of differences exist. While as self-absorbed as any other creator, Einstein seems to have directed little overtly negative behavior toward others; he wanted chiefly to be left alone. Picasso represents the opposite extreme: He seems to have obtained sadistic pleasure, if not creative inspiration, from inducing discomfort in others. The remaining five creators can be placed somewhere in between these two extremes, perhaps somewhat like this:

Disregarding others		*Difficult toward others*	*Frankly sadistic*
Einstein	Eliot	Gandhi	Picasso
		Stravinsky	
		Graham	
		Freud	

A related dimension concerns the degree of effort attached to self promotion. One can be quite distanced from, or even sadistic toward, others and still devote considerable energy to self-promotion. All seven creators recognized the importance of bringing their work to the attention of others; and in the absence of a parent, spouse, or aide who could accomplish this task on their behalf, they were expected to do so themselves. Much of the self-promotion was dedicated to the work; as far as I can determine, Gandhi was much more interested in bringing attention to his program than to his own person, but his efforts at self-promotion were still striking. I would array our creators in this approximate order:

Ordinary self-promotion				*Extraordinary self-promotion*		
Einstein	Picasso	Eliot	Stravinsky	Gandhi		Freud
		Graham				

A notable characteristic of creativity, I have argued, is its special amalgam of the childlike and the adultlike. This amalgam can occur both in the sphere of personality and in the sphere of ideas. It can be more positively tinged (when the childlike feature is innocence or freshness) or more negatively tinged (when the childlike feature is selfishness or retaliation). A brief comment is in order on the relation between the child and adult "faces" in these seven creators:

- In the realm of personality, the adult Freud showed few childlike features; if anything, he sought to present himself as mature and judicious. However, his interests in the unconscious, in the stream of consciousness, and in childhood wishes, fantasies, dreams, and sexual preoccupations underscore the extent to which the consciousness of the child remained crucial to his thinking.

- Einstein prided himself on the preservation of certain childlike features, such as curiosity and a defiance of convention. Like other creators, he placed the mind and spirit of the young child on a pedestal. While his concepts were technical contributions, they represented attempts to answer the same kinds of questions that preoccupy young children—Piaget-style questions about the basic nature of the universe and of experience.

- Like Einstein, Picasso cultivated certain childlike personality features. Besides his clowning for the media, his enormous quest for possessions (human as well as material) and his desire to control all aspects of his life (and others' lives) can be fairly described as infantile. In his artwork, he cultivated the fragmentation of forms, searched for the simplest underlying shapes, and strove to capture all aspects of a visual experience simultaneously on paper—all characteristics of the art of young children.

- Stravinsky was interested in the world of children, but certainly did not dote on his own childhood and took no special pleasure in appearing to act like a child. He probably was most reminiscent of a child in his extraordinarily litigious nature—his desire to pick, and then to win, every fight and, if possible, humiliate "the enemy" in the process. Like other modern artists, he anchored his work in the most basic elements of the medium—primitive rhythms and harmonies of the sort that had so impressed him when he was a young child.

- In Freudian terms, Eliot was the most rigid and repressed of the seven personalities, in his life, if not in his work. He had seemed old when young, and he enjoyed the role of the elder statesman. Yet even he retained a certain childlike nature, loving puzzles, producing bawdy doggerel, as well as verses designed for children; and he did not let his conservative political views undermine his appreciation of the novel and the offbeat. In the fragmented nature of his verse and its concern with unconscious and symbolic themes, he inhabits the same childlike universe of the artist Stravinsky and the scientist Freud.

- Martha Graham sought to remain forever young in her person and her work; any sign of aging terrified her. Her self-centeredness, furious temper, and single-minded passion speak to the preservation of certain behavioral patterns of the young. Both the art form that she selected (use of the body

339

for expression) and the kind of elemental expressions that she favored draw on the reservoir of the child's imagination.

• We think of Gandhi as elderly and wise, yet in many ways he was very childish and even cultivated the look of the young child—naked to the world, proud of his body, excessively interested in its functions. Moreover, his major conceptual breakthrough—*satyagraha*—can be seen as childlike in the best sense of that term: individuals confronting one another in terms of actual equality and ensuring that they each feel renewed by a mutually satisfactory arrangement. Of course, it takes a most mature individual to bring this vision to fruition.

Having touched on the childlike component of each creator, I shall conclude this discussion of personality features by considering two remaining elements. The extent to which each of the creators engaged in public display of emotions, particularly the powerful ones of passion and rage, raises a complex issue. While each creator no doubt experienced very powerful feelings, some expressed them directly, while others preferred to "speak" through their works. There are few works of the period more powerful than *The Waste Land*, yet Eliot struck many observers as lifeless, without affect, frighteningly shy and reserved. Picasso and Graham, on the other hand, were as dramatic in their bedrooms and their working spaces as they were expressive in their works of art. The same contrast can be observed among our more scholarly creators. Einstein kept his feelings under wraps but wrote compellingly of the aesthetic elements of doing science; Freud took a clinical approach to emotion in his writings, but he was not afraid to confront his emotions, to express his feelings directly, and to mastermind and lead an intellectual revolution.

Particularly at the time of greatest creative tension, these creators felt under siege. So far as I can tell, all of them experienced periods of despondency when work was not going well, and virtually all had some kind of documented breakdown. The only possible exception here is Gandhi, who appears to have experienced two significant periods of depression. As described in chapter 9, these preceded his decisions to return to a far simpler life: in his South African period (1906 to 1910) and in his return to India in the 1930s following the disastrous London conference.

Social-Psychological Aspects. Though each of these creators seems to have come from a reasonably supportive household, unconditional intimacy and warmth may have been in short supply, except perhaps in the care of a nanny. When there was a close tie to the mother (Freud, Einstein, Eliot) or the father (Gandhi, Graham, Picasso), it seems to have been conditioned on achievement. Perhaps these contexts of early life stimulated the creators to regard work as the area where they would feel most whole. The French writer Gustave Flaubert once declared, "I love my work with a love that is frantic and perverted, as an ascetic loves the hairshirt that scratches his belly."

Strictness also marked most of the households. A disciplined "Protestant ethic"–style regimen led to children who were able to stick to tasks and to advance quickly in their studies or in their area of expertise. Ultimately, each of the creators rebelled against control: Freud, by calling explicit attention to the various motivational forces that had been covert in Vienna; Einstein by reveling in the permissiveness of the Aarau school and by confronting his teachers; Picasso, by rejecting his family, and particularly his father; Stravinsky, by spurning a legal career and finding a new father figure (Rimsky-Korsakov); Eliot, in similar fashion, by choosing a non-professorial career and leaving his native land for good; Gandhi, by rejecting aspects of his Hindu heritage and leaving home for over twenty years; and Graham, by choosing a dance profession, pursuing a unique lifestyle, and offering explicitly erotic performances.

I do not think that such a rebellion would have been possible without two factors: (1) sufficient skill and talent to allow one the option of a life different from one's forbears, and (2) positive models in childhood of a creative life. The homes of these seven creators may have been strict and conservative, but hints were given, either inside or around the home, that it was permissible to strike out on one's own, so long as one gave a good account of oneself. Freud's parents ultimately approved of his pursuing whichever career he liked; Einstein's uncle Jakob and his older friend Max Talmey promoted curiosity and scholarly pursuit; Picasso's uncle funded his study trips abroad; Stravinsky's family home served as a congregating place for artists of the time; Eliot's mother was herself a poet; Graham's father had an artistic side, nicely revealed in his decision to escort young Martha to a performance by Ruth St. Denis; and Gandhi's family was judicious in personal matters and permissive in religious matters.

Despite the support that the young creators received from their families, the theme of marginality pervades this work. Some of the creators were distinctly marginal by accident of their birth: Einstein and Freud as Jews in German-speaking countries, Graham as a woman in a male-oriented world. Others were marginal by virtue of where they came to live, by choice or by necessity: the Indian Gandhi abroad in the British empire; the Russian Stravinsky in Western Europe and the United States; the American Eliot in London; the Spaniard Picasso in Paris.

In addition to their demographic marginality, each of our creators used his or her marginality as a leverage in work. Not only did they exploit their marginality in what they worked on and how they worked on it; more important, whenever they risked becoming members of "the establishment," they would again shift course to attain at least intellectual marginality. Freud became suspicious whenever his work was too readily accepted; Einstein labored for thirty years on the unpopular side of the quantum-mechanical enterprise; Picasso and Stravinsky renounced first the mainstream artistic heritage and, in later decades, their own unrelenting departures from it; Eliot embraced unfashionable political and social ideas and then attempted in midlife to become a playwright; Graham took on new and challenging genres throughout her life, finally making the shift to choreography successfully (if reluctantly) in her eighties; and Gandhi constantly embraced unpopular causes and controversial groups.

While each creator was determinedly marginal and was willing to give up much to retain this marginality, it is too simple to say that each was simply aloof from the world of other people. At least two further patterns were at work. First, these creators often moved from a period of life in which they were comfortable with many persons to a period of maximum isolation, at the moment of their major discovery, only to return to a larger, and perhaps more accepting, world in the later years of life. Second, at the time of their greatest isolation, these creators needed, and benefited from, a special relation to one or more supportive individuals.

I return to this special relation at the conclusion of this chapter, but let me comment now on the general shift in the texture of interpersonal relations across the life span. Freud is the prototypical figure here—popular and engaged as a youth, increasingly isolated as he seeks his own domain, and then a firm leader of an ever-expanding crusade in the last decades of his life. Einstein's life pattern is somewhat similar, but in his case, the ultimate relationship to the wider world took a more distanced form, as he concerned himself with weighty issues of war, peace, philosophy, and religion. Gandhi's trajectory reflects aspects of both of these models: the need to organize a small group of loyal lieutenants, in the spirit of Freud's psychoanalytic circle, coupled with the capacity to relate to larger segments of humanity in a more distanced way, through writings, through the media of mass communication, and through his inspiring personal example.

With respect to the four artists, I discern a somewhat different pattern. Whatever their configuration of childhood relations, all experienced a period of isolation when they were working on their pioneering compositions. Once their work began to be accepted, they found themselves necessarily enmeshed in a political network, of the sort I have detailed in chapter 6. Stravinsky seems to have been most energized by this political world; Eliot accepted it as part of the territory and negotiated it surprisingly well; Picasso and Graham, in their different ways, left as much as possible to other individuals, and yet rose effectively to the occasion when their own presence was wanted.

If I can risk another generalization, none of these individuals had a particular need for friends who could be treated as equals. Rather, they used others to advance their professional work, being charming, seductive, and at least superficially loyal, while dropping these peers quietly or dramatically when the usefulness was judged to be at an end. The carnage around a great creator is not a pretty sight, and this destructiveness occurs whether the individual is engaged in a solitary pursuit or ostensibly working for the betterment of humankind.

Life Patterns: The Shape of Productivity. Without wishing to invest more magic in a numeral than is warranted, I have been struck throughout this study by the operation of the ten-year rule. These seven creators can be well described in terms of careers in which important events and breakthroughs occurred at approximately ten-year intervals, with the number of such ten-year periods allotted to the creators differing across domains. As has already been well documented in studies of cognitive psychology, it takes about ten years for an individual to gain initial mastery of a domain. Should one begin at age four, like Picasso, one can be a master by the teenage years; composers like Stravinsky and dancers like Graham, who did not begin their creative endeavors until later adolescence, did not hit their stride until their late twenties.

The decade of an apprenticeship heightens the likelihood of a major breakthrough. Such a breakthrough generally follows a series of tentative steps, but when it occurs, it represents a decisive break from the past. In this vein I have described Freud's "Project," Einstein's special theory of relativity, Picasso's *Les demoiselles d'Avignon*, Stravinsky's *Le sacre du printemps*, Eliot's *The Waste Land*, Graham's *Frontier*, and Gandhi's strike at Ahmedabad as breakthrough events.

In the years that follow, the creator comes to terms with his or her breakthrough. The appeal of innovation rarely atrophies, but generally speaking, the subsequent breakthrough is of a broader and more integrative sort, with the creator proceeding in a more nuanced way, tying innovations more directly to what has gone on in the past of the domain and to what others have been executing in the domain. Freud's *Interpretation of Dreams* (or perhaps *Totem and Taboo*), Einstein's general theory of relativity, Picasso's *Guernica*, Stravinsky's *Les noces*, Eliot's *Four Quartets*, Graham's *Appalachian Spring*, and Gandhi's salt march are candidates for a second, culminating breakthrough.

What happens after the second breakthrough is more a reflection of the nature of the domain than of the skills and aspirations of the creator. If the domain is wide open, freshly charted, and graced with relatively little competition, the creator retains the opportunity to continue to be innovative for as long as he or she remains active. This is what happened to Graham, Freud, Stravinsky, Gandhi, and Picasso. (Freud, in fact, thought that he had a breakthrough every seven years.) If, however, the domain is already well delineated, there are many other younger individuals working in the domain, or the creator's energies are sapped, then the possibility of further innovation is reduced. Neither Eliot nor Einstein was able to continue his innovations beyond the second decade of efforts, though Eliot wrote plays and Einstein worked on theoretical and philosophical issues until his death. The varying permeability of the domains cuts across the arts and the sciences: Lyric poetry ends up as closer to physics than to painting.

After the second decade, a different kind of opportunity arises. The individual may begin to look back on the relevant domain in a historical or reflective way. Picasso, Stravinsky, and Graham each pursued an impressive neoclassical period; Eliot tried to do so, but with less pronounced success. One can also become a metacommentator on one's field, as did Einstein and Eliot. When there exists a respected role within the domain, as there has been in literary criticism, one can continue in this reflective vein indefinitely. In the sciences, however, people who become philosophers of science are considered to have left their domain; thus, in the final decades of his life, Einstein was not considered central to the discussions pursued by the most innovative scientists.

In table 10.1 I summarize the trajectories of creativity across the decades. The radical breakthrough is indicated by a single asterisk, and the comprehensive work by two asterisks. Note that no two creators exhibit exactly the same trajectory, but that the "ten-year rule" still proves suggestive. The issue of productivity may appear to be a confounding factor. In some domains one can produce works at an enormous rate. Picasso produced on the average of one work a day throughout his adult life; Freud produced dozens of books and hundreds of papers. On the other hand, Eliot wrote less than fifty poems, some of them very brief, and Einstein's published scientific works were far fewer than Freud's. Yet there are first-class poets (like W. H. Auden) and major scientists (like the chemist Carl Djerassi) who prove that creativity in literature and in science is also compatible with fecundity.

While even creative individuals can differ enormously in terms of energy, I think it is important not to dwell on the actual number of products listed in an encyclopedia. What strikes me about our subjects is that they were each productive every day. Eliot may not have written poetry, but he wrote hundreds of reviews, edited major publications, and issued books on a wide range of subjects. Gandhi's literary output fills ninety volumes. Einstein worked on questions of physics until the last years of his life, even though his publication output lagged. It would be more opportune to monitor the number of new ideas or separate projects than to count the number of "final products" by a painter, poet, or physicist. Picasso may have made a thousand paintings over a five-year period, but in his own mind one or two of them were far more important (in my terms, "defining works") than the others. Freud may have written a dozen papers a year, but he could be repetitive in these essays, and he stressed his own need to search actively for new ideas. The ten-year period is revealing in this respect because it suggests that, independent of the number of discrete works issued by an individual, there may be a limit to the number of genuinely innovative works or ideas that an individual can produce in a finite period of time.

TABLE 10.1 THE TEN-YEAR RULE AT WORK

	Origin	10 Years	20 Years	30 Years and Beyond
Freud	Charcot	"Project"* *Interpretation of Dreams***	*The Three Contributions to the Theory of Sex*	Social works
Einstein	Light-beam thought experiment	Special theory of relativity*	General theory of relativity*	Philosophical works
Picasso	Barcelona circle	*Les demoiselles d'Avignon** Cubism	Neoclassical style	*Guernica**
Stravinsky	Rimsky-Korsakov influenced works	*Le sacre du printemps**	*Les noces**	Later styles
Eliot	"Prufrock" Juvenilia	*The Waste Land**	*Four Quartets***	Playwright/critic
Graham	St. Denis troupe	First recital	*Frontier**	*Appalachian Spring*** Neoclassical style
Gandhi	Natal	So. Africa Satyagraha	Ahmedabad*	Salt march**

*Radical breakthrough
**Comprehensive work

Domain Level

If the level of the individual reveals as many similarities as differences, the domain level constitutes the crucial location for the most telling differences across the creators. Youths interested in some aspect of their world evolve into young adults who choose (or are chosen) to work in a recognized domain or discipline within the culture. Each will be working in a domain for decades, and so the nature of that domain becomes crucial.

When I embarked on this study, I believed one could describe the steps of creativity across domains in a relatively comprehensive way. Building on Graham Wallas's well-known fourfold scheme—from preparation and incubation, to illumination and revision—I described in chapter 2 the nature of local disturbances, the initial surgery undertaken to fix them, the unsatisfactory nature of this stop-gap measure, and the gradual emergence of the need to create some kind of a new language or symbol system adequate to the problem at hand. And I posited an "afterlife" to this emerging scheme, where other knowledgeable individuals attempted to understand the new symbolic scheme and to promulgate it, and where the new invention, as understood by others, gradually became accepted by the field and even contributed to the reformulation of the domain.

I still stand by this scheme, in a general sense, but I now believe that it leaves out two very important, related dimensions. First, the kinds of symbol systems with which individuals work in different domains vary strikingly; one cannot simply lump them together under the broad rubric of symbol systems. Freud worked with words as shorthand for scientific concepts about human dreams and behaviors, and with simple diagrammatic schemes. Einstein thought in terms of complex spatial schemes, bodily imagery, and mathematical equations, with words entering as afterthoughts at the end of the process. Picasso dealt with colors, textures, lines, and forms as they relate to objects in the world and, increasingly, in terms of their own intrinsic features. Stravinsky treated analogous elements in the world of sound (timbre, rhythm, pitch, color); though these bear some relation to the world of experience, they gain significance chiefly in terms of intramusical associations. The verbal elements, allusions, and sounds used by Eliot lead in a wholly different direction from the words used by a Freud or an Einstein. Graham worked chiefly with the materials of the human body, attempting to capture plot, emotion, and formal relations in explicit gestures, and integrating them with the accompanying music and decor. And, finally, Gandhi's texts and talks represented an effort to paint a convincing picture of the experiences of a group of people: He built a model of current beliefs and behaviors within a group, as well as a model of how to change them, through the mounting of certain pivotal performances ranging from ritualistic to high-stake.

Not only do these symbols and symbol systems differ dramatically from one another, but the kinds of mental skills needed to work with them, and to communicate discoveries to others, are distinctly different—so much so, that grouping them all together as symbol systems obscures as much as it clarifies. Indeed, as described in the Interludes, these creative individuals were involved, respectively, in at least five distinct kinds of activities:

1. *Solving a particular problem* (usually a scientific one). Einstein's early papers, for example, on Brownian motion, reflect such a practice. Particular assignments tackled in the course of artistic training, such as Stravinsky's reorchestration of classical pieces, constitute examples from a different domain.

2. *Putting forth a general conceptual scheme.* Whatever their original missions, Einstein and Freud are most remembered for the broad schemes they

developed—relativity theory, by Einstein, and the psychoanalytic theory of unconscious processes, by Freud.

3. *Creating a product.* Artists create small-scale products, such as preliminary sketches or brief poems, or larger-scale ones, such as murals, operas, or novels. These works embody ideas, emotions, and concepts, but they are not well described, overall, as efforts to solve problems or to create conceptual schemes. Rather, they are often highly original instances of works within a genre, or attempts to initiate a new genre. Picasso, Stravinsky, and Eliot fall into this scheme, as does Graham in her guise as a choreographer.

4. *A stylized kind of performance.* In forms like dance or drama, an individual creator may embody the art form; in this case the "autographic" work does not exist apart from a particular realization by one person at a specific historical moment. The performance may be prescribed in various ways, but opportunities always exist for innovation, improvisation, and interpretation. The condition of the body and the exigencies of the historical moment circumscribe such performances.

5. *A performance for high stakes.* When one enters the political or spiritual realm, an individual's own public words and actions become the terrain in which the creativity unfolds. Gandhi may have had brilliant or scatterbrained ideas; but in the end it was his capacity to appear credible to his followers, and to the rest of the world, by virtue of his example at specific historical moments, that constituted the central aspects of his creation. Unlike the ritualistic dancer or dramatic artist, the high-stakes performer is risking security, health, and even life in the service of a mission. In Clifford Geertz's famous phrase, it is a form of very "deep play."

In light of the new distinctions introduced—a consideration of the nature of specific symbol systems and a position of five different kinds of activities that merit the term *creative*—it is necessary to conceptualize a more complicated scheme, which entails three components:

1. The particular symbol system (s) employed

2. The nature of the creative activity

3. Particular moments in the course of a creative breakthrough or performance

Rather than simply speaking generically of incubation, one needs to configure such dimensions in light of whether one is dealing (1) with words, gestures, or mathematical concepts, for example; (2) with the solving of a problem, the creating of a work, or the influencing of behavior among individuals living in one's community; and (3) with the period of conceptualization, the execution of the work, or the actual time when a performance is unfolding.

So far I have focused on the kinds of symbol systems and activities that characterize each domain. But domains differ as well in terms of their *structures* at a given historical moment. One key structural aspect is the extent to which a domain may be considered paradigmatic. In the way in which this word is usually employed, only physics can lay claim to the status of a *scientific paradigm*—a domain in which established practices and norms are accepted by all members. Psychology in Freud's time and even psychology today are preparadigmatic: The principal issues differ between rival schools, not between rival interpretations of mutually agreed upon phenomena and findings. But the physics paradigms associated with Newton, with Maxwell and Faraday, or with Mach and Helmholtz were not adequate either; the uneasiness with current concepts expressed by Lorentz and Poincaré around 1900 pointed to the possibility that a new paradigm might soon be needed.

The term *paradigm* can be stretched, or analogized, outside of science. When one performs such stretching, it becomes clear that there are times in other domains where there is also a dominant paradigm. In the late eighteenth century, Western classical music embraced a paradigm of composition; by the same token, in British law courts today, there are accepted paradigms for handling disputes.

At the beginning of the twentieth century, it seems fair to say, there were no equivalently entrenched paradigms in the major art forms. The romantic approach in music and literature and the academic and impressionist movements in the visual arts were in their waning phases; dance was not taken seriously as an art form. Thus, these domains can be considered "paradigmless" and, hence, open for new and competing approaches. If one can use the term *paradigm* for relationships between geographical entities within the British empire, it can perhaps be said that the British still believed that they knew what was best for their colonies and colonists, while the indigenous residents themselves were becoming increasingly restive.

The hegemony of a single paradigm is probably the best prognosticator of the rapidity with which a new approach can be broadly accepted. Despite some initial skepticism, the merit of Einstein's breakthrough could be readily and rapidly appreciated within the physics community. On the other hand, the very centripetal nature of a paradigmatic domain also means that younger individuals can soon make contributions that build on the new paradigm, and they thereupon become competitive with the originator of the new paradigm. What happened to Einstein has happened to other paradigm makers at work in established domains; he was soon overtaken by the younger scientists, who readily mastered his contributions and were able to build on them.

None of the other six creators had to confront this situation; in a sense, all of them had enough work to last a lifetime. And, indeed, the creators did continue to work and make innovations throughout the rest of their lengthy lives. The chief exception is Eliot. Of the various possibilities, the explanation that I have favored here pertains to the nature of the domain of lyric poetry, as I mentioned earlier. While other kinds of writing seem relatively resistant to the processes of aging, lyric poetry is a domain where talent is discovered early, burns brightly, and then peters out at an early age. There are few exceptions to this meteoric pattern. Eliot attempted to lengthen his creative life by becoming a playwright and critic: His accomplishments, particularly in the critical sphere, are notable, but they do represent a lifeline different from that pursued by other creators; the closest analogy would be Graham after she had been forced to stop dancing.

Field Level

Once the creators have begun to advance within a chosen domain, they inevitably encounter other individuals with whom they must interact. Typically, each creator will have one or more mentors; if reasonably successful, she will also spawn colleagues, rivals, and followers, and she will become involved in political battles, to at least a limited degree. How do each of the creators fare on these field dimensions?

It would be virtually inconceivable to envision any mature expert devoid of at least some competent mentoring. Still, our creators differ noticeably in the kind and degree of mentoring to which they were exposed. Freud had perhaps the most traditional mentoring picture: a number of strong father figures, ranging from Bruecke to Charcot to Breuer, who introduced him to important disciplines, problems, and methods. Einstein was unusual for a scientist in the relative lack of personal mentors. His mentoring seems to have occurred at a greater distance, by virtue of the reading that he did, first in popular accounts, and then in the writings of Mach, Poincaré, Maxwell, and other major professional figures. In this way he is reminiscent of certain artistic titans of the past, like Shakespeare or Beethoven, who are not considered to have had major personal mentoring figures.

Each of the four artists benefited from mentors. Stravinsky had the most traditional experience, with Rimsky-Korsakov serving as a primary musical mentor and Diaghilev introducing him to other aspects of the theatrical world. Eliot was influenced by several of his teachers at Harvard, by the slightly older and more daring Pound, and by the writings of Laforgue and Symons. Picasso benefited most directly from his father; thereafter, he was exposed to a multiplicity of figures from the recent and distant past, no one of whom seems to have dominated his artistic formation. Graham had relatively demarcated mentoring from the team of Ruth St. Denis and Ted Shawn; thereafter, as we have seen, her close confidant Louis Horst doubled as her best teacher.

Again, Gandhi seems anomalous here. One can point to some individuals whom he knew well, like Gokhale and Rajchandra, and to those whose writings he admired, such as Tolstoy and Ruskin. He also had a circle of confidants, such as Polak and Schleslin. Yet to a greater extent than with most other creative figures, Gandhi seems to have invented himself. Perhaps this explains why he wrote so much about his own experiments on himself and why, in certain senses, he felt on an equal footing with the religious innovators of the past and with his own God.

Earlier I touched on the generally dismal relationships between the creators and other human beings. What can be said more specifically about the creators' relationships with others in their chosen field? The paper trail can be a confounding factor here, for those who liked to write, like Freud, Stravinsky, and Eliot, have provided much more evidence about their relations to others in the field than have those who did not conventionally resort to correspondence, memos, and diaries.

With respect to the seven creators, I would locate Freud and Stravinsky at one end of the continuum, with most other practitioners close by, and Einstein at the other end. Freud and Stravinsky were both intensely competitive individuals who saw—and labeled—many others as rivals. They doggedly protected their territory, divided the world into supporters and enemies, proved quick to reward loyalty and to punish apparent disloyalty. They perceived the sociopolitical scene in zero-sum terms: If you were not for them, you were against them. Each had an ensemble of followers who did their bidding, and neither welcomed close colleagues, perhaps because they felt that few of their domain peers were their equals. It is no accident that their closest, least charged friendships occurred with individuals outside of their own domain.

Picasso was at least as competitive as Freud and Stravinsky, and probably far more ruthless toward men and women than either of the others. But perhaps because, from a fairly early age, he could afford to have others negotiate for him, and because he did not have as ready a recourse to writing, he does not qualify as quite so politically embroiled as Freud and Stravinsky. Eliot became involved in political relations as well and proved quite skilled at them; but I perceive little relish on his part about this sphere of life, and he seems to have been happy to metamorphose into the role of an elder statesman. Additionally, in comparison with other figures, Eliot seems to have devoted less effort to discovering enemies, labeling them as such, and seeking to destroy them. Graham's relationships with colleagues were charged, and she gave at least as well as she received; but her focus on her work and on the performance was primary; and she was pleased to leave financial and logistic arrangements to other people.

Gandhi and Einstein constitute the exceptions here. Einstein was simply less interested in the personal and political sphere surrounding his work, rarely taking a chance to defend himself or to attack others, unless controversy hinged on a nonscientific issue. More so than the other creators, Einstein was content to allow the work to speak for itself, though he occasionally encouraged his supporters. As for Gandhi, he often affirmed the essential political nature of what he was doing and was endlessly enterprising in promoting his own work. Yet, as the central portion of his message had to do with maintaining peace with his adversaries, he could not afford to be openly jealous or competitive with them. Still, his sorry record with his own family indicates that Gandhi could be very difficult interpersonally, especially when he could not prevail over those who were closest to him. The master at large-scale politics proved a disaster in more intimate relations.

A final dimension of the field points to the complementary concept of the domain. I refer here to the extent to which a field is organized hierarchically and the extent to which one's position in the hierarchy influences one's behavior. Here the differences across fields are again striking. Early on, Einstein was placed at the summit of the hierarchy of physics, and his position remained secure even after his death; but he himself took little interest in this position, except perhaps insofar as it permitted him to focus on his scientific work and to bring attention to his nonscientific interests. Freud was never highly ranked in any internationally acknowledged field, so he created his own. Thenceforth, controlling the hierarchical structure of psychoanalysis became a chief preoccupation.

Picasso's position as the outstanding painter of the century was also widely acknowledged from the start of his middle age. Picasso was far more interested in his own work and his own success than he was in influencing the behavior of other painters; and while he retained relationships with some artists, only Matisse seems to have occupied a significant part of his consciousness. For Stravinsky, the hierarchy was more complex, inasmuch as the rival schools of music continued throughout his life. For many years he saw himself as locked in competition with Schönberg, the leader of the rival school; only after Schönberg's death did Stravinsky himself feel free to grapple with twelve-tone music. Graham found herself similarly embattled with other leaders of modern dance; and though Graham had trained most of the best of the next generation, she desired ardently to remain the figure emblematic of dance, even after her long-delayed retirement.

A major part of Eliot's assignment as an editor and critic was precisely to attend to the cultivation of the domain of literature, spanning fiction and poetry. Seldom has an honored practitioner also served as its chief evaluator (Freud might be considered another exemplar of this dual role within psychoanalysis). Eliot performed this task with more generosity of spirit than one might have predicted; he believed in a hierarchy of literary quality, susceptible to judgment apart from political and social attitudes. At the same time, particularly with respect to predecessors, he enjoyed playing an iconoclastic role; and in his rewriting of literary history, he was, whether or not consciously, attempting to boost his own stock.

The model of a field, with a set of judges operating consensually, is perhaps least germane in a consideration of Gandhi. In one way, his domain, politics, was the oldest and broadest; but in another way, satyagraha, like psychoanalysis, was a domain of his devising, and as the founder, he was by definition in the best position in the field to render judgments about its practitioners and practices. But Gandhi was not in the business of making evaluations; he was in the business of bringing about change. Here he was playing for very high stakes; and his competitors were the major political figures of his time, like Lenin and Churchill, and the major religious thinkers of other times, like Christ and Buddha. Gandhi was surely aware of these considerations, though he seems to have been genuinely humbled by such lofty comparisons. In the end, Gandhi's creativity is closely linked to the success of his reform efforts: He was amazingly successful in the middle years of his life, but far less successful thereafter.

With this discussion of individual, domain, and field considerations, I conclude a review of the "data" relevant to the theoretical and empirical issues that guided my study. Definitive answers about these issues remain elusive. At the same time, however, one may speak of strong trends with regard to certain issues—for example, the prevalence of marginality or the unimportance of prodigiousness; and of striking domain differences with respect to other issues, such as the possibility of lifelong creativity or the inevitability of political strife. As others add data points from these and other domains, and from this and other eras, we should be able to facilitate the transition from an idiographic, Gruber-like effort to a nomothetic, Simonton-style research enterprise.

ASYNCHRONIES ASSESSED

In part I of this book, I introduced an organizing framework, designed in response to the question, Where is creativity? The essential burden of the "triangle of creativity" has been to investigate the dialectics among the *individual* person, or talent; the *domain* in which the individual is working; and the *field* of knowledgeable experts who evaluate works in the domain. No matter how talented the individual is, in some abstract sense, unless he or she can connect with a domain and produce works that are valued by the relevant field, it is not possible to ascertain whether that person in fact merits the epithet "creative." In some cases, of course, there may not be a fit initially between the nodes of the triangle; but unless some kind of rapprochement can be arranged among individual, domain, and field, an ultimate decision about an individual's creativity cannot be made.

Occasionally, an almost perfect fit among individual, domain, and field will exist: This is the textbook example of a prodigy. Indeed, in some societies, such alignment is all that is ever wanted. In our modern world, however, few if any prodigies make a ready transition to the world of the creative adult. That is because we seek from adults a kind of innovation, a departure from the norm, that not even the most talented youth can fathom. The closest that one comes to the adult prodigy is an individual like Mozart or Picasso, who was blessed with a stunning talent and who eventually became an acknowledged master; but as is well known, both men experienced anything but a smooth transition from youthful to adult practice. More commonly, as I have shown, the individuals who made the most remarkable breakthroughs could not have been considered prodigies by any strict definition of that term.

What seems defining in the creative individual is the capacity to exploit, or profit from, an apparent *misfit* or *lack of smooth* connections within the triangle of creativity. From an analytic point of view, there are six possible areas of asynchrony: within the individual, within the domain, within the field; between individual and domain; between individual and field; and between domain and field. Individuals who avoid any kind of asynchrony may well be prodigies or experts, but they are unlikely to become creative people; those who experience asynchrony at all points may be overwhelmed. I have hypothesized that an individual will be judged creative to the extent that he or she exhibits several asynchronies and yet can withstand the concomitant strain.

In the previous pages I have provided evidence of each of these kinds of asynchronies. It would be possible to look for each kind of asynchrony in each case study, but such a quest for forty-two or more asynchronies would be forced and, therefore, not revealing. Instead, I want to recall some of the more striking asynchronies that have emerged.

At the level of the individual, I noted the asynchrony between Picasso's excellent spatial and bodily capacities, on the one hand, and his meager scholastic capacities, on the other. Within the domain of physics, the strains at work in the years before Einstein's cutting of the Gordian knot were apparent. Within the field of clinical psychiatry, there were the deep divisions between those who valued Freud's work and those who felt that it was errant nonsense; and, of course, the tension between mass (low) and elite (high) fields characterizes the several art forms.

One can with equal readiness amass instances of asynchronies across nodes. Freud exhibited a set of intelligences that were unusual for a natural scientist but that finely tuned the newly formed domain of psychodynamic practice. Graham's early dances were remote from the tastes of regular concertgoers, but they excited pivotal newspaper reviewers, who helped create a new field for modern dance. Finally, with Gandhi, the domain of legalistic or militaristic conflict resolution functioned adequately within a British setting but made less sense to the field of twentieth-century Indians seeking to build their own society.

A problem with the hypothesis of fruitful asynchrony is that one can all too readily find instances of asynchronies. Do creative individuals really experience or exhibit more asynchronies, or are they simply better at exploiting them? Here the example of marginality is useful. By definition, most individuals are not marginal within their community; hence, to the extent that there is a larger proportion of marginal individuals within the ranks of the creative, one has evidence that asynchronies may actually be associated with creative output in a statistically verifiable way. But it seems equally true that creative individuals, once they have felt the pain and pleasure of asynchrony, often continue to seek asynchrony, even as many other individuals "escape from freedom" and rush to the comfort of majority status.

I maintain that each of our individuals stands out in the extent to which he or she *sought* conditions of asynchrony, receiving a kind of thrill or flow experience from being "at the edge" and eventually finding it difficult to understand why anyone would *not* wish to experience the fruits of asynchrony. Such a pattern clearly characterizes each of the creators, independent of the degree of asynchrony or marginality with which each began. Though highly gifted in many areas, Freud was riven with asynchronies within himself and with respect to other individuals and to the several domains in which he worked. And whenever it looked like he might be moving somewhat closer to the establishment, Freud made the move typical of creators of traveling closer to the edge, confronting yet more complex issues, making even stronger demands on those around him.

Though by any definition as creative and successful as Freud, Einstein may not have had the same drive to be asynchronous in his life or his work. At an early age he had already identified the issues on which he wished to work, and like the bear of which his secretary spoke, he would have continued to work on them for the next millennium. Einstein's personality and gifts had suited him well for his revolutionary discoveries between 1905 and 1920. In that sense, the world of physics interacted perfectly with his particular strengths and style. Thereafter, his asynchrony with physics became too great. He distinguished himself as a commentator on the world scene and on the enterprise of science.

Each of the four artists was characterized by considerable asynchrony, but this lack of fit affected them in different ways. From an established family and with strong academic and professional credentials, Eliot had to stretch the most to induce asynchrony. But the combination of his strange personality, difficult marriage, and decision to live abroad on his own finances, yielded an almost desperately asynchronous individual by the end of the First World War. Thereafter, the asynchronies lessened somewhat, perhaps mirroring a muting of his creative talent. Graham can be seen as an opposite to Eliot, since she had neither the family connections nor the social advantage, as then defined, of being male. By personality, she was strengthened by challenges (as Eliot was probably not), and so she was able to thrive under asynchronies of her own devising, as well as those inherent in her situation.

Stravinsky and Picasso can be seen as similar on some dimensions, opposite on others. Stravinsky came from a family that was centrally involved in the musical arts, while Picasso's father, though an artist, possessed distinctly limited talent. Whereas Stravinsky wished to escape from bourgeois complacency, Picasso wanted to escape from provinciality. Both men could have remained with their early successes—the music of *Firebird* and *Petrouchka,* the art of the blue and the pink periods—but both were impelled to strike out in more radical directions. And despite the eventual success that greeted their frankly iconoclastic works, they were stimulated for the rest of their lives to search for asynchronies in both their professional efforts and in their relations to other individuals.

Once again, Gandhi presents a complex picture. On demographic grounds, both within his country and as a citizen of the world, Gandhi was most asynchronous with his surroundings. In addition, his decidedly odd personality and philosophy helped ensure that he would always stand out from those around him. At the same time, however, his work was predicated on the assumption that he could illustrate, indeed exemplify, deep connections to the rest of his society: that he could appear as a typical representative of the larger Indian community. Thus, he had to cultivate, to embody, a being who was at once synchronous with the rest of society and humanity and distinctly marginal, someone positioned to bring about radical social change. Perhaps this dual assignment can itself be seen as a form of asynchrony; that is, it may be even more anomalous for the creator to retain one foot in the camp of the ordinary than to be completely dissociated from one's society.

Those in search of asynchronies can feel rewarded by these case studies, which document considerable initial asynchrony as well as a decided taste for creating more. In this way, the creative individual certainly differs from the individual who does not seek to stand out in any way. But is the creator also different from members of a reasonable control group—a group composed of individuals in a related domain who are equally ambitious but perhaps less successful, such as Wilhelm Fliess in the domain of medicine or Pierre Janet in the domain of psychiatry (see chapter 3)? My own hunch is that our seven model creators *are* different, and that the degree and type of asynchrony they represent is somehow more fruitful—more fruitful, say, than Wilhelm Fliess's scheme, which was too bizarre, or Pierre Janet's scheme, which was less sharply delineated and less expertly disseminated. But in the absence of convincing methods for evaluating both asynchrony and fruitfulness, this must remain a speculation.

THE TWO EMERGING THEMES

Any researcher embarking on a large-scale study must be guided by certain assumptions. In the absence of such a rough roadmap, his or her journey is almost impossible to envision. Yet, at the same time, most researchers remain open to the possibility of surprises or of new discoveries: After all, if one knew exactly what one expected to find, then the journey would hardly be worth undertaking. Here, as elsewhere, the favorable degree of asynchrony is delectable.

Two themes I had not anticipated emerged during my work on these case studies. Consistent with the developmental perspective, one theme represents a relatively brief period of time, during which the creator made a major breakthrough; the second theme covers a significant portion of the creator's adult life.

The Matrix of Support at the Time of Breakthrough

Because of my familiarity with Freud's life, I had known that, during the time of his greatest loneliness, he had gained sustenance from his relationship with Fliess. While few scholars have felt that Freud obtained indispensable ideas from Fliess, Freud clearly needed the latter's support and listening ear. And since Freud destroyed Fliess's correspondence, we cannot determine the extent to which he gave Freud either valuable ideas or acute criticisms.

When I began to learn about other creators, I gradually became struck by the fact that, far from being an isolated case, the Freud-Fliess confidant relationship represented the norm. As we have seen in the preceding chapters, Braque played much this role for Picasso; Horst, for Graham; Pound (along with Vivien Eliot), for Eliot; and the Diaghilev circle (along with special figures like Roerich and Ramuz), for Stravinsky.

It is possible to stretch the facts to cover the remaining two creators as well. For Einstein, the first rank of support came from the members of the Olympiad, with whom he had such regular and intimate contact during the years before his epochal discoveries. More proximal support for his relativity theory came from his close friend Besso and, with reasonable likelihood, from his wife Mileva as well.

I feel less secure in invoking the name of one or a few individuals who served as confidants for Gandhi. But, as suggested in chapter 8, the defining moment at Ahmedabad might not have been possible if Gandhi had not been aligned with one family member (Anasyra Sarabhai) and against another (Ambalal Sarabhai), to both of whom he felt powerfully connected. Perhaps, in this sense, I can legitimately include the Gandhi example within the pattern of confidants as well.

There is more to be said about this confidant relationship. First, under ideal circumstances, it ought to have two dimensions: an affective dimension, in which the creator is buoyed with unconditional support; and a cognitive dimension, where the supporter seeks to understand, and to provide useful feedback on, the nature of the breakthrough. The prototypical supporters—Fliess, Horst, and Braque—apparently assumed both roles. Certainly, between them, Pound and Vivien Eliot cover the waterfront, too. With Einstein, the need for affective support may have been less pronounced; with Stravinsky, the various kinds of support may well have been distributed among diverse figures, including Fokine, Nijinsky, Benois, Monteux, Roerich, and Diaghilev himself. Again, I have the least to say about Gandhi, though it might be relevant to remark that the support of large multitudes of strikers, workers, religious individuals, and his own loyal team of intimates must have meant a great deal to the Indian leader.

As I have perceived it, these relationships hearken back to important associations early in life. One model entails the exchanges that take place between the mother and the infant, as the mother attempts to teach the child the language and rules of the culture in which they will both live. Thanks to the mother's constant efforts at interpretation, the infant passes here from a state of ignorance to a state of knowledge. Another model entails the exchanges that take place between close friends—siblings or peers—as they explore the unfamiliar world together and relate to one another what they have discovered.

Such processes must be replayed at the time of the breakthrough. The difference is that the language being forged is new not just for a single child but for the rest of humankind. The creator is in the throes of discovering this language, as a means of solving certain issues—often personal as well as discipline-based—and, perhaps, of illuminating others as well. The creator must be able to devise and understand the language well enough to use it and then gain sufficient mastery to communicate it to others (lest it be autistic). In so doing, the creator draws on earlier models of teaching a new language to an ignorant but willing pupil.

It would be bizarre and unnecessary to maintain that this form of communication represents any kind of a conscious replay of the mother-child dialogue, or of the kind of intimate conversations that take place in early life, for instance, between siblings, twins, or close friends. Yet, I find these analogies helpful in conveying the phenomenal flavor of this exchange. I would submit, further, that a creator who has not gone through an earlier effective communication process, such as the mother-child or nurse-infant dialogue or the conversation between close friends, would have difficulties in effecting this most radical kind of adult communication. It is notable that this support in adult life is related specifically to the creation of new work—a replaying of an earlier situation in which elders rewarded a gifted young child's achievements.

My claim, then, is that the time of creative breakthrough is highly charged, both affectively and cognitively. Support is needed at this time, more so than at any other time in life since early infancy. The kind of communication that takes place is unique and uniquely important, bearing closer resemblance to the introduction of a new language early in life, than to the routine kinds of conversations between individuals who already share the same language. The often inarticulate and still struggling conversation also represents a way for the creator to test that he or she is still sane, still understandable by a sympathetic member of the species.

The Faustian Bargain and a Creative Life

I have alluded in previous chapters to a Faustian bargain struck by each creator. The Faust legend is but the best-known exemplar of a widely held belief that creative individuals are special by virtue of their gift, and that they must pay some kind of a price or adhere to some kind of an agreement to sustain that gift. In a trivial sense, of course, this proposition has to be true: One cannot remain an expert writer or performer unless one practices one's craft regularly. In its more dramatic sense, however, this claim has the air of fancy: After all, why should we think that a creator need be in communication, or in cahoots, with a personal god or a private devil?

I was quite surprised to find that the creators, in order to maintain their gifts, went through behaviors or practices of a fundamentally superstitious, irrational, or compulsive nature. Usually, as a means of being able to continue work, the creator sacrificed normal relationships in the personal sphere. The kind of bargain may vary, but the tenacity with which it is maintained seems consistent. These arrangements are typically not described as pacts with anyone, but at least to me, they resemble that kind of semimagical, semimystical arrangement in the West we have come to associate with Dr. Faustus and Mephistopheles. Equally, they have a religious flavor, as if each creator had, so to speak, struck a deal with a personal god.

The only allusion I found to a deliberate experience of this sort was in the biography of Picasso, the most overtly superstitious of the seven creators, regarding his sworn oath to stop painting if one of his sisters recovered from a critical illness (indeed, a singular bargain). As noted in chapter 5, Picasso elected to devote his life to painting when she died, and I argue that Picasso took this commitment as a license to sacrifice not only himself but others in the service of his painting; the bargain that had not been honored gave rise to a "counterbargain" that sanctioned his otherwise outrageous behavior toward so many other people over the course of his life.

The extreme asceticism associated with the lives of Freud, Gandhi, and Eliot represent another variant of the bargain. Freud and Gandhi both renounced sexual relations at a very young age and subjected themselves to many kinds of seemingly unnecessary (and perhaps unwise) deprivations. In addition to being virtually celibate, Eliot endured a miserable marriage for many years, seemingly as part of the bargain that he had forged to lead the poetic life. Revealingly, following his divorce, and especially following his second marriage, he became much happier, but also much less productive.

And what of the other figures? Graham did not renounce the pleasures of the flesh, but she was wary of maintaining an intimate love relationship for many years, denying herself not only a spouse but also children. She seems to have been celibate once her marriage broke down. Stravinsky seems not to have been notably ascetic or abstemious, but, like Picasso, he retained a singularly cruel attitude toward other people—possibly, in his mind, a necessary part of his creative personality. The fact that he wrote legal documents on the very day that he completed *Le sacre du printemps* underscores to me the close relationship that he discerned between the committed innovator and the embattled litigant. (Cf. Picasso's reported equation of Rape and Work). Of these seven figures, Einstein seems the least likely to have made any kind of conscious or unconscious pact with respect to his own creativity; yet he commented so often on his distance from other individuals, and his inability to relate to them, as to suggest that he viewed this disjunction from the human sphere as part of the price he had to pay for being able to think originally about the physical world.

As with the many empirical issues just discussed, it would be an exaggeration to maintain that either of my emerging themes constitutes a prerequisite for a creative breakthrough. I have tried to introduce various nuances while considering each creative figure. But I believe that the two emerging themes do open a unique window into the experience of being highly creative. If one feels in possession of (or possessed by) an enormous talent, one may well feel that the talent comes with a price; and one may seek to make that covenant as explicit and unmistakable as possible. By the same token, when one is working at the edge of one's creative powers, invading territories never touched before, the need for help and support is unprecedently great: At least in some ways, the best model for this is the time, shortly after birth, when the caretaker helps the infant make initial sense of a new world.

REMAINING QUESTIONS

Even if this study is convincing in its major lines, it raises a host of questions. I will comment on five raised frequently by those who have become familiar with my point of view.

- *Did I select the right people?* My original intention in this study was to choose individuals representing each of the several intelligences. I then decided to add the condition that the individuals must have lived in the shadow of the twentieth century. By dealing with individuals who lived at roughly the same time and were exposed to the same general international currents, I could gain some control over at least one source of variation. At the same time, I would have to restrict the emerging portrait of creativity to its practice during a specific historical era.

Of course, within any domain, or set of domains, many individuals could have been chosen. In addition to the significance of the figures I was considering, my major criteria were individuals on whom considerable information existed, and individuals to whose work I was personally attracted. While I sympathize with those who might have preferred a cohort with fewer white males or more non-Europeans, I hope that the study will be judged on its power for explaining the work of these seven individuals, rather than on the costs of not including subjects who represent other populations.

- *Did I select the proper domains?* Again, the decision to deal with a manageable number of domains meant that many areas had to go unsampled. Dealing with a poet meant that I could not deal with a novelist; dealing with a physicist left no room for a biologist, mathematician, chemist, or astronomer; a focus on high art involved a neglect of popular art; and neither inventors nor business people nor athletes have infiltrated my sample. Again, I hope that readers can focus on the insights gained from the present analysis and, if so motivated, extend this study to other individuals, other domains, or other populations. Only additional studies will reveal whether the generalizations offered in this chapter can withstand extension to domains, eras, or individuals that I failed to sample.

- *Has my focus been too cognitive?* Without question, a study of equal length could have been carried out with an entirely different focus: a focus on personality, on conscious or unconscious motivation, on social supports. Similarly, instead of focusing on individuals, I could have focused on the field, as a sociologist might have done, or on the domain, as a historian, a philosopher of science, or a philosopher of art might have done. I focus on the cognitive area both because it is the one that I know the best and because I think it is the one that can currently provide the most illumination. At the same time, I am well aware that the cognitive story is not the whole story; I hope that I have at least construed the cognitive domain in a relatively broad way, ultimately reaching to affect, religion, and spirituality. Even so, determined cognitivists will note that I do not probe as deeply as I might have into the specific mental processes used by the creators, nor do I propose any model of information processing by the creative person.

- *Have I really focused on creativity?* While most readers will accept my list of individuals as creative, some will balk at my criteria. For example, acceptance by the field indicates to some that I am looking at popularity or worldly success rather than sheer creativity. And many will point to individuals whom they consider to be at least as creative as the members of the cohort I selected.

I do not insist on the notion of acceptance by the field because I believe that creativity is a popularity contest, but rather because I know of no other criterion that is reliable in the long run. The phrase "in the long run" is critical here; probably at any historical moment during the first half of the twentieth century, the individuals listed here would not have been considered the best by the relevant fields. Certainly, the number of negative, and even outraged, reviews of the works of Freud, Graham, or Stravinsky would give anyone pause. But I believe that, with the passage of time, individuals of merit do come to stand out; and I underscore my belief that there *is* such a thing as merit within a domain. Of course, this by no means denies the existence of many other meritorious individuals who happen to have been missed by the field. It is just that we have not even heard of these people, or, if we have, we do not (at least yet) quite know what to make of them.

- *To what extent are the results of this study limited to the modern era?* Having deliberately selected individuals who are roughly contemporaneous, I am not in a position to say whether my findings about creativity would apply to another time. My own guess is that certain findings are time-bound, while others would have recurred whether I was looking at ancient Athens, Renaissance Italy, France in the Enlightenment, or China in the T'ang dynasty. However, it is clear to me that other factors, such as the nearly instant availability of information about one's own domain and about events in the world, color the picture I have presented. Moreover, I think that the variety of creativity on which I have focused, with its radical, revolutionary breakthroughs, is characteristic of our own era in the West, rather than a generic property of all creativities in all societies. Whether the picture that I have presented, in turn, helps us understand the nature of the era and the culture in which we *do* live is a question to which I turn in the epilogue.

EPILOGUE:
THE MODERN ERA AND BEYOND

IN CHARACTERIZING AN ERA with a specific label—indeed, in defining an era altogether—one runs the risk of making a claim that cannot be substantiated. It is perhaps least contentious simply to sort on the basis of numerical units, contrasting the seventeenth century with the eighteenth or, perhaps in *Time* magazine fashion, treating each decade of the century as a separate entity. However, purely chronological division has its costs. In drawing an arbitrary line at 1800, for example, one may ignore political watersheds like 1776 or 1789 or 1815, each of which seems far more important for the understanding of historical trends. And one misses the opportunity to define epochs in meaningful terms—for example, the period from 1815 to 1914 (a period of relative peace on the European continent) or the period of 1914 to 1989 (the two world wars and the cold war).

THE PROBLEM

In writing of the modern era, I am clearly transcending a purely chronological metric and averting a political delineation of eras. The term *modern era* is put forth in the same spirit as one might speak of the Renaissance, the Reformation, the Enlightenment, or the romantic era. And just as each of these refers, roughly, to the centuries following 1500, the term *modern era* is designed to refer to the personalities, events, and, above all, the ideas that have dominated the twentieth century in the West. At the same time, I intend no slavish adherence to the span 1900 to 2000. At least two of these seven modern masters, Freud and Gandhi, accomplished considerable work before 1900; and the last of the surviving titans, Graham, died in 1991.

My focus, here, has been on breakthroughs that occurred, as I have phrased it, in the shadow of 1900—roughly speaking, the era from 1900 to 1939, from publication of *The Interpretation of Dreams* to the death of Freud. If Picasso, Stravinsky, Eliot, Einstein, Graham, and Gandhi had stopped their activities around the beginning of the Second World War, their contributions would still be considered epoch-making. And most authorities would agree that, after a high watermark at mid-century, the most definitive aspects of the modern era have gradually receded; perhaps we are already well into a postmodern era or some new, yet-to-be-codified period of human history.

As mentioned in the Introduction, I do not want in any sense to suggest the existence of some kind of invisible hand that guides the course of history. At the same time, I do believe that certain themes cut across the accomplishments of these individuals and help to confer a common spirit on the age. These commonalities can be accounted for entirely in nonmystical terms. To have lived at a time when certain experiences (such as access to mass communications) pervaded the Western world, to be conscious of the activities of one another, and to have experienced cataclysmic events like the First World War suffice to encourage, even if not ensure, the stimulation and dissemination of certain ideas and practices. There may even be a productive sense of rivalry. The novelist Norman Mailer comments: "Look at who Picasso's contemporaries were—Freud, Einstein. Picasso wanted to compete with the greatest men alive and he really had some competition."

One terminological point: In using *modern era*, I intend to span the contributions of each of the seven creators. The term *modernism* has a different connotation, because that term refers specifically to a certain formally oriented artistic movement at this time, of which Picasso, Stravinsky, Graham, and Eliot are each representatives. There is little question that these artists have a considerable amount in common, more than they share with the other masters of the modern era. In what follows, I use *modern era* as a generic term, spanning all of the domains; the expression *modernism* will be reserved for an artistic movement of the time. I contrast it with *postmodernism*, the term usually applied to the artistic and cultural movement that has succeeded modernism.

THE BACKGROUND

According to some historians of ideas, the modern era actually began around 1500. At this time, a wedge began to be driven between religious and secular society, various movements of protest gelled, the spirit of scientific inquiry and philosophical critique, and a more open, pluralistic, and tolerant tone could be discerned in the West. I have no objection to this formulation per se: The ideas that animated the twentieth century may well have been initially planted several centuries earlier. Yet the society that existed around 1600 or even 1800 was quite different from the one at the turn of the nineteenth century into the twentieth, and the artistic and scientific forms that had become common coin by the late nineteenth century could not even have been envisaged at the height of the Renaissance.

From one vantage point, the nineteenth century was a period of unparalleled calm and prosperity. In contrast to the Europe of the Napoleonic era or of the future world wars, nineteenth-century Europe was mostly at peace. Progress seemed the order of the day, as an increasingly prosperous, confident, and powerful middle class asserted itself in various societal and cultural domains. The United States experienced its cataclysmic Civil War, but it also grew astonishingly in size, ambition, and power. The world beyond Europe and the United States was more turbulent, but trends toward independence, democracy, urbanization, and industrialization could be detected in pockets of Latin America, Africa, and Asia as well.

One does not have to be a revisionist historian to call this progressivist or Panglossian view into question. Urbanization and industrialization entailed severe costs in many quarters. The exploitative practices of colonization exacted costs from the colonizers and enervated those whose land was occupied and whose culture was suppressed by alien forces. Slavery and serfdom may have ended as legal matters but their consequences were not as readily attenuated. Less visibly, even those lands and populations that were "surviving most fitly" experienced doubts about purpose, direction, and the quality of life.

It was into an increasingly uncertain and unsettled world that the seven modern masters were born in the latter years of the nineteenth century. Interestingly, and perhaps revealingly, most were born in smaller communities outside the great urban centers and led youthful lives that were at least somewhat sheltered from the most punishing aspects of the Industrial Revolution. Their families were at least moderately comfortable, and, while in some instances personally religious, they generally showed tolerance toward "free thinking." They embodied and passed on to their children the bourgeois values of hard work and high achievement. At a time of much exploitation of young children, the future modern masters were spared the most awful traumas, though they were not necessarily happy during the formative period of their lives.

As if under the control of the same strong magnetic force, our modern masters all found themselves drawn, as young adults, to the major cities of Europe or North America. They were the "youths" of Milosz's poem, which appears at the front of this book. In these cities, the masters found like-minded youths; set up their study groups, or artistic or scientific circles; launched iconoclastic journals and performances, and experienced a period of intellectual gestation that ultimately led, in each case, to important creative breakthroughs.

A spate of books, exhibitions, and studies of the major cities of Europe around 1900 have appeared recently. Discounting a certain retrospective halo effect, there does seem to have been something magical about London, Paris, Berlin, Zurich, Budapest, and other metropolises around the turn of the century—the pride of accomplishment and a sense of possibility, accompanied perhaps by equally potent glimmerings of anxiety and foreboding. Even if Freud loathed aspects of Vienna, even if Picasso felt estranged in Paris, they would not have easily been persuaded to move away from these centers of activity.

Each of these sites had its peculiarities. St. Petersburg (where Stravinsky lived) harbored a remarkable, free-thinking intelligentsia living in the midst of a reactionary regime. The seat of the waning Hapsburg empire, Vienna looked backward in history and forward in the arts, prompting the wit Karl Krauss to quip that it was a "test lab for the end of the world." Budapest spawned an impressively meritocratic educational system, leading to a highly educated elite that seemed out of step in this somewhat removed corner of the Austro-Hungarian empire. London remained smug in the wake of the Victorian era, but the decline of the long-dominant liberal middle class had already begun.

Perhaps the most interesting contrast involved Paris and Berlin. Paris featured an enviable balance between the old and the new, between the national and the foreign: Individuals from all over the world were attracted to the beautiful French capital with its influential avant-garde, while Parisians, in turn, were fascinated by the foreign cultures of Italy, Russia, Germany, and the Far East. Berlin was the most determinedly modern of European cities—energetic, dynamic, improvised. The population grew at a staggering rate and the military was becoming powerful, while the recently formed German state was attempting to find its center and to define its mission. Open to new ideas, opposed to French hyperrationality, uncomfortable with Anglo-Saxon materialism, Berlin saw itself as a harbinger and a blueprint for a new kind of spiritual existence. Whether it would ultimately veer in the direction of Kantian ethical balance, the Wagnerian mystique, the Nietzschian Superman, or the military-industrial might of the Prussian state was still open.

The talented young individuals who had gravitated to these cities were aware of these potent pulls. The start of a new century signified a time of opportunity, a time to reject the weight of the past and to forge a future world along their own lines, a time to address the tensions and uncertainties that lingered beneath the surface. These sensibilities were most apparent in the arts. Diaghilev's Ballets Russes treated the arts as a unique form of deliverance and regeneration. Convinced that the English literary art of the time was third-rate, Eliot believed that poetry had to be completely reconstituted to capture the rhythm and tone of the modern era. Picasso, Braque, and Matisse rejected impressionism no less than academicism: They were attracted to the art of primitive societies, which they found refreshingly uncontaminated by the smugness of bourgeois society or the pseudo-scientism of the intelligentsia. Einstein concluded that all attempts to rescue traditional physics were fundamentally flawed. The Austrian novelist Robert Musil described the feeling of the era in his *Man Without Qualities*:

> Out of the oil smooth spirit of the last two decades of the nineteenth century, suddenly throughout Europe, there rose a kindling fever. Nobody knew exactly what was on the way; nobody was able to say whether it was to be a new art, a New Man, a new morality, or perhaps a reshuffling of

society. So everyone made of it what he liked. But people were standing up on all sides to fight against the old way of life. . . . Talents developed that had previously been choked or had taken no part at all in public life.

Historians confirm that the feeling of dramatic change, and of possible apocalypse, was in the air. While the First World War did not erupt until August 1914, it was incipient throughout the preceding decade. Virginia Woolf declared that "in or about December 1910, human nature changed." The French writer Charles Peguy declared in 1913 that "the world has changed less since Jesus Christ than it has in the last thirty years." The forms of *Les demoiselles d'Avignon* depicted a social fabric being torn apart. And, perhaps most symbolically, *Le sacre du printemps* occasioned virtual warfare in the performance halls, as the established audience signaled its dismay at a form of expression that it could not understand and that seemed to threaten its spiritual core. This radical work seemed to argue that the meaning of life could only be found in death, that there was a tie between death and ecstasy, and that destruction was a necessary way station en route to creation.

Once the First World War had erupted, all the intimations seemed but a pale foretaste of what was actually happening. The web of optimistic assumptions that had survived the nineteenth century came crumbling down, as the part of the world that had considered itself the most advanced was tearing apart its own vitals. At the time, the British social commentator Leonard Hobhouse said: "It turns out to be in sober truth a different world from that which we knew, a world in which force had a greater part to play than we had allowed, a world in which the ultimate securities were gone, in which we seemed to see all of a sudden through a thick crust of civilization the setting forces of barbaric lust for power and indifference to life." More decisively, the historian William Pfaff comments that "it is now clear that the First World War was the most important event of contemporary Western history, one whose effects are still not exhausted."

Except for Graham, still a student at the time of the war, each of the seven masters was severely shaken in a personal sense. Even if their own lives were not physically dislocated, the lives of those around them were; and most lost family members or close friends. More pointedly, the assumptions that underlay their own professional work were now subjected to greater scrutiny. Absent the First World War and its aftermath, it is not possible to think of Eliot's *Waste Land*, Picasso's *Guernica*, Stravinsky's *Histoire du soldat*, Einstein's application of atomic theory, Gandhi's changing stance toward the British, or Freud's agonized reflections on *Civilization and Its Discontents*.

And here we come to the central reason why we may speak of individuals responding to the same general milieu. For the first time in human history, individuals drawn from cultures as disparate as India and America, from European communities as diverse as Spain and Russia, were—and considered themselves to be—members of the same world community. The rise of industrialism, the growth of cities, the ever-improving access to information about events occurring around the world, the growing feeling of uneasiness before the war, and the unbelievable carnage of war itself—all of these factors ineluctably influenced the consciousness of these modern masters.

DEFINING THE MODERN ERA ACROSS DOMAINS

As early as the 1840s, the French poet Baudelaire was already beginning to define the modern temper in the arts. Writing about the painter Constantin Guy, he said, "[Modernity] is the experience of life lived in fragments, the swift pace of change in our time, fragmenting experience." And he added, "By modernity I mean the ephemeral, the contingent, the half of art whose other half is eternal and immutable." It is not a mere coincidence that Baudelaire lived in Paris: As the critic Walter Benjamin was to note a century later, it was in Parisian urban life that the peculiar rhythm and quality, the time and the space, of modern existence first appeared.

As captured in the works of Picasso, Stravinsky, Eliot, and Graham, the modern element in art displays certain features. To begin with, in contrast to the art of earlier centuries, it disdains plot, melody, linearity, virtuosity, canonical forms, an explicit moral stance, a full-blown treatment of personalities and scenes—all features one might expect in the works of traditional academic artists. In its place is a determined effort to capture the feeling of everyday experience, through snapshots, fragments, pulsing rhythms, and sharp accents. This artistic work may find its roots in common experience, but it tends toward abstraction, toward capturing the formal elements that help define the universal aspects of experience. Yet, unlike the art of a slightly later era—for instance, that of the abstract expressionists or of serial composers at mid-century—it avoids complete abstraction, pulling back as it nears the void of nonrepresentationality, and it spurns the elements of complete randomness.

Modern art arises in a context of constant change. It makes a determined effort to question tradition altogether, and, as the critic Harold Rosenberg puts it, to create a "tradition of the new." It deliberately crosses the once firm line between high art and low art, as epitomized by the juxtaposition of idiosyncratic contemporary elements and instantly identifiable traditional elements in Picasso's canvases, Stravinsky's scores, Graham's dances, and Eliot's verses. It challenges a particular face of the human condition—the mass, bureaucratic, faceless or hollow person. And yet, such art is not conceivable except in relation to the traditions that have preceded it, and the mixing that it permits is achieved unambiguously from the stance of elite art. In that sense, it is pointedly opposed to the self-stylized nihilism and "ignorabimus" of postmodern art.

Discerning common elements in artistic modernism is a relatively straight-forward task. But a search for parallel features proves more problematic once one ventures beyond the arts. In the political sphere, the modern aspects are most obviously evident in what might be called, paraphrasing Bertold Brecht, the "theater of politics." In the propagandistic rallies and films put forth in totalitarian (and, one must add, some nontotalitarian) societies, aspects of modern presentation are certainly exploited. There was little innovative about Hitler's or Stalin's or Mao's totalitarian programs: A long history of tyrants preceded them in model, if not in scale. But their advertising techniques, staging of demonstrations, and propaganda methods reflect the sensibility that was defined by the modern masters I have described. It is no accident that Hitler's architect Albert Speer was interested in the dance theories of Mary Wigman; that Stalin had a love-hate relationship with the movie director Sergei Eisenstein; and that Franco craved the artworks of his nemesis, Pablo Picasso. Yesterday's heresy becomes the sound bite of today's politician.

The religious and political innovations of Mahatma Gandhi are remote from the Nazi's "aesthetics of murder" in which one can discern traces of Nietzsche, Wagner, and Stravinsky. This is not only because Gandhi came from India, but also because his methods are in many ways a throwback to forms of human contact that characterized an earlier, more intimate, and less complex era. To be sure, Gandhi made use of the most advanced forms of communication, and his efforts could not have been successful without telegraph, printing, and other of the contemporary mass inventions that he generally decried.

Gandhi can be thought of as a modern personage in at least one sense: He questioned the assumptions of traditional power politics and sought to strip down the political encounter into its most basic elements—a human being nakedly confronting another human being. This effort to strip down a domain to its quintessential components is a telling symptom of a modern temperament. And, I would submit, Gandhi probably would not have arrived at his method if he had not lived for many years in disparate corners of the British empire and absorbed the challenging writings of Thoreau, Ruskin, and Tolstoy.

Arts and politics, as parts of the contemporary world, can be expected to reflect secular events. It is not surprising when a poet or a politician reacts to an event as dramatic as a famine, a depression, or a world war. Science and mathematics are different kinds of domains. Workers in such domains seek principles, rules, and patterns, which presumably operate in a universal fashion, irrespective of what is happening in the next community or halfway around the world.

In that sense, scientific advances, like Darwin's evolutionary analysis or Einstein's theory of relativity could have been made at any historical moment, just as the works of the ancient Greek philosopher and mathematician Pythagoras, the nineteenth-century works of the German mathematician and astronomer Carl Gauss, or the twentieth-century works of the Austrian-American mathematician

Kurt Gödel did not await a specific historical event or epoch. Of course, in a literal sense, this claim is not sensible. All scholars build on the work of their predecessors, and Einstein was unthinkable without Newton, Maxwell, Lorentz, and others. Also, technical advances, mathematical discoveries, and experimental results often underlie a discovery. And it is not mere metaphor to note that the world described by quantum mechanics resembles the programmed world captured by modern artists. Still, I believe that scientific and mathematical advances are more insulated from the effects of the daily world, and that work in the social or behavioral sciences, like that of Freud, falls somewhere in between the pristine principles of the physicist and the practical purview of the poet.

Even in the cases of Freud and Einstein, however, the influence of the modern world proves more than negligible. It was burgeoning knowledge about the microscopic world of the atom and the macroscopic worlds of space and time that animated Einstein's curiosity; it was the advent of certain kinds of neuroses in the urban-layered society of Vienna that caught Freud's attention. Both working in a post-Newtonian (but still Helmholtzian) era, they sought the basic principles that could illuminate the worlds in which they had become interested: the physical domain and the psychological domain, respectively. And, in each case, they ended up identifying a network of elementary principles or assumptions that could organize an enormous amount of data in their realms of interest. In doing so, of course, they were merely carrying out the mission of any scientist and reflecting a quest that had been vigorously pursued at least since the seventeenth century; but as I have argued with respect to Gandhi, the interest in basic underlying forms and structures is perhaps especially characteristic of the modern era.

Whatever the influence of their time on the specific scientific discoveries, Freud and Einstein unquestionably felt very much a part of their political and social era. When young, they were part of the youthful culture that was systematically questioning the religious, scientific, and political verities of the past: They were freethinkers in search of new syntheses to replace those that had failed. As they grew older, both spent an increasing amount of time pondering issues of war and peace, of community, and of religious and philosophical issues. As described earlier, their own personal acquaintance in fact derived from their mutual interests in those realms. And I think it is fair to argue as well that the particular scientific and philosophical questions in which they became interested also reflected their times: in Freud's case, his increasing concern with the impulses toward destructiveness; in Einstein's case, his lending of support to the effort to construct an atomic bomb and, thereafter, his efforts to make sure that his fearful discoveries were no longer used in the service of destruction. Clearly, then, even those who work in domains more remote from daily life are affected by the events and the "figures of thought" of the world in which they live; when a group of brilliant innovators live in the "same" world, they are likely to display certain common features. But can one go beyond this vague generalization?

In my view, any creative breakthrough involves a linkage between two seemingly disparate realms: (1) a thorough, often precocious mastery of the relevant domains of practice; and (2) a form of understanding, a variety of intuition, that is properly associated with the consciousness of human beings at an earlier point in their lives. The creative breakthrough inheres in the successful wedding of these two realms, and this fusion allows other people to apprehend the breakthrough.

This clearly is a most speculative point of view. As a first approximation, I claim that during the Renaissance (roughly 1400 to 1600), domain mastery became linked with the adolescent's capacity for hypothetic-deductive thinking and interest in "possible worlds"; this linking led to the consolidation of the scientific method and to more specific inventions and methods, such as the use of a linear perspective in drawing. The Enlightenment (1700 to 1800) fixated on the mind of the young adult, as it strained to reach the ultimate level of rationality. In the romantic era, both the sentimentality of the preadolescent and the primordial, often preverbal, emotions of the young child were captured in the arts; the sciences continued their investigative program with even greater faith in the future, while the political sphere, shaken by the revolutions at the end of the nineteenth century, remained in a sober holding pattern.

Assuming at least a surface plausibility to this speculative line of thinking, one may ask what is characteristic of the modern era. As I have drawn out in the portraits in part II, each of the modern masters uniquely exemplified a link between the childlike, in general, and the most advanced thinking in his or her domain. I specify now that the *defining characteristic* of the modern way of thinking involves a revisiting of the mind of the child at the cusp of formal schooling—the child aged four to seven. By this age, children have available a range of human symbol systems and can already use them in a generative way; they have already evolved robust theories about the physical and the psychological worlds, and they are willing to test whether these are adequate; and they manifest an ability to express themselves in terms of the elemental forms of experience, a capacity as impressive as it is often short-lived. At the same time, while already aware of convention, they are not yet unduly constrained by rules, norms, or expectations.

New insights in the nineteenth century had legitimized an interest in childhood. Taking full advantage of this legitimation, the modern masters were fascinated by childhood and looked to children—and to their own childhoods—as catalysts for their own work. We have discerned this tendency in each of the masters. Freud studied the dreams and associations of young children and located the basis of the adult personality in the Oedipal strivings and rivalries of the young child. The free associations that fascinated him most closely resemble those of a child who is in the process of mastering the symbolic systems of the culture. Einstein, the other scientist, traced his own scientific curiosity to events in his early childhood; he often resorted to the kinds of experiments that fascinate the mind of the young child.

The links to early childhood abound in the artistic realm. When but children, all of the modern masters were already fascinated with the domains of their artistry. As adults, they continued to examine the productions of young children and of populations that seemed to them primitive and childlike, and they often sought to capture such aspects in their own work. And perhaps most fundamentally, the modern masters centered their own work around the elements that are salient for the young child: Picasso, around the rough scribbles and collagelike juxtaposition of the toddler; Stravinsky, around primordial rhythms and repetitive tonal clusters reminiscent of singing in the nursery; Eliot, around the remembered images and semations of earliest childhood, the pungent phrases, and the fragmentary flavor of youthful experience; and Graham, around the stripped-down versions of movement and the basic dualities of contraction and relaxation that permeate early physical experience. In each case, these childlike fragmentary features occur in works that simultaneously call on the audience members to achieve some kind of integration—a childlike sense of the whole.

I contend, finally, that a childlike quality also exists in the political innovations devised by Gandhi. Spurning complex rhetoric or advanced weaponry, Gandhi sought to institute (or to recapture) an unadorned situation that would define a problem clearly for all who have a stake in its resolution. Acting honestly and straightforwardly, confronting one's antagonist directly, and exhibiting a willingness to sacrifice for what one believes in are all notions that a child can understand and that may prove easier to endorse if one is not yet jaded. Gandhi's own mien and appearance were also childlike. Religious ideas take root when individuals are still young; and satyagraha would be most likely to work successfully if it became part of an individual's rearing from the earliest years of life. Alas, other, less palatable features of the child's world often undermine this process before it can become established.

In arguing for significant links between the world of the young school child and the world of the accomplished master, I have sought to capture something that is special about the achievements of the modern era. Moreover, I suggest tentatively that other eras in human history may have been characterized by an analogous link between adult mastery, on the one hand, and some earlier period of individual human consciousness, on the other. I do not want to overstate the link: Without question, the modern masters retained links to periods of their childhood; and in all probability, masters in some other eras have also had a special affinity with early childhood. To put it differently, there have been "modern" persons in many historical eras. Still, if the modern era differs from others, one basis for this difference may be its privileged links to one particularly pregnant moment in the life of the young child.

By drawing these analogies between complex breakthroughs and the mind of the young child, I do not in any way wish to demean the quality of the achievement. None of these achievements would have been possible had not the creator already reached early in adulthood the uppermost limits of his or her domain as then currently constituted. Their creative breakthroughs represent the mature products of mature persons. Yet, it may well be part of the birthright of the most creative individuals that they retain a privileged access to sensations and points of their earlier development, including the years of early childhood. As Baudelaire once remarked, genius is the ability to recapture one's childhood at will. My point is not to denigrate creative masters: It is to extol the amazing power of childhood, as well as its startling endurance, at least in certain individuals.

BEYOND THE MODERN ERA

If my account is designed to illuminate the events and breakthroughs that occurred in the shadow of 1900, it should not be equally applicable to all other eras. For this reason, I conclude by touching on the period beyond the modern era, often called "postmodernism" or "beyond modernity," and determining the extent to which it may differ from that finite period I have considered in this book. This discussion applies especially to the arts; I am not prepared to speculate about events in the sciences or politics in the period after 1950.

As we are much closer to postmodernism (and do not know its dates!), describing it succinctly and confidently is more difficult. One characterization assumes that postmodernism is simply a heightening of modernism: If modernism is ironic, postmodernism is even more so. A second stance is to see postmodernism as a reaction: If modernism spurns tradition, postmodernism revels in it. There are also more positive characterizations of postmodernism—for example, as an effort to reintroduce personal, cultural, historical, subjective, and political features into human creations. No doubt the story varies within and across domain: The composer John Cage differs from the composer John Adams; Italo Calvino is a literary figure quite different from Joseph Brodsky; and Andy Warhol, Robert Rauschenberg, Frank Stella, Anselm Kiefer, and Julian Schnabel inhabit five distinctive graphic universes.

But for me, the definitive feature of the so-called postmodern era is a deliberate blurring of genres; a defiant ignoring of historical precedents and orderings; a challenge to any effort to be deadly serious; a ready shift among styles, surfaces, and identities; a relinquishing of the effort to find meaning or structure beneath the surface chaos; and a license of "anything goes" in the worlds of creation and interpretation. There is no moral mooring, and if *The Waste Land* achieves this conclusion with regret, the postmodern version accepts it as a given, if not a virtue. In one way, these traits can be seen as a continuation of modernism, for modernism certainly challenged many established forms and practices. But modernism's challenge was mounted from within a community where these forms, practices, and values were taken very seriously, and so any reaction was seen as—and *had* to be seen as—an earnest dialogue with earlier practices.

In light of the triumph of modernism itself, however, the memory of preceding eras has dimmed; and now an effort to challenge history, tradition, and established canonical forms enjoys virtually free run. Instead of one truth challenging another, the whole notion of truth is abandoned; instead of the contemporary challenging the traditional, any sense of history is ignored; instead of high art being provocatively mingled with low art, no sense of separate arts, cultures, or traditions exists. There is impatience with the idea of change or of an avant-garde: All has already been tried; everyone has already been shocked to exhaustion; today's novelties are tomorrow's packaged goods. All authority is called into question; there are no reliable moorings; indeed once one assumes the stance of the "deconstructionist" critic, every object or text harbors within it the potential to destroy or undermine itself.

It is probably evident that my own sentiments are closer to those of the modern era than to those of the postmodern era. Even though I have lived in the postmodern era, my sensibility was formed by the modernist canon, and I am unlikely ever to embrace fully the motifs and the spirit of a postmodern age. Possibly, that is because I have my doubts as to whether the postmodern spirit can ever be sustainable in a positive sense. A revolt against tradition makes sense within a world still governed by tradition; a questioning of the typical modes of interpretation is reasonable if the traditional modes have not been adequately scrutinized or challenged; but once the memory of these older forms becomes attenuated, a continuing protest against them becomes obscure. Mannerism, spectacle, effects become all.

Of course, this critique does not apply with equal measure to the full range of forms created after the passing of the canonical Modern era: for instance, those that hearken back to a time of belief, those that find virtue in the repetition of simple forms, those that deal with political themes, or those that feature the conscious teasing of a knowledgeable audience. These "versions" can be considered comfortably in terms of the practices and standards of earlier eras, including the modernist era.

One can view postmodernism in a much more positive spirit than I have. From one vantage point, the modernists are seen as still caught in a traditional bind: Though critical of progressivist or materialistic or rationalistic or deterministic modes of discourse, they know no others, and so they must ultimately adhere to them, if against their better judgment. Only the postmodern spirits have been able to throw off these intellectual carapaces and contemplate experience with an unjaundiced eye. From another vantage point, postmodernism can be seen as liberating. For perhaps the first time in human history, the various biases and prejudices that have dictated consciousness can be seen as such; and it may be possible to pursue practices and to cultivate a sensibility in which no person or work is considered more privileged than any other, or in which each vantage point is seen at once for what it highlights and what it obscures. In the philosopher Stephen Toulmin's revisionist version, a postmodern sensibility harkens back to the humane and tolerant atmosphere of the early Renaissance: From this perspective, much of the modern era is seen as hierarchical, elitist, authoritarian, and accepting of untenable dualities, such as that between reason and emotion or that between humanity and nature. In my view, however, the politics of postmodernism remains up for grabs: To some, it appears more liberal, democratic, and multicultural; to some, it simply acknowledges the political aspects of all ideas and work; while to others, it seems more faith-oriented and authoritarian.

What stance does the postmodern era assume vis-à-vis the childhood of human beings? As I have characterized it, full-blown postmodernism poses a direct challenge to childhood as a set of distinct stages. Rather, the assumption seems to be that, just as one can wander freely across eras and cultures, so, too, one can ramble indifferently across the ages and stages of childhood, sometimes dipping back into infancy, sometimes perseverating in early childhood, sometimes adopting the rigidity of the child in middle years, or the studied openness of the adolescent. The hybrid vehicles of postmodernism—MTV, "designer" shopping malls, theme parks, computer culture—belong equally to all groups and all ages. During an era in which all cultures are in touch with one another, and in which children are exposed from early on to the knowledge, mysteries, wonders, and horrors of the world, perhaps childlike innocence must forever be sacrificed. No wonder that many pundits have spoken nostalgically, or with alarm, about the "disappearance of childhood."

And here, I think, we encounter the way in which the modern era was not as modern as its apostles believed. While desperately hoping to overthrow the burdens of the past, the seven modern masters were in fact steeped in that past—be it religious, historical, traditional, academic, or some combination thereof. They stepped to the brink of the abyss but were called back: partly because of increasing age, but partly because they continued to see a reason for tradition—in the case of Eliot or Stravinsky, quite vividly so; in the case of Freud or Picasso or Graham or Gandhi, with marked ambivalence. By the same token, they retained a respect for their own childhood and for childhood in general, thereby paying tribute to their own pasts. In a sense they were rooted in two locations: their (and their peers') early childhood and their most sophisticated mastery. Modernism is freeing, but only if this freedom is purchased in terms of a recognition of history and of prior constraints. One consequence of postmodernism is to deny that past and that childhood, to question the reasons for any form of constraint, to expunge any rootedness.

Such a denial of childhood, or such an annihilation of the past, could possibly work; but signs are already mounting that a more virulent counter-reaction may ensue in the aesthetic domain, and perhaps in others as well. Rather than proceeding indefinitely in the direction of greater openness, tolerance, or blurring of genres, human beings may be condemned to oscillate back and forth between periods of innovation and tradition, between modernism and historicism, between creative breakthroughs and periods of stasis or regression that may result in tribal destructiveness. A modern spirit à la 1900 may be as far as human beings are capable of venturing, and perhaps they can only reside in that bracing region of the mind for brief intervals of history.

NOTES

CHAPTER 1. CHANCE ENCOUNTERS IN WARTIME ZURICH

3   The following quotations are from Stoppard, 1975: "Zurich during the war . . . " is on p. 98; "Literature must become party literature . . . " is on pp. 85–87; "Doing the things . . . " is on p. 38; "You are an overexcited . . . " is on p. 62.

4   For more on Malraux's museum without walls, see Malraux, 1963.

4   For more on McLuhan's global village, see McLuhan, 1964.

6   For a discussion of different dates for modernism and the modern era, see Johnson, 1991; Lutz, 1991; and Toulmin, 1990.

11   For more on Eliot's borderline mental disturbance, see Lutz, 1991.

13   On challenges to the notion of a zeitgeist, see Gombrich, 1979, 1991; and Popper, 1964. For a defense of this notion, see Boring, 1950; and Kroeber, 1944.

14   For a discussion of eras characterized by underlying assumptions about the nature of knowledge, see Foucault, 1970.

15   For more on standard stories of the end of the century and the rise of the modern era, see Berman, 1988; Dangerfield, 1935; Eksteins, 1989; Schorske, 1979; Strachey, 1988; and Varnedoe, 1986. For more on revisionist views, see Gay, 1984; Showalter, 1990; Toulmin, 1990; and Wilson, 1990.

15   For more on Vienna from 1890 to 1920, see Janik and Toulmin, 1973; Schorske, 1979; and Varnedoe, 1986. For views of other cities, see Gyongyi and Jobbagyi, 1989; and Lukacs, 1991.

16   For details on Eksteins's argument, see Eksteins, 1989.

17   On the necessity of challenging convention, see Martindale, 1990.

CHAPTER 2. APPROACHES TO CREATIVITY

19   For more on studies of intelligence, see Block and Dworkin, 1976; Gardner, 1983; and Sternberg, 1985.

19   For more on intelligence testing, see Sternberg, 1985.

19   On Guilford's call for a scientific focus on creativity, see Guilford, 1950.

20   For a more complete discussion of psychometric measures of creativity, see Guilford, 1950; Torrance, 1988; Vernon, 1970; and Wallach, 1971. For a critical review, see Gardner, 1988b.

20   On the intelligent person and the creative person compared, see Getzels and Jackson, 1962; and Wallach, 1971.

20   Suggestive findings about the validity of psychometric measures of creativity are discussed in Torrance, 1988.

21   For more on Gestalt psychology, see Koehler, 1969.

21   On Gestalt investigations of creativity, see Duncker, 1945; and Wertheimer, 1959.

21   For more details on cognitive science approaches to creativity, see Boden, 1990; and Perkins, 1981, 1991.

21   Artificial intelligence investigations of creativity are discussed in Langley et al., 1987.

21   For Csikszentmihalyi's critique of artificial intelligence efforts, see Csikszentmihalyi, 1988.

21   For more on cognitive studies of specific domains, see Johnson-Laird, 1988.

22   Holistic studies on Darwin are discussed in Gruber, 1982.

22   For a more complete discussion of the evolving systems approach, see Gruber, 1982; Gruber and Davis, 1988; and Wallace and Gruber, 1990.

23   On personality studies of creative individuals, see Barron, 1969; and MacKinnon, 1962. For a critique, see Weisberg, 1986.

23   For details on Freud's psychoanalytic approach, see Freud, 1959, pp. 43–44; and Nelson, 1958.

23   Freud, "before creativity . . . " is from Freud, 1961b, p. 177.

23   Freud, "the nature of artistic attainment . . . " is from Freud, 1967, p. 119.

24   Freud, "Might we not say . . . " is from Freud, 1959, pp. 143–144.

24   For more on the Freudian tradition of creativity analyses, see J. Gedo, 1983; and Murray, 1981.

24   Skinner's behaviorist view of creativity is described in Skinner, 1953.

25   For details about experimental social psychological demonstrations, see Amabile, 1983.

25   The flow experience is described in Csikszentmihalyi, 1990; and Csikszentmihalyi and Csikszentmihalyi, 1988.

25   For more about historiometric studies, see Kroeber, 1944; Martindale, 1990; and Simonton, 1984, 1988, 1989, 1990.

31   For more on initial romance with a subject, situation, or person, see Bloom (with Sosniak), 1985; and Whitehead, 1929.

31   Crystallizing experiences are discussed in Walters and Gardner, 1986.

31   Evidence on the need for ten years of practice is discussed in Hayes, 1981.

31   For a full discussion of the microgenesis of creative works, see Gardner and Nemirovsky, 1991; and Wallas, 1926.

35   Medawar, "Creativity is beyond analysis . . . " is from Medawar, 1969, p. 46.

35   For more on the subpersonal analysis of creativity, see Findlay and Lumsden, 1988; and Gardner and Dudai, 1985.

35   For more on the impersonal analysis of creativity, in the context of domains, see Feldman, 1980.

35   For more on the multipersonal analysis of creativity, see Cole, 1979; Csikszent-mihalyi, 1988; Martindale, 1990; Merton, 1961, 1968; Simonton, 1989, 1990; and Zuckerman and Merton, 1972.

36   The three nodes of creativity are described in Csikszentmihalyi, 1988.

37   For further discussion of the structures of a domain, see Kuhn, 1970.

38   Fruitful asynchrony is explored in Gardner and Wolf, 1988.

CHAPTER 3. SIGMUND FREUD: ALONE WITH THE WORLD

48   For more on the Wednesday evening group's activities, see Clark, 1980, p. 214; Jones, 1961; and Nunberg, 1962. See also Sulloway, 1983.

48   Jones, "I was not highly impressed . . . " is quoted in Clark, 1980, p. 215.

49   For details on Freud's early childhood, see Clark, 1980; and Jones, 1961.

49   Freud, "At the Gymnasium I was . . . " is from Freud, 1935, p. 13.

49   For more on Freud's being given his own room, see Clark, 1980, p. 19.

49   Details about the piano's removal are in Jones, 1961, p. 15.

50   J. Freud, "My Sigmund's little toe . . . " is quoted in Jones, 1961, p. 16.

50   For Goethe's essay "On Nature," see Gay, 1988, p. 24.

50   Freud, "felt no particular partiality for the position . . . " is quoted in Gay, 1988, p. 25.

50   Freud's wide readings are discussed in Gay, 1988; and Sulloway, 1983, p. 468.

51   Freud, "more towards human concerns . . . " is in Freud, 1935, p. 13.

51   Freud's comments about his future biography appear in E. Freud, 1960, p. 140.

51   Freud's comments on the merits and demerits of a paper on which he was working, the route to greatness, and the key to fame appear in E. Freud, 1960, pp. 98, 57, and 73, respectively.

52   Freud, "Strange creatures are billetted . . . " is quoted in E. Freud, 1960, p. 68.

52   Freud's work in the laboratory of Bruecke is described in Jones, 1961, pp. 33ff.

53   Freud's virtual discovery of the neuron is described in Clark, 1980, p. 44.

53   Freud, "not so sure . . . " is quoted in Clark, 1980, p. 44.

53   Freud's positive discovery of cocaine is discussed in Jones, 1961, pp. 52–67.

54   Freud, "succeeded in proving . . . " is quoted in Clark, 1980, p. 72.

54   Freud, "I do nothing . . . " is quoted in E. Freud, 1960, p. 188.

54   Freud, "possibly reach . . . " is quoted in E. Freud, 1960, p. 202.

55   For Erikson's description of Freud's twenties as a "psychosocial moratorium," see Erikson, 1959.

56   On Pappenheim's waking up from hypnosis, see Clark, 1980, p. 102.

56   For details about Pappenheim's symbolic relationship, see Freud, 1962, p. 34.

56   Freud, "[The cure] brings to an end . . . " is quoted in Clark, 1980, p. 131.

56   Freud, "Now comes Dr. B's child." is quoted in Gay, 1988, p. 67.

56   For Freud's comments on his "unpretentious theory," see Freud, 1935, p. 39.

57   Freud, "The development of psychoanalysis . . . " is in Freud, 1935, p. 34.

57   On Breuer's preference for purely physiological explanations, see E. Freud, 1960, p. 413.

57   For more on the presentation "The Etiology of Hysteria," see Jones, 1961, p. 171.

57   Freud, "Whatever cause and whatever symptom . . . " is from Freud, 1962, p. 119.

57   Krafft-Ebing, "It sounds like a scientific . . . " is quoted in Clark, 1980, p. 158, and in Jones, 1961, p. 171.

57   Freud, "an icy reception . . . " is quoted in Clark, 1980, p. 148.

57   Freud, "a solution to a more than . . . " is quoted in Clark, 1980, p. 158.

58   Freud, "absurd, wildly conjectural . . . " is quoted in Clark, 1980, p. 148.

58   For more on the letters to Eduard Silberstein, see Boelich, 1990.

59   For details about Fliess's influence on Freud, see Sulloway, 1983, pp. 135–237.

59   Freud, "One finds scientific support . . . " is quoted in Gay, 1990, p. 169.

59   Freud's letters to Fliess are in Masson, 1985: "I am pretty much alone . . . " is on p. 74; "If both of us . . . " is on p. 180; "Oh, how glad . . . " is on p. 243; "I have finished nothing . . . " is on p. 261; "I have resigned myself . . . " is on p. 430.

60   Freud, "At that time . . . " is quoted in Clark, 1980, p. 141.

60   Freud, "When I look back . . . " is from Freud, 1963, p. 304.

61   For a discussion of pre-Freudian work on the unconscious, see Clark, 1980; and Whyte, 1978.

61   On repression as the central theme of Freud's work, see Holton, 1988.

61   Freud, "The doctrine of repression . . . " is from Freud, 1949, p. 297.

62   Freud, "Here, on July 24, 1895 . . . " is quoted in Masson, 1985, p. 417.

62   Freud, "the royal road . . . " is from Freud, 1938a, p. 181.

62   On Freud's discussion of neuroses of repression and anxiety, see Clark, 1980, pp. 156–157; and Jones, 1961, p. 167.

62   For further discussion of the five different categories of neuroses, see Masson, 1985, p. 111.

63   "Project for a Scientific Psychology" appears in Freud, 1954, pp. 355–445; see also Pribram and Gill, 1976.

63   Freud, "is to furnish a psychology that shall . . . " is quoted in Clark, 1980, p. 152.

63   Freud, "A man like me . . . " is quoted in Masson, 1985, p. 129.

63   For Freud's stringing together of predicates, see Freud, 1966, p. 297; and Masson, 1985, pp. 103, 248.

64   Freud, "In the phi neurones . . . " is from Freud, 1954, p. 314.

64   On psychoanalysis as Newtonian in scope, see Gay, 1988, pp. 79–80.

64   On Freud's clinical breakthroughs discernible in scattered passages on dreams, see Freud, 1953, p. 291.

64   Freud, "the barriers suddenly lifted . . . " is quoted in Masson, 1985, p. 146; Clark, 1980, p. 154.

64   Freud, "concoct the scheme . . . " is quoted in Masson, 1985, p. 152.

65   Breuer, "Freud's intellect is . . . " is quoted in Jones, 1961, p. 157.

66   Freud, "Whenever I began . . . " is quoted in Clark, 1980, p. 147.

67 Freud, "The most important patient for me . . . " is quoted in Gay, 1988, p. 96.

68 For a discussion of Freud's recognition of the importance of sexual factors, see Gay, 1988, p. 91.

68 For more on Freud's mistake about sexual exploitation, see Masson, 1985, p. 264.

69 Freud, "the most valuable of all the discoveries . . . " is from Freud, 1938a, p. 181.

70 For details on Freud's view about all psychoneurotic symptoms as wish fulfillments, see Freud, 1938a, p. 511.

70 Freud, "I have very restricted capacities . . . " is quoted in Jones, 1961, p. 366.

70 Freud, "I have an infamously . . . " is quoted in E. Freud, 1960, p. 292.

72 Freud, "Not a leaf . . . " is quoted in Clark, 1980, p. 181.

72 Freud, "It seems to be my fate . . . " is quoted in Jones, 1961, p. 228.

72 Freud, "Klimt and Wagner . . . " is quoted in Varnedoe, 1986, p. 20. See also Janik and Toulmin, 1973; and Schorske, 1979.

73 Freud, "Vienna has done everything . . . " is from Freud, 1949, p. 326.

75 For details about Freud's professorial appointment, see E. Freud, 1960, p. 242; and Gay, 1988, p. 137. See also Schorske, 1979.

76 Freud, "honorable but painful isolation . . . " is quoted in E. Freud, 1960, p. 256.

77 For more on the Vienna Psychoanalytic Society, see Gay, 1988, p. 174.

77 For more on the International Psychoanalytic Association, see Freud, 1935, p. 96.

77 For a discussion of Freud's feud with Janet, see Sulloway, 1983.

77 For details about Freud's vigilant monitoring of those who wrote about him, see Groddeck, 1961; and Wittels, 1924.

78 For more on the suicides among Freud's followers, see Sulloway, 1983, p. 482.

79 Freud's comment on this phase as a "regressive development" is from Freud, 1935, p. 137.

80 For more on questions about Freud's integrity, see Goleman, 1990; Waldholz, 1991; and Raymond, 1991.

CHAPTER 4. ALBERT EINSTEIN: THE PERENNIAL CHILD

84 On Einstein's characterization of his autobiography as his "obituary," see Schilpp, 1949, p. 9.

84 For more about Einstein's puzzle of a box falling freely down a long shaft, see Clark, 1971, p. 118.

84 For details about Piaget's questions to young children, see Matthews, 1980; and Piaget, 1965.

85 Einstein, "How did it come to pass . . . " is quoted in Clark, 1971, p. 10.

85 For Einstein's suggestions to Piaget about studying children's notions of speed and time, see Flavell, 1963, p. 256.

85 On Baudelaire's characterization of the child as "the painter of modern life," see Gardner, 1980, p. 8.

85   Rabi, "I think that physicists . . . " is quoted in Bernstein, 1980, p. 78.

86   On the characterization of Einstein's mother as having had "a touch of the ruthlessness," see Clark, 1971, p. 22.

86   For more on Einstein's father manufacturing gadgets, see Swenson, 1979, p. 2.

86   For more on Einstein's building of houses of cards, see Hoffmann, 1975, p. 18.

86   For more on Einstein's walks through the streets of Munich, see A. Miller, in press.

86   For more on Einstein's playing alone as a child, see Hoffmann, 1975, p. 18.

86   For more on Einstein's temper tantrums, see Pais, 1982, pp. 35, 37.

87   For more on Einstein's student life, see Pyenson, 1985, chap. 1.

87   For more about Einstein's religious strain, see Hoffmann, 1975, p. 16; and Stachel, 1990.

87   For more on Einstein's arguments with geometry books, see Frank, 1953, p. 14; and Hoffmann, 1975, p. 23.

87   On Talmey giving Einstein books, see Clark, 1971, p. 15.

87   For details about the series of volumes by Bernstein, see Gregory, 1990.

87   Talmey, "the flight of (Einstein's) mathematical genius . . . " is quoted in Clark, 1971, p. 16.

87   For details on the effect of Einstein's studies on his early intoxication with formal religion, see Hoffmann, 1975, p. 24.

87   For more on Einstein's difficult adolescence, see Holton and Elkana, 1982, p. 284.

88   Einstein, "[The school] made an unforgettable impression . . . " is quoted in Holton, 1988, p. 391; and Pyenson, 1985, pp. 12–14.

88   Einstein, "Here are the causes . . . " is quoted in a *New York Times* article on Einstein's school record.

88   The following quotations from Einstein's paper sent to his uncle are quoted in Clark, 1971: "the elastic deformation . . . " is on p. 23; "I believe that the . . . " is on p. 23.

89   For details on courses Einstein attended at the Zurich Polytechnic Institute, see Clark, 1971, p. 32

89   For more on Weber ignoring the work of Maxwell, see Clark, 1971, p. 34.

89   For details about Einstein's contact with Föppl, see Clark, 1971, p. 98; and Holton, 1988, pp. 218–224.

89   Föppl, "There can be no recourse . . . " is quoted in Holton, 1988, p. 221.

90   Einstein, "What made the greatest impression . . . " is quoted in Schilpp, 1949, p. 19.

90   For more on Einstein's desire to build an apparatus to measure the earth's motion, see Pais, 1982, p. 131; and Swenson, 1979, p. 101.

90   For Einstein's ideas about a new and simpler method for the investigation of matter and light ether, see Pais, 1982, p. 132.

90   Cohen, "Like so many . . . " is quoted in Clark, 1971, p. 32.

91   For more on Kuhn's definition of preparadigmatic domains, see Kuhn, 1970.

91   Einstein, "it was as if the ground . . . " is quoted in Schilpp, 1949, p. 45.

91   For further discussion of Galilean and Newtonian laws, see Einstein, 1921, chap. 4; Frank, 1953, pp. 25ff; Swenson, 1979, p. 12.

92   The following phrases are from Newton's work: "absolute motion" is referred to in Barnett, 1948, p. 40; Clark, 1971, p. 75; Frank, 1953, p. 31; and Swenson, 1979, pp. 13–14; "'absolute time' flowed uniformly" is quoted in Nordmann, 1922, p. 22; "absolute space" is referred to in Clark, 1971, p. 75.

92   For more about Newton's doubts, see Barnett, 1948, p. 40.

92   For more on strict causality and the "relativity principle," see Clark, 1971, p. 34.

92   For Frank's assessment of Einstein's achievement, see Frank, 1953.

92   For a discussion of the question about whether ether impedes the progress of objects, see Frank, 1953, p. 22.

92   For more on Faraday's work on electromagnetic induction, see Holton, 1988, p. 10.

92   For a discussion of Faraday's concept of the field as the medium, see Clark, 1971, p. 77

92   For Faraday's discussion of the field as a change in physical state, see Schilpp, 1949, pp. 517–518.

93   For more on Maxwell's abstract, mathematically oriented work, see Pais, 1982, p. 119.

93   Maxwell, "all our knowledge . . . " is quoted in Swenson, 1979, pp. 33–34.

93   For Einstein's characterization of Maxwell's discoveries as "a revelation," see Schilpp, 1949, p. 33.

93   Einstein, "mechanics as the basis of physics . . . " is quoted in Schilpp, 1949, p. 27.

93   For Hertz's point of crucial importance, see Frank, 1953, p. 38.

94   For more on Mach's statements regarding observable phenomena, see Frank, 1953, p. 39.

94   Mach, "All masses and all . . . " is quoted in Swenson, 1979, p. 156.

94   Mach, "Every body maintains . . . " is quoted in Swenson, 1979, p. 156.

94   On the notion of the Earth moving through the ether without dragging it along, see Frank, 1953, p. 33.

94   For more on the Michelson-Morley experiments regarding the ether, see Clark, 1971, pp. 78–80; and Swenson, 1979, p. 22.

95   On arguments about the inadequacy of a mechanical explanation, see Frank, 1953, p. 45.

95   For more on the Lorentz equations, see Hoffmann, 1975, p. 67.

95   For more on Poincaré's "principle of relativity," see Pais, 1982, p. 126.

95   Poincaré, "we have no direct intuition . . . " is quoted in Pais, 1982, pp. 127–128.

96   Poincaré, "as demanded by the relativity principle . . . " is quoted in Pais, 1982, p. 128.

96   Poincaré, "perhaps we must construct . . . " is quoted in Pais, 1982, p. 128.

96   For Einstein's comment about having sold himself body and soul to success, see Rhodes, 1986, p. 113.

97   Solovine, "[Poincaré's *La science et l'hypothèse*] "profoundly impressed us . . . " is quoted in Pais, 1982, p. 134.

98   Einstein, "I lived in solitude . . . There are certain callings . . . " is quoted in Holton and Elkana, 1982, p. 268.

98   For more on Einstein's relaxation techniques, see Clark, 1971, p. 106.

98   Einstein, "For the rest of my life," is quoted in Schilpp, 1949, p. 225.

98   Einstein, "The fact that I . . . " is quoted in Hoffmann, 1975, p. 8.

98   Einstein, "In a man of my type . . . " is quoted in Schilpp, 1949, p. 230.

99   For more on Einstein's attraction to theoretical physics and to the simplest possible mathematical formulas, see Frank, 1953, p. 17; and A. Miller, in press.

99   Einstein, "The words of the language . . . " is quoted in Ghiselin, 1952, p. 43.

99   For more on Einstein's view of his thinking, see A. Miller, 1986b, pp. 43–44.

99   Einstein, "When I examine myself . . . " is quoted in Clark, 1971, p. 87.

100   Frank, "When Einstein had thought through . . . " is from Frank, 1949, p. 90.

100   Miller, "It cannot be overstated . . . " is from A. Miller, 1986b, p. 86.

100   Einstein, "When God created the ass . . . " is quoted in Clark, 1971, p. 46.

100   On Besso's rescuing of Einstein from insulting Planck, see Bernstein, 1989.

100   For more on Einstein's critique of Boltzmann, see A. Miller, in press.

100   For more on Einstein's pride in not knowing the literature of the field, see Pais, 1982, p. 165.

100   Einstein, "I have little patience . . . " is quoted in Clark, 1971, frontispiece; see also Frank, 1949.

100   For Einstein's comment about scientific discoverers as monomaniacs, see Bernstein, 1989.

100   For Millikan's characterization of Einstein as "reckless," see Holton and Elkana, 1982, p. 63.

101   For Einstein's comments on Langevin, see Goldberg, 1968, p. 219.

101   Newton, "In those days I was . . . " is quoted in a *Times Literary Supplement* review entitled, "Let Newton Be."

102   Einstein, "A practical profession . . . " is quoted in Clark, 1971, p. 51.

102   For more on the perception of young Einstein as somewhat of a failure, see Pyenson, 1985, pp. 58–59.

102   For Einstein's comment about his early papers as "worthless," see Clark, 1971, p. 52.

103   Einstein, "It is a wonderful feeling . . . " is quoted in Pais, 1982, p. 57.

103   Einstein, "Between the conception . . . " is quoted in Holton, 1988, p. 192.

103   For comments about Einstein's lack of a reference to a "fundamental decision," see Hoffmann, 1975, p. 74; and Pais, 1982, p. 139.

103   Einstein, "there is an inseparable connection . . . " is quoted in Hoffmann, 1975, p. 74.

104   Bondi, "is to leave . . . " is quoted in Clark, 1971, p. 85; see also Pais, 1982, p. 139.

104   Einstein, "From the beginning it appeared . . . " is quoted in Schilpp, 1949, p. 53.

104   For more on Einstein's pondering of a paradoxical incompatibility, see Schlipp, 1949, p. 57.

104   For a discussion of the two assumptions underlying Einstein's theory, see Infeld, 1950, pp. 23–24; and A. Miller, 1986b, p. 52.

105   For more on the reduction of Lorentz's equations, see Swenson, 1979, p. 175; see also A. Miller, 1986b, pp. 114–115; and Pais, 1982, p. 139.

105   Einstein, "We cannot attach . . . " is quoted in Clark, 1971, p. 88.

105   For more about rods appearing to contract in the direction of their motion, see Clark, 1971, p. 89.

105   For more about electric and magnetic field strengths, see Frank, 1953, pp. 63–64.

105   For a discussion of why the central notions of physics must be rethought, see Hoffmann, 1975, p. 77.

105   For details about how the mass of a moving body increases with its velocity, see Barnett, 1948, p. 61.

105   For a discussion of $E = mc^2$, see Frank, 1953, p. 65; and Swenson, 1979, p. 176.

105   Einstein, "an amazingly simple summary . . . " is quoted in Holton, 1988, p. 195.

106   Einstein, "The introduction of a luminiferous ether" is quoted in Holton, 1988, p. 311.

106   For a discussion of how mechanical and field theory could be joined, see Infeld, 1950, p. 35.

106   Einstein, "Shortly after 1900 . . . " is quoted in Holton, 1988, p. 252.

106   Miller, "resolving problems in a Gordian manner . . . " is from A. Miller, 1986a, p. 207.

108   For Cassirer's comments on the conception of constancy and absoluteness, see Holton and Elkana, 1982, pp. 187–188.

108   For more on Einstein's reaction to there being no mention of his paper in *Annalen der Physik*, see Pais, 1982, p. 150.

108   For more on Einstein's theory being confused with Lorentz's, see Holton and Elkana, 1982, p. ix.

108   For more about the pace of the discussion in Germany on relativity theory, see Goldberg, 1968.

108   For more about the young physicists who read Einstein's paper, see Clark, 1971, p. 118.

109   Minkowski, "Gentlemen! The ideas on space . . . " is quoted in Clark, 1971, p. 123.

109   For more on Einstein's Nobel Prize nomination in 1912, see Pais, 1982, p. 153.

109   For more on Mach's posthumous document, see Clark, 1971, p. 161.

109   For more on Planck's muted enthusiasm, see Goldberg, 1968, pp. 155–158.

109   For more on Poincaré's stance on relativity as an empirical assertion, see Holton, 1988, p. 204.

110   For a discussion of how Einstein proceeded from postulates, see Goldberg, 1968, p. 28.

110   Holton, "Lorentz' work can be seen . . . " is from Holton, 1988, p. 201.

110   For more on the University of Göttingen seminar in 1905, see Pyenson, 1985, p. 101.

110   For a discussion of Einstein's lack of response to empirically based attacks, see Goldberg, 1968.

111 For more on Einstein, Freud, and others' endorsement of the statement, see Clark, 1971, p. 154.

112 For more on Einstein's unsystematic collecting of his papers until 1920, see A. Miller, in press.

113 The following quotes are from Clark, 1971: Dyson, "The results of the expedition . . . " is on p. 232; the society's president, "One of the greatest . . . " is on p. 232; Whitehead, "A great adventure . . . " is on p. 232.

113 For more about the prize of thousands of dollars, see Infeld, 1950, p. 121.

113 Hoffmann, "Einstein was world-famous . . . " is from Hoffmann, 1975, p. 133.

114 Einstein, "If you will not take the answer . . . " is quoted in Frank, 1953, p. 179.

114 For a discussion of Einstein's enduring commitments, see Clark, 1971, p. 179.

114 For more on Szilard's letter, see Clark, 1971, pp. 581–583.

115 For more on the exchange between Freud and Einstein on "Why War?," see Jones, 1961, p. 462.

115 Freud, "He understands . . . " is quoted in Jones, 1961, p. 462.

115 For more on Einstein's nervous breakdown, see Clark, 1971, pp. 191–193.

115 For Infeld's description of Einstein's work on the general theory of relativity, see Infeld, 1950, pp. 48–89.

116 Einstein, "almost inertial" system is quoted in Einstein, 1921.

116 For more about the experiment on the extension of a spring, see Holton and Elkana, 1982, p. 124.

116 For a discussion of space as curved, see Swenson, 1979, p. 201.

116 Barnett, "The universe is not a rigid and immutable edifice . . . " is from Barnett, 1948, p. 85.

117 For more on the astronomers' 1919 measurement of light rays during the solar eclipse, see Clark, 1971, p. 207; and Infeld, 1950, pp. 51–54.

117 For Einstein's comment to a student about the eclipse studies, see Clark, 1971, p. 230.

117 For more on the recently discovered correspondence, see Wilford, 1992, p. C1.

117 For more on Bohr's explanation of the consensus, see Swenson, 1979, p. 228.

118 For a discussion of Einstein's view of quantum-mechanical claims, see Pyenson, 1985, p. 75; Schilpp, 1949, p. 81; and Swenson, 1979, p. 226.

118 Einstein, "God does not play . . . " is quoted in Pais, 1982, p. 440.

118 Einstein, "I find the idea . . . " is quoted in Clark, 1971, p. 211.

119 Einstein, "It is not improbable that Mach . . . " is quoted in Schilpp, 1949, p. 272.

119 Einstein, "I have thought a hundred times . . . " is quoted in Pais, 1982, p. 9.

120 Hoffmann, "No matter what he was doing . . . " is from Hoffmann, 1975, p. 249.

120 For more on Einstein's conservative streak, see Holton, 1988, p. 207.

121 For more on Einstein's last writings, see Pais, 1982, p. 171.

121 Einstein, "I want to know how God . . . " is quoted in Clark, 1971, p. 19.

121 Einstein, "What has perhaps been overlooked . . . " is quoted in Holton and Elkana, 1982, p. 247.

122   For Dukas's commentary on Einstein, see Dukas and Hoffman, 1979.

122   Berlin, "in the case of seminal discoveries . . . " is quoted in Holton and Elkana, 1982, p. 291.

123   Einstein "My passionate interest . . . " is quoted in Infeld, 1950, pp. 118–119.

CHAPTER 5. PABLO PICASSO:
PRODIGIOUSNESS AND BEYOND

128   Neurobiological aspects of prodigiousness are discussed in Gardner and Dudai, 1985, pp. 1–6.

129   For more on Wang Yani's artistry, see Ho, 1989.

129   On prodigiousness and "co-incidence," see Feldman, 1986.

130   For more about prodigies' "mid-life crises," see Bamberger, 1982, pp. 17, 61–78.

130   Berlioz, "He knows everything . . . " is quoted in Schönberg, 1969, p. 17.

130   Stein, "Picasso wrote painting . . . " is from Stein, 1970, p. 4.

131   For a discussion of Picasso's early facility as an artist, see Richardson, 1991, chap. 3.

131   For more on Picasso as a poor student, see M. Gedo, 1980, p. 15.

131   Gedo, "he anthropomorphized . . . abilities . . . " is from M. Gedo, 1980, p. 15.

132   For more about Picasso's youthful notebooks, see Glaesemer, 1984.

132   On the relation between youthful experimentation and cubism, see Staller, 1986.

132   For more on Picasso's father as a painter, see Penrose, 1981.

133   The traumatic experiences of Picasso's early childhood are described in Huffington, 1988, p. 20; and Penrose, 1981, p. 14.

134   Picasso's relation to his mother is discussed in Huffington, 1988, p. 36.

134   Richardson, "It would seem that Picasso . . . " is from Richardson, 1991, p. 29.

134   Picasso, "In contrast to music . . . " is quoted in Glaesemer, 1984, p. 30.

135   Picasso, "When I was their age . . . " is quoted in Penrose, 1981, p. 307.

135   On Picasso's facility at art academies, see Huffington, 1988, p. 32.

135   For more on Picasso's meeting other artists in Barcelona, see Barr, 1974; and Hilton, 1975, p. 10.

136   For more on the disparate subjects of Picasso's early works, see M. Gedo, 1980, p. 25; Huffington, 1988, p. 64; and Penrose, 1981, p. 50.

136   The Parisian scene, particularly as an artistic capital, is described in Richardson, 1991, p. 168.

137   For more on Picasso's contemplated suicide, see Huffington, 1988, p. 68.

138   Fagus, "they say that he is not . . . " is quoted in Richardson, 1991, pp. 198–199.

138   Fagus, "Danger lies for him . . . " is quoted in Penrose, 1981, p. 68.

138   Critiques of Picasso's blue period: "Beauty of the horrible" is quoted in Penrose, 1981, p. 63; "this sterile sadness" is quoted in Huffington, 1988, p. 69; and "a negative sense of life" is quoted in Richardson, 1991, p. 263.

139   Richardson, "served as each other's catalyst . . . " is from Richardson, 1991, p. 329.

139   Picasso, "Thinking of Casagemas . . . " is quoted in Conrad, 1986, p. 14.

139   Richardson, "whether or not Picasso . . . " is from Richardson, 1991, p. 181.

140   For M. Gedo's study of *La vie*, see M. Gedo, 1983.

142   Richardson, "Like all the major arcana in the Tarot . . . " is from Richardson, 1991, p. 274.

142   On the nature of defining works, see Barr, 1974, p. 36.

143   Apollinaire, "Never has there been . . . " is in Apollinaire, 1949, p. 22.

143   Picasso, "If I can draw as well as Raphael . . . " is quoted in Stein, 1970, p. 23.

143   For more about the Stein portrait, see Richardson, 1991, p. 36.

144   Picasso, "Don't worry . . . " is quoted in Goodman, 1976, p. 33.

144   For more on general considerations concerning *Les demoiselles d'Avignon*, see Hilton, 1975.

145   Picasso's competition with Matisse and Derain is discussed in Leighten, 1989.

145   For more on Picasso having seen Degas's private drawings, see Glimcher and Glimcher, 1986.

145   Cézanne, "You must see in nature . . . " is quoted in Barr, 1986, p. 30.

145   Picasso, "Do I know Cézanne? . . . " is quoted in Conrad, 1986, p. 15.

145   For details on the discovery of the only existing preparatory oil sketch, see Kimmelman, 1992, p. C13.

146   Picasso, "My work is like a diary . . . " is quoted in Richardson, 1991, p. 3.

147   Hilton, "there was never before . . . " is from Hilton, 1975, p. 79.

147   For more on the only two dealers who showed interest in *Les demoiselles*, see Penrose, 1981, p. 131.

147   Kahnweiler, "What I'd like . . . " is quoted in Leighten, 1989, p. 90.

147   Picasso, "Painting is freedom . . . " is quoted in Huffington, 1988, p. 109.

148   For more about the association between Picasso and Braque, see Museum of Modern Art, 1989.

148   Braque, "It made me feel . . . " is quoted in Huffington, 1988, p. 93.

149   On Picasso and Braque as Orville and Wilbur Wright, see Penrose, 1981 p. 171.

149   Braque, "We lived in Montmartre . . . " is quoted in M. Gedo, 1980, p. 85.

150   For more on cubism and the elevation of low art to high art, see Gopnik, 1983.

150   For more on cubism and children's art, see Staller, 1986; and Stein, 1970, p. 26.

150   For more on cubism and psychology, see Teuber, 1982.

150   Barr, "Mathematics, trigonometry, chemistry . . . " is from Barr, 1946, p. 74.

151   Picasso, "When we invented cubism . . . " is quoted in Barr, 1946, p. 74.

151   For more on the source of the name "cubism," see Barr, 1974, p. 63.

151   Picasso's and Braque's approaches to painting are described in Museum of Modern Art, 1989, p. 26.

153   On Picasso seeing a squirrel in a Braque painting, see Huffington, 1988, p. 84.

153   On collage and papier collé, see M. Gedo, 1980, p. 97; and Museum of Modern Art, 1989, pp. 30ff.

154   On the later phases of cubism, see Barr, 1974, pp. 88, 90, 133.

155   On Picasso's first trip to Horta de San Juan, see Penrose, 1981, p. 42.

155   On Picasso's trip to Gosol, see Richardson, 1991, p. 441; and Penrose, 1981, pp. 119–123.

155 On Picasso's later summer trips, see Penrose, 1981, chap. 5.

156 On the summer Picasso and Braque spent together in Ceret, see M. Gedo, 1980, p. 94.

156 For more on the denunciations of cubism, see Huffington, 1988, p. 117; Leighten, 1989, p. 98; and Penrose, 1981, pp. 186–187.

156 For more about Apollinaire on cubism and a wholly new conception of beauty, see Penrose, 1981, p. 188.

156 For more on the defenders of cubism, see Hilton, 1975, p. 109.

156 For Murry on cubism as great art, see Huffington, 1988, pp. 117–118.

157 For more on cubism as part of camouflage, see Blunt, 1969, p. 2.

157 For more on Picasso as an internationally renowned prodigy, see Huffington, 1988, p. 174.

158 Picasso, "Don't expect me . . . " is quoted in Huffington, 1988, p. 166.

159 Picasso's typical output is described in M. Gedo, 1980, p. 140.

159 For a discussion of Picasso staving off a collapse of inspiration, see Hilton, 1975, p. 164.

159 For more about Picasso raising the ante in his life, see Richardson, 1991, p. 227.

159 For comments on Picasso's cool ("icy") style, see Hilton, 1975, p. 126; and Huffington, 1988, p. 136.

159 Picasso, "A work of art . . . " is quoted in Huffington, 1988, p. 290.

160 Picasso, "After all, you can only work . . . " is quoted in Huffington, 1988, p. 261.

160 For more on the *Franco* series, see M. Gedo, 1979, p. 199.

160 For more on *Minotauromachie* as a child's view of sex, see Arnheim, 1962, p. 17.

161 For a general discussion of *Guernica*, see Arnheim, 1962; Blunt, 1969; and Russell, 1980.

161 Picasso: "If it were possible . . . " is quoted in Hilton, 1971, p. 261.

162 Picasso, "[All of my paintings] are researches . . . " is quoted in Arnheim, 1962, p. 14.

162 Picasso, " . . . it's not sufficient . . . " is quoted in M. Gedo, 1980, p. 3.

163 Picasso, "Basically . . . a picture . . . " is quoted in Arnheim, 1962, p. 30.

165 For more on what the bull in *Guernica* represents, see Hilton, 1975, p. 241.

166 Picasso, "I have always believed . . . What do you think . . . " is quoted in Blunt, 1969, pp. 56–57.

166 Blunt, "It combines the emotional intensity . . . " is from Blunt, 1969, p. 56.

167 For details on Picasso's range of output and activities, see Huffington, 1988, pp. 257, 454.

168 For more on Picasso's visit to the Louvre, see Conrad, 1986, p. 11.

168 Berger, "Picasso is only happy when working . . . " is quoted in Huffington, 1988, p. 421.

168 For more about Picasso not knowing whether his works were any good, see Huffington, 1988, p. 422.

168 For more on *Le mystère de Picasso*, see Clouzot, 1955.

168   On Gilot's negative portrayal in *Life with Picasso*, see Gilot and Lake, 1981.

169   For more on Picasso's abuse of others, see Huffington, 1988, pp. 203, 351, 375.

170   On Gedo's depiction of Picasso as a "tragedy addict," see M. Gedo, 1980, pp. 9, 235.

170   Picasso, "When I die . . . " is quoted in Huffington, 1988, p. 470.

170   For more on the career of Gris, see Huffington, 1988, p. 127.

170   Sabartés, "Picasso chooses friends . . . " is quoted in Richardson, 1991, p. 115.

170   Richardson, "[Picasso] never outgrew . . . " is from Richardson, 1991, p. 97.

170   For more on Picasso's relation to Matisse, see Huffington, 1988, pp. 236, 403.

170   Picasso, "I have mastered drawing . . . " is quoted in Richardson, 1988, p. 417.

170   Picasso, "in the end, there is . . . " is quoted in Gilot, 1990, p. 316.

170   For more about the two Matisse works that stimulated *Les demoiselles*, see Richardson, 1988, pp. 414–416.

171   For more on Matisse's friendship with Gilot, see Gilot, 1990.

171   On Picasso as deeply superstitious, see M. Gedo, 1983.

171   On Picasso's fear of death, see M. Gedo, 1980, p. 101.

172   For more on Picasso's time with Braque as the happiest period of his life, see Huffington, 1988, p. 97.

CHAPTER 6. IGOR STRAVINSKY:
THE POETICS AND POLITICS OF MUSIC

174   Stravinsky, "Music is by its very nature . . . " is quoted in Druskin, 1983, p. 70.

175   Craft, "Whether or not Stravinsky's letters . . . " is from Craft, 1984, p. 261.

175   Craft, "The correspondence does not include . . . " is from Craft, 1984, p. 276.

176   For more about Stravinsky's childhood memories, see Stravinsky, 1962, pp. 3–4.

176   For more on Stravinsky's early interest in improvisation, see Boucourechliev, 1987, p. 29.

177   Stravinsky, "I never came across . . . " is from Stravinsky, 1962, p. 8.

177   For more on Stravinsky's lack of interest in formal schooling, see White and Noble, 1980, p. 240.

178   Stravinsky, "No matter what the subject may be . . . " is quoted in White, 1947, p. 17.

178   Rimsky-Korsakov, "Igor Stravinsky may be my pupil . . . " is quoted in Craft, 1982, frontispiece.

179   Noble, "the distance Stravinsky . . . " is quoted in White and Noble, 1980, p. 243.

179   The early years of Stravinsky's career are described in Boucourechliev, 1987, p. 35.

180   Diaghilev, "I am firstly . . . " is quoted in Eksteins, 1989, p. 21.

180   Stravinsky on Diaghilev, "He had a wonderful flair . . . " is quoted in Boucourechliev, 1987, p. 39.

181   Benois's remarks about the young Stravinsky are quoted in White, 1947, p. 25.

181   Stravinsky: "I worked hard . . . " is from Stravinsky, 1936, p. 42.

182   Diaghilev, "Take a good look at him . . . " is quoted in Boucourechliev, 1987, p. 31; see also White, 1947, p. 27.

182   "The success of *The Firebird*" is from White and Noble, 1980, p. 244.

182   Stravinsky, "It is more vigorous . . . " is quoted in Van den Toorn, 1983, p. 2.

183   "a picture of a puppet . . . " is from White and Noble, 1980, p. 244; see also Druskin, 1983, p. 40.

183   A description of the score of *Petrouchka* is in Tansman, 1949, p. 170.

184   Boucourechliev, "It is impossible to exaggerate . . . " is from Boucourechliev, 1987, p. 52.

184   For more on the dispute about characterization, see White, 1947, p. 36.

184   Stravinsky, "As Petrouchka he was . . . " is quoted in Van den Toorn, 1983, p. 98.

184   Stravinsky, "The success of Petrouchka . . . " is quoted in Van den Toorn, 1983, p. 98.

184   For Simonton on the number of good and bad works produced by creative individuals, see Simonton, 1988.

185   Stravinsky, "There arose a picture . . . " is from Stravinsky, 1989, p. vii.

185   For more on the sources used for the early drafts of *Le sacre*, see Stravinsky, 1969; and Van den Toorn, 1987.

185   Roerich, "The new ballet . . . " is quoted in Van den Toorn, 1987, p. 34.

185   Roerich, "The action begins . . . " is quoted in Druskin, 1983, p. 64.

185   For more on there being no prospect of a performance for *Le sacre*, see Van den Toorn, 1987, p. 34.

186   Stravinsky, "accepted it with joy . . . " is quoted in Van den Toorn, 1983, p. 139.

186   Stravinsky, "I had imagined . . . " is from Stravinsky, 1962, p. 28.

187   For more on the problems associated with the composing of *Le sacre*, see Van den Toorn, 1987, p. 34.

187   For more on the reordering of scenes of *Le sacre*, see Van den Toorn, 1983, p. 31.

188   Stravinsky, "Today November 17, 1912 . . . " is quoted in Van den Toorn, 1987, p. 24.

189   Laloy, "as though by a hurricane . . . " is quoted in Boucourechliev, 1987, p. 64.

189   Hugo, "It was as if the theater . . . " is quoted in Riding, 1990, p. 17.

189   Van Vechten, "Cat-calls and hisses . . . " is quoted in Eksteins, 1989, p. 13.

189   For more on the early criticisms, see Lesure, 1980; and White, 1947, p. 44.

190   Newman, "the work is dead" and other comments, see Lesure, 1980, p. 75.

190   On the method of melodic development in *Le sacre* as a shock to ears nurtured on nineteenth-century symphonic forms, see Tansman, 1949, p. 39.

191   On the superimposition of thematic material, see White, 1947, p. 41.

191   On Nijinsky's choreography, see Eksteins, 1989, pp. 50–51. For a revisionist view, see Riding, 1990.

191   For Ravel on *Le sacre* as a novel entity, see Stravinsky, 1970.

191   Stravinsky, "What I was trying to convey . . . " is quoted in Vlad, 1967, p. 29.

191   On the listener being called on to carry out a creative, integrating function, see Boucourechliev, 1987, p. 73.

192   Debussy, "An extraordinary ferocious thing . . . " is quoted in Eksteins, 1989, p. 51.

192 Tansman, "It is difficult to tell . . . " is from Tansman, 1949, p. 17.

192 On Diaghilev's delight in the scandal surrounding *Le sacre*, see Horgan, 1989, p. 20.

192 For more on revisions of *Le sacre*, see Craft, 1982, p. 398.

192 For details on the version performed by Graham in New York in 1930, see Graham, 1991, pp. 127–133.

193 For more on the *Montjoie!* incident, see Van den Toorn, 1987, p. 5.

193 "In the *Prelude* . . . " is quoted in Van den Toorn, 1987, p. 5.

193 Stravinsky, "It is highly inaccurate . . . " is quoted in Craft, 1982, p. 55.

193 On Stravinsky's many disavowals of the *Montjoie!* article, see Van den Toorn, 1987, pp. 5–6.

194 Stravinsky, "His 'moral integrity' . . . " is quoted in Craft, 1982, p. 134.

194 Stravinsky, "Two words in response . . . " is quoted in Craft, 1982, p. 226.

194 For more on the exchanges with Monteux, see Craft, 1982, p. 210; and Craft, 1984, pp. 66–67.

194 Stravinsky: "I hold firmly to my argument . . . " is quoted in Craft, 1985, p. 55.

195 On Stravinsky's letters to his family, see Craft, 1992, chaps. 8 and 9.

195 On Stravinsky's relation to Reinhart, see Craft, 1985, p. 139.

195 On Stravinsky's wish to enter the French Academy, see Craft, 1982, p. 5.

195 On Stravinsky as unabashedly ingratiating, see Craft, 1982, p. 94; and White, 1947, p. 61.

195 On how Stravinsky reveled in negotiating, see Libman, 1972.

196 Stravinsky, "The impression which [Japanese lyrics] . . . " is quoted in Druskin, 1983, p. 126.

196 For more on the theme of a Russian peasant wedding, see Van den Toorn, 1983, p. 155.

197 On *Les noces* as Stravinsky's favorite piece, see Libman, 1972, p. 227.

197 For a description of *Les noces*, see Van den Toorn, 1983, p. 155. For more on the exposition and development of *Les noces*, see Van den Toorn, 1983, pp. 130–134.

197 Stravinsky, "so many instrumental metamorphoses . . . " is quoted in Van den Toorn, 1983, p. 156.

197 On the evolution of the composition of *Les noces*, see Vlad, 1967, p. 70.

197 Stravinsky, "I suddenly realized . . . " is quoted in Vlad, 1967, p. 70.

197 Components of the traditional wedding ceremony are described in Vlad, 1967, p. 69; and White, 1947, p. 71.

198 On *Les noces* coming from a small number of melodic scraps, see White, 1947, p. 73.

198 For more about the reactions to *Les noces*, see Tansman, 1949, pp. 186–187; and White, 1947, pp. 75–76.

199 Boucourechliev, "He was determined to make . . . " is from Boucourechliev, 1987, p. 18.

199 Stravinsky, "Picasso accepted the commission . . . " is quoted in Druskin, 1983, p. 88.

199 Stravinsky, "[Picasso] worked miracles . . . " is from Stravinsky, 1962, p. 81.

200   Stravinsky, "Pulcinella was my discovery . . . " is quoted in Boucourechliev, 1987, p. 141.

200   Stravinsky, "It is in the nature of things . . . " is from Stravinsky, 1962, p. 75.

201   For more on Stravinsky's relation to Balanchine, see Druskin, 1983, p. 62.

201   Stravinsky, "Did not Eliot and I . . . " is quoted in Druskin, 1983, p. 79.

202   Stravinsky, "I regard my talent . . . " is quoted in Boucourechliev, 1987, p. 158.

202   Stravinsky, "not only believe in the symbolic sense . . . " is quoted in Boucourechliev, 1987, p. 158.

202   Stravinsky, "I was born out of . . . " is quoted in Druskin, 1983, p. 4.

203   Druskin, "Stravinsky's work table . . . " is from Druskin, 1983, p. 11.

203   Stravinsky, "For me as a creative musician . . . " is from Stravinsky, 1962, p. 174.

203   Freud, "When inspiration does not come to me . . . " is quoted in Jones, 1961, p. 225.

203   Stravinsky, "I stumble upon something . . . " is from Stravinsky, 1970, p. 55.

203   Stravinsky, "What fascinated me . . . " is quoted in Vlad, 1967, p. 14.

203   For more on Stravinsky's tendency toward obsessiveness, see Tansman, 1949, p. 9.

203   Stravinsky, "I would go on eternally . . . " is quoted in Stravinsky and Craft, 1962, p. 197.

203   Stravinsky, "They think I write like Verdi . . . " is quoted in White, 1947, p. 126.

204   Stravinsky, "My freedom will be . . . " is from Stravinsky, 1970, p. 87.

204   Stravinsky, "I loathe all communism . . . " is quoted in Craft, 1984, p. 236.

204   Stravinsky, "I am the first . . . " is from Stravinsky, 1970, p. 15.

205   For more on Stravinsky's encounter with Rose, see Craft, 1982, p. 211.

206   Stravinsky, "this brilliant instrumental masterpiece . . . " is quoted in Vlad, 1967, p. 39.

206   Stravinsky, "Whatever opinion one may have . . . " is from Stravinsky, 1970, p. 17.

206   Schönberg, "I have made a discovery . . . " is quoted in Boucourechliev, 1987, p. 210.

207   For more on Stravinsky's brand of serial music, see Tansman, 1949, p. 58; and White and Noble, 1980.

207   Stravinsky, "The general public . . . " is from Stravinsky, 1962, p. 26.

207   "the two authors . . . " is quoted in White and Noble, 1980, p. 258; see also Libman, 1972.

207   Stravinsky, "It is not a question of simple ghost-writing . . . " is quoted in Boucourechliev, 1987, p. 251.

CHAPTER 7. T. S. ELIOT: THE MARGINAL MASTER

212   For details about the manuscript recovered in 1968, see V. Eliot, 1988, p. xv.

212   Gardner, "Pound turned a jumble . . . " is from Helen Gardner, 1973, p. 83.

213   For more details about Eliot's background, see Ackroyd, 1984; Gordon, 1977; and Sencourt, 1979.

213   Eliot, "I feel that there is something . . . " is quoted in Ackroyd, 1984, p. 23.

213   For more on Eliot's early sensory impressions, see Gordon, 1977, p. 3.

213   For more on Eliot's early poetry, see Ackroyd, 1984, pp. 28–29.

214   For details about Eliot's wide readings and efforts as a student, see V. Eliot, 1988, pp. 6–7.

214   For more on Snow's two academic cultures, see Snow, 1959.

215   For more on Eliot's Harvard experience, see Ackroyd, 1984, chap. 2.

215   For more on Eliot's use of Symons's *The Symbolist Movement in Literature*, see Gordon, 1977, p. 28.

215   Eliot, "I do feel more grateful . . . " is quoted in V. Eliot, 1988, p. 191.

215   For details about Eliot's perception of the streets shrinking and dividing, see Gordon, 1977, p. 15.

216   Eliot, "the awful daring of a moment's surrender . . . " is quoted in Gordon, 1977, p. 2.

216   For more on streets as a recurring symbol in Eliot's writings, see Clampitt, 1988.

216   Eliot, "The kind of poetry . . . " is quoted in Frye, 1963, p. 1.

217   For more on Eliot's nightly panics, see Gordon, 1977, p. 42.

217   For more on Eliot's changing handwriting, see Gordon, 1977, p. 43.

217   Eliot, "a maddeningly brief visionary moment . . . " is quoted in Gordon, 1977, p. 51.

218   Eliot, "How much more self-conscious . . . " is quoted in V. Eliot, 1988, p. 74.

218   Pound, "the last intelligent man . . . " is quoted in Gordon, 1977, p. 67.

218   Eliot, "In 1914 my meeting . . . " is quoted in V. Eliot, 1988, p. xvii.

218   Eliot, "the situation of poetry . . . " is quoted in V. Eliot, 1988, p. 33.

220   On the comparison of Eliot with Keats, see Bate, 1963.

221   For Eliot's comment on "Prufrock" as his swan song, see V. Eliot, 1988, p. 151.

221   Eliot, "I should be better off . . . " is quoted in V. Eliot, 1988, p. 75.

222   For more on Russell's affair with Vivien, see Sencourt, 1979, p. 62.

222   Eliot, "I have lived enough . . . " is quoted in V. Eliot, 1988, p. 126.

223   Eliot, "to secure introductions . . . " and "to have these magazines . . . " are quoted in V. Eliot, 1988, pp. 104–106.

223   Eliot, "There are only two ways . . . " is quoted in V. Eliot, 1988, p. 285.

224   Eliot, "It is time I had . . . " is quoted in V. Eliot, 1988, p. 245.

224   For Eliot's reference to a "long poem," see V. Eliot, 1988, p. 44; and Gordon, 1977, p. 87.

225   For Eliot's reference to his own declining health, see letter to Henry Eliot in V. Eliot, 1988, p. 471.

225   For a discussion of the complete story of the composition of *The Waste Land*, see, for example, Litz, 1973.

226   For details about Pound's editing, see V. Eliot, 1988; and Litz, 1973. I am also indebted to Mindy Kornhaber for her careful reading of the editing by the three individuals.

226   On the two verses added by Vivien, see H. Gardner, 1973, p. 88.

227   Eliot, "a personal and wholly insignificant grouse . . . " is quoted in V. Eliot, 1971, p. 1.

228 The comments from the early reviews of *The Waste Land* are quoted in Cox and Hinchliffe, 1968, unless otherwise noted. Lowell, "a piece of tripe," p. 1; *Manchester Guardian*, "so much waste paper," p. 29; Aiken, "a series of brilliant . . . ," p. 12; Wilson, "enhanced and devastated . . . ," p. 12; and Schapiro, "the most important poem . . . ," p. 12. *Times Literary Supplement*, "We have here . . . " is quoted in Medcalf, 1992. Williams, "I felt at once . . . " is quoted in Ozick, 1989, p. 153.

228 The comments from the early reviews of *The Waste Land* are quoted in Cox and Hinchliffe; 1968: Richards, "music of ideas," p. 12.

228 For Brooks on the stance of irony, see Cox and Hinchliffe, 1968.

229 Eliot's characterization of *The Waste Land* as "a good one" appears in V. Eliot, 1988, p. 519.

229 Eliot's assessment of *The Waste Land* as "the best I have ever done" appears in V. Eliot, 1988, p. 530.

230 Lewis, "[*The Waste Land*] gives an authentic . . . " is quoted in Cox and Hinchliffe, 1968, p. 58.

231 Eliot, "I am about ready . . . " is quoted in V. Eliot, 1988, p. 522.

231 Eliot, "bringing together the best . . . " is quoted in Frye, 1963, p. 4.

232 Eliot, "classicist in literature . . . " is quoted in E. Wilson, 1959, p. 105.

233 For more on Eliot's spurning of sweets, see Ackroyd, 1984, p. 314.

233 For more on Eliot's tempering of his opinions, see Cox and Hinchliffe, 1968, p. 50.

234 For Eliot's comments on an American becoming a European, see Sencourt, 1979, p. 72.

234 For Eliot's views on poetry as an escape, see E. Wilson, 1959, p. 103.

234 Eliot, "The more perfect . . . " is quoted in Ozick, 1989, p. 123.

234 For Eliot's comment on the distinction between immature and mature poets, see Cox and Hinchliffe, 1968, p. 51.

234 For Eliot's comment on the poet's mind as a receptacle, see Williamson, 1953, p. 34.

234 For Eliot's comment on a "logic of the imagination" and his view of the reading of poetry as an emotional experience, see Cox and Hinchliffe, 1968, p. 222.

235 Eliot, "a set of objects . . . " is quoted in Williamson, 1953, p. 35.

235 Eliot, "unless we have those few . . . " is quoted in Frye, 1963, p. 24.

235 For Eliot's comments on the objective correlative, see Williamson, 1953, p. 35.

236 Eliot, "and they are what we know" is quoted in Frye, 1963.

237 Eliot, "Joyce I admire . . . " is quoted in V. Eliot, 1988, p. 450.

237 For more on Eliot's admiration for Virginia Woolf, see Gordon, 1977, p. 84.

237 Eliot, "Whether Stravinsky's music . . . " is quoted in Litz, 1973, p. 19.

238 On Eliot's letters to children, see Whitney, 1991, p. 13.

238 For more on Eliot's friendship with Marx, see Gordon, 1977, p. 32.

239 Eliot, "we may perceive . . . " is quoted in Frye, 1963, p. 31.

239 For Ozick's attack on Eliot, see Ozick, 1989, p. 152.

240 Ozick, "Not since Dr. Johnson . . . " is from Ozick, 1989, p. 121.

241 For Davie's comments about Eliot as a Londoner, see Davie, 1973, p. 43.

241 For Eliot's self-depiction as "a metic, a foreigner," see V. Eliot, 1988, p. 318.

241 Eliot, "The arts insist . . . " is quoted in Gordon, 1977, p. 11.

CHAPTER 8. MARTHA GRAHAM:
DISCOVERING THE DANCE OF AMERICA

249 On dance as obsessed with mimicking plants and animals, see J. Martin, 1936, p. 15.

249 De Mille, "Isadora cleared away . . . " is quoted in Terry, 1960, p. 48.

250 Terry, "Martha turned out to be . . . " is from Terry, 1975, p. 5.

250 Graham, "Man, I'm Doctor Graham's daughter . . . " is quoted in Stodelle, 1984, p. 4.

250 George Graham, "Don't you know . . . " is quoted in Terry, 1975, p. 57.

251 Graham, "From that moment on . . . " is quoted in McDonough, 1973, p. 16.

251 On the death of Graham's father, see de Mille, 1991b, p. 24.

252 For more on Graham's enormous demands on herself, see McDonough, 1973, p. 25.

252 Shawn, "She would look . . . " is quoted in Graham, 1991, p. 68.

252 Graham, "But I do know it," is quoted in Terry, 1975, p. 32.

252 The review from the *Tacoma New Tribune*, "a brilliant young dancer . . . " is quoted in Stodelle, 1984, p. 32.

253 Graham, "I'm going to the top . . . " is from Graham, 1991, p. 86.

253 Graham, "I will dance like that." is quoted in Kisselgoff, 1991, sec. B, p. 7.

253 "white dress, a flaxen wig . . . " is quoted in Terry, 1975, p. 48.

253 A critic from the *Morning Telegraph*, "she is in appearance . . . " is quoted in J. Martin, 1936, p. 189.

253 Graham, "childish things, dreadful" is quoted in McDonough, 1973, p. 50.

253 Sabin, "the idiom was still . . . " is quoted in Kisselgoff, 1991, sec. B, p. 7.

253 For more on Stravinsky's first concerts of 1907 and 1908, see Boucourechliev, 1987, p. 35; see also this book, chap. 6, p. 193.

254 For more on *Dance* of 1929, see Stodelle, 1984, p. 51.

254 For more on *Heretic*, see McDonough, 1973, p. 63; and Stodelle, 1984, p. 59.

254 For more on *Lamentation,* see Stodelle, 1984, p. 64; and Siegel, 1979, p. 37.

255 For more descriptions of the dances, see, for example, Jowitt, 1988; de Mille, 1951; Siegel, 1979; and Stodelle, 1984.

255 De Mille, "This was a stirring period . . . " is from de Mille, 1951, p. 114.

255 De Mille, "I am glad I participated . . . " is from de Mille, 1950, p. 26.

256 Graham, "Life today is nervous . . . " is quoted in Cohen, n.d., p. 6.

256 Graham, "Once we strove to imitate . . . " is quoted in Armitage, 1978, p. 84.

257 Martin, "The actual number . . . " is from J. Martin, Jan. 5, 1930. Many quotes have been gleaned from a newspaper clippings file, and page numbers are not available.

257 Martin, "Whatever the ultimate . . . " is from J. Martin, Oct. 22, 1929.

257 Martin, "A distinguished American audience . . . " is from J. Martin, Feb. 8, 1931.

257   Martin is quoted in Stodelle, 1984: "Audiences who come to be amused . . . "
      p. 85; "When the definitive history . . . " p. 95.

258   Watkins, "Dancing is no longer . . . " is from Watkins, Jan. 25, 1931.

258   Terry, "innocent people . . . " is from Terry, 1975, p. 78.

258   Watkins, "the cerebral . . . " is quoted in Stodelle, 1984, p. 85.

258   Young, "If Martha Graham ever gave birth . . . " is quoted in McDonough,
      1973, p. 103.

258   Denby, "violent, distorted . . . " is quoted in Steinberg, 1973, p. 11.

258   Kirstein, "When I first saw Graham . . . " is quoted in Armitage, 1978, pp. 25–27.

258   Fokine, "ugly in form . . . " is quoted in Stodelle, 1984, p. 87.

258   Graham, "We shall never . . . " is quoted in Stodelle, 1984, p. 88.

259   St. Denis, "the open crotch school of music" is quoted in Terry, 1975, p. 77.

259   Humphrey, "I haven't much faith . . . " is quoted in McDonough, 1973, p. 57.

260   Graham, "You're breaking me . . . " is quoted in de Mille, 1950, p. 26.

260   Horst, "Every young artist . . . " is quoted in de Mille, 1950, p. 26.

260   For more on the relationship between Graham and Horst, see Stodelle, 1984,
      p. 50.

260   Graham, "The work is no good . . . " is quoted in de Mille, 1950, p. 29.

260   Horst, "When you get down to it . . . " is quoted in de Mille, 1950, p. 29.

262   For more on *Primitive Mysteries*, see Siegel, 1979, p. 50; and Stodelle, 1984,
      pp. 73–77.

262   Morgan, "Each member seems to walk . . . " is from Morgan, 1941, p. 61.

262   Watkins, "The most significant . . . " is from Watkins, Feb. 22, 1931.

263   Martin, "probably the finest . . . " is quoted in de Mille, 1950, p. 28.

263   For more on Graham's considerable versatility, see Armitage, 1978, p. 114.

263   For more on Graham's ability to leap like a prima donna, see McDonough,
      1973, p. 71.

263   Martin, "she ran furiously . . . " is from a J. Martin article in the *New York
      Times*, 1933.

263   Stodelle, "To remember Martha Graham . . . " is from Stodelle, 1984, p. 264.

263   Martin, "her spacious extension . . . " is from a J. Martin article in the *New York
      Times*, 1953.

264   For more on *American Provincials*, see a J. Martin article in the *New York
      Times*, 1934.

264   For more on *Frontier*, see Stodelle, 1984, p. 97; and Siegel, 1979, p. 142.

266   Graham, "The answer to the problem . . . " is quoted in *New York Times*,
      November 10, 1930.

266   Graham, "The American dancer . . . " is quoted in Armitage, 1978, p. 107.

266   Kirstein, "Martha Graham has a specifically . . . " is quoted in Armitage, 1978,
      p. 32.

266   Graham, "I'm afraid . . . " is quoted in McDonough, 1973, p. 148.

266   Graham, "Now that we moderns . . . " is quoted in McDonough, 1973, p. 148.

269   McDonough, "whenever Graham took . . . " is from McDonough, 1973,
      p. 137.

270   For more on *Appalachian Spring*, see Siegel, 1979, p. 145.

270 Graham, "behind the structure is the emotion" is quoted in Stodelle, 1984, p. 125.

272 Hawkins, "I was her equal . . . " is quoted in Kaye, 1991, p. 46.

273 For more on *Cave of the Heart*, see Jowitt, 1988, p. 223.

273 For more on *Errand into the Maze*, see Siegel, 1979, p. 198.

273 The program notes for *Night Journey*, "Jocasta sees with double insight . . . " are quoted in Stodelle, 1984, p. 148.

273 Graham, "I felt that when . . . " is quoted in Terry, 1975, p. 107.

274 For more on *The Notebooks of Martha Graham*, see Graham, 1973.

275 de Mille, "elliptical . . . " is from de Mille, 1991b, p. 14.

275 Graham, "Runs with tip 3X . . . " is from Graham, 1973, p. 157.

275 Hawkins, "What was most important for Martha . . . " is quoted in Kaye, 1991, p. 44.

276 Garafola, "Graham *was* her body . . . " is from Garafola, 1992, p. 16.

276 Mishnun, "a deadend road . . . " is from a *Nation* article by V. Mishnun, January 22, 1944.

276 Simon, "*Deaths and Entrances* is a long . . . " is from Simon, 1944.

276 A *Detroit News* writer, "Miss Graham is the most perplexing . . . " is quoted in McDonough, 1973, p. 115.

277 Graham, "It would be a criminal waste . . . " is quoted in Terry, 1975, p. 60.

277 For more on Graham's technique, see Cohen, n.d.; and Terry, 1975, p. 54.

278 Graham, "The body must be tempered . . . " is quoted in de Mille, 1991a, p. 22.

278 Graham, "it took years . . . " is quoted in McCosh, n.d.

278 Graham, "The difference between the artist . . . " is quoted in Armitage, 1978, p. 109.

278 Elisa Monte, "Nothing was ever 'just fine' . . . " is quoted in Kaye, 1991, p. 44.

278 On Graham's perfectionism, see McDonough, 1973, p. 50.

279 Graham, "I don't want to be understandable . . . " is quoted in Kisselgoff, 1984, p. 51.

279 Graham, "I would put a typewriter . . . " is quoted in Kisselgoff, 1984, p. 52.

279 Graham, "I get the ideas going . . . " is quoted in McDonough, 1973, p. 162.

279 Graham, "I am a thief . . . " is from Graham, 1973, p. xi.

279 Graham, "You draw from memory . . . " is quoted in *Christian Science Monitor*, November 15, 1962.

279 Graham, "a time of great misery . . . " is quoted in Tobias, 1984, p. 64.

279 Varèse, "Everyone is born with genius . . . " is quoted in Kisselgoff, 1984, p. 46.

280 Graham, "I didn't choose to be a dancer . . . " is from Graham, 1965, p. 54.

280 Graham, "Only if there is . . . " is quoted in Stodelle, 1984, p. 180.

280 Graham, "I want to know what . . . " is quoted in Kaye, 1991, p. 44.

280 de Mille, "Martha felt that she must . . . " is from de Mille, 1991a, p. 22.

280 On Graham being vindictive like Picasso, see McDonough, 1973, p. 224.

280 Graham, "I know I am arrogant . . . " McDonough, 1973, p. 196.

280 Lang, "An element of destructive savagery . . . " is quoted in de Mille, 1991b, p. 303.

280 Cohan, "she had to live out . . . " is quoted in Kaye, 1991, p. 46.

280   Graham, "Everyone has the right to fail . . . " is quoted in *Playbill*, 1969, p. 39.

281   Cohan, "While you were in the company . . . " is quoted in Kaye, 1991, p. 46.

281   de Mille, "whenever she was done . . . " is from de Mille, 1991b, p. ix.

281   Graham, "I want people to think . . . " is quoted in Terry, 1978, p. 6.

281   Graham, "an angry wild wicked woman" is quoted in Stodelle, 1984, p. 184.

282   On the two categories of Graham's later works, see Barnes, *New York Times*, May 4, 1972.

282   On the neoclassical flavor of Graham's later works, see Kisselgoff, 1984.

282   On Graham not giving her dancers sufficient credit, see McDonough, 1973, p. 174.

283   Graham, "when you get Nureyev . . . " is quoted in Siegel, 1973, p. B21.

283   Graham, "A dancer's instrument . . . " is quoted in Morgan, 1941, p. 11.

283   On Angelica Gibbs's biography of Graham, see McDonough, 1973, p. 75.

283   On members of the company demanding the right to perform without Graham, see McDonough, 1973, p. 189.

283   For the *New York Times* announcement of Graham's retirement, see Saal, 1973.

283   Graham, "The decision made me physically ill . . . " is quoted in Saal, 1973, p. 87.

284   Graham, "I would much rather be dancing . . . " is quoted in Kaye, 1991, p. 44.

286   Stodelle, "Here is the largest, most awe-inspiring . . . " is from Stodelle, 1984, p. xiii.

286   Barnes, "It is Miss Graham's . . . " is from Barnes, Oct. 31, 1965.

286   For Virgil Thomson's comment on Graham, see de Mille, 1950, p. 26.

286   de Mille, "In this hundred years . . . " is from de Mille, 1991a, p. 1.

286   de Mille, "It is probably the greatest addition . . . " is from de Mille, 1951, p. 117.

286   On St. Denis not taking Graham seriously, see Terry, p. 85.

287   Graham, "If I were to take that step . . . " is quoted in de Mille, 1991b, p. 239.

CHAPTER 9. MAHATMA GANDHI: A HOLD UPON OTHERS

290   Shirer, "It was the only instance . . . " is from Shirer, 1979, p. 18.

290   For more on the early history of British-controlled India, see Fischer, 1950; and Harris and Levy, pp. 1325–1326.

291   On Gandhi's family background, see Fischer, 1950, p. 12.

291   "to keep two or three . . . " is quoted in Fischer, 1983, p. 6.

291   Gandhi, "I am an average man . . . " is quoted in Nanda, 1985, p. 133.

292   On Gandhi and Mehtab stealing money, see Fischer, 1983, p. 8.

292   Orwell, "a few cigarettes . . . " is quoted in Nanda, 1985, p. 9.

292   Gandhi, "the cruel custom . . . " is quoted in Nanda, 1985, p. 9.

292   On Gandhi never forgiving himself for leaving his father's deathbed, see Mehta, 1976, p. 82; and Payne, 1990, p. 42.

293   On Gandhi barely passing his college entrance exam, see Mehta, 1976, p. 83.

293   The community headman, "In the opinion . . . " is quoted in Fischer, 1983, pp. 20–21.

293   The community headman, "This boy shall be treated . . . " is quoted in Fischer, 1983, pp. 20–21.

293   On Gandhi dressing like a dandy, see Fischer, 1983, p. 24.

293   On Gandhi's early taste for eccentricities, see Mehta, 1976, p. 91.

294   Fischer, "mediocre, unimpressive . . . " is from Fischer, 1950, p. 28.

294   Gandhi, "No one else has ever . . . " is quoted in Fischer, 1950, p. 39.

295   On Gandhi's belief in the effectiveness of compromise and reconciliation, see Mehta, 1976, p. 101.

295   For more on the meeting in Pretoria, see Erikson, 1969, pp. 165–166; Mehta, 1976, p. 100; and Payne, 1990, p. 92–93.

295   Gandhi, "I thus made . . . " is quoted in Brown, 1989, p. 32.

296   On Indians technically considered as equal to citizens of European descent, see Brown, 1989, p. 32.

296   On Indians described as "semi-barbarous Asiatics," see Nanda, 1985, p. 28.

296   On Indians in South Africa, see Brown, 1989, p. 57.

296   On Indians having to be "properly dressed" to ride first- or second-class, see Fischer, 1950, p. 43.

296   For more on Gandhi attaining concessions from the British Secretary of State, see Payne, 1990, p. 170.

297   On Gandhi characteristically feeling sorry for those who attacked him, see Fischer, 1983, pp. 49, 52.

297   For more on Gandhi's first time in jail, see Payne, 1990, pp. 113–114 and 177.

297   On Gandhi's extraordinary shyness, see Fischer, 1983, p. 27.

297   On Gandhi belonging to many organizations, see Payne, 1990, p. 105.

298   On Tolstoy Farm, see Mehta, 1976, p. 124.

298   On Gandhi's fasting when two boys engaged in sodomy, see Payne, 1990, p. 234.

298   Gandhi, "It became my conviction . . . " is quoted in Fischer, 1983, p. 69.

299   Gandhi, "The true remedy lies . . . " is quoted in Fischer, 1983, p. 118.

299   On Gandhi leading a series of protest marches, see Mehta, 1976, p. 127.

299   On Gandhi asking his followers to resist laws, see Jack, 1956, p. 59; and Polak et al., 1949, p. 57.

300   On the relations between Smuts and Gandhi, see Payne, 1990, p. 272.

300   Smuts, "The saint has left . . . " is quoted in Fischer, 1950, p. 117.

301   Gandhi, "I wanted to acquaint India . . . " is quoted in Fischer, 1983, p. 126.

301   For more on Gandhi calling attention to unfair practices, see Fischer, 1983, pp. 138–141.

301   Gandhi, "I had to disobey . . . " is quoted in Fischer, 1983, p. 140.

302   Payne, "For Anasuyabehn . . . " is from Payne, 1990, p. 324.

303   Gandhi, "In my opinion I would have been . . . " is quoted in Erikson, 1969, p. 51

303   Gandhi, "The words came . . . " is quoted in Mehta, 1976, p. 137.

303   On Gandhi's first fast for a public cause, see Fischer, 1950, p. 155.

303   On the drama, potency, and simplicity of fasting, see Payne, 1990, p. 324.

303   Gandhi, "[They] received my words . . . " is quoted in Bondurant, 1958, p. 71.

303   On reaching a settlement, see Fischer, 1983, pp. 143–145.

303   For more on the method of arbitration, see Fischer, 1950, p. 157.

304   Brown, "[Gandhi's] Ahmedabad campaign . . . " is from Brown, 1989, p. 121.

305   Gandhi, "the greatest battle . . . " is quoted in Brown, 1989, p. 128.

305   On Gandhi seeking to unite the Muslim and Hindu forces, see Brown, 1989, pp. 124–125.

305   On Gandhi being particularly distraught by a murderous Indian mob in Bardoli, see Fischer, 1950, p. 197.

305   On Gandhi conceding his errors, see Fischer, 1950, p. 179.

305   On Gandhi being placed on trial for sedition, see Payne, 1990, p. 361.

305   Broomfield, "high ideals . . . " is quoted in Payne, 1990, p. 361.

306   Gandhi, "The only course open . . . " is quoted in Payne, 1990, p. 364.

306   Gandhi, "a law designed . . . " is quoted in Payne, 1990, p. 364.

306   On the British connection making India helpless, see Brown, 1989, p. 129; and Payne, 1990, p. 227.

306   Gandhi, "I have no doubt . . . " is quoted in Payne, 1990, pp. 365–367.

306   Gandhi, "I hold it to be a virtue . . . " is quoted in Payne, 1990, pp. 365–367.

306   Broomfield, "It will be impossible . . . " is quoted in Payne, 1990, p. 367.

306   Broomfield, "So far as the sentence is concerned . . . " is quoted in Payne, 1990, p. 367.

307   Brown, "Like many great visionaries . . . " is from Brown, 1989, p. 177.

307   On Thoreau's influence on Gandhi, see Fischer, 1950, p. 87.

307   Tolstoy, "a question of the greatest importance . . . " is quoted in Brown, 1989, p. 79.

308   Erikson, "why certain men of genius . . . " is from Erikson, 1969, p. 128.

308   Gandhi, "I regard untouchability . . . " is quoted in Fischer, 1950, p. 134.

309   Gandhi, "For me politics bereft of religion . . . " is quoted in Fischer, 1983, p. 217.

309   On Gandhi's dismissiveness of religion that ignores practical problems, see Bondurant, 1958, p. 110; and Brown, 1989, p. 124.

309   Gandhi, "I take part . . . " is quoted in Brown, 1989, p. 133.

309   Gandhi, "I can say without the slightest humility . . . " is quoted in Brown, 1989, p. 123.

309   Gandhi, "What I want to achieve . . . " is quoted in Fischer, 1983, p. 4.

309   On Gandhi's reference to satyagraha, see Fischer, 1983, p. 6.

309   On the satyagrahis hoping to convert the opponent, see Gandhi, 1951, p. 111.

309   Gandhi, "No matter how badly . . . " is quoted in Fischer, 1983, p. 90.

309   Gandhi, "If by using violence . . . " is quoted in Fischer, 1983, p. 17.

310   For more on the wider political significance of the struggle, see Brown, 1989, p. 117.

310   On the specific codes of satyagraha, see Bondurant, 1958, p. 39.

310   Gandhi, "A satyagrahi differs . . . " is quoted in Fischer, 1983, p. 85.

311   For more on the specific details of satyagraha, see Fischer, 1983, pp. 139–140.

311   Gandhi, "The rope dancer . . . " is quoted in Fischer, 1983, pp. 108–109.

311   Gandhi, "I have the unquenchable faith . . . " is quoted in Fischer, 1983, p. 327.

312   On the satyagraha as a new form of ritualization, see Erikson, 1969, p. 395.

312   Nanda, "In every case . . . " is quoted in Erikson, 1969, p. 97.

312   Gandhi, "Such experiments . . . " is quoted in Brown, 1989, p. 230.

313   Gandhi, "I own no property . . . " is quoted in Fischer, 1983, p. 308.

213   An observer, "it cost a great deal . . . " is quoted in Mehta, 1976, p. 56.

313   For more on Gandhi's commitment to spinning, see Fischer, 1983, pp. 223 and 232.

313   Gandhi, "Fasting cannot be undertaken . . . " is quoted in Bondurant, 1958, p. 37.

314   On Gandhi's meticulous concern for every step of organizing a protest, see Brown, 1989, p. 113.

314   For more on Gandhi's routine and sleeping patterns, see Shirer, 1979, p. 132.

314   Followers of Gandhi, "In his presence . . . " is quoted in Erikson, 1969, p. 63.

314   Nehru, "It is nearly three and a half decades . . . " is quoted in Bondurant, 1958, p. xvii.

315   On Gandhi's denial of having special or divine powers, see Bondurant, 1958, p. 124.

315   Gandhi, "They should leave it to me alone" is quoted in Nambodiripod, 1981, p. 59.

315   A follower, "Of course when we were living . . . " is quoted in Mehta, 1976, p. 6.

315   Shirer, "They felt in the presence . . . " is from Shirer, 1979, p. 76.

316   Sarabhai, "There was something in Bapuji . . . " is quoted in Mehta, 1976, p. 56.

316   For more on Gandhi disinheriting his son, see Erikson, 1969, pp. 369–370.

317   Gandhi, "Certainly not . . . " is quoted in Fischer, 1950, p. 206.

317   Lord Irwin, "rather remote . . . " is quoted in Brown, 1989, p. 144.

317   Lord Birkenhead, "Poor Gandhi . . . " is quoted in Brown, 1989, p. 177.

317   Gilbert Murray, "A dangerous and uncomfortable . . . " is quoted in Fischer, 1950, p. 118.

317   Churchill, "seditious fakir . . . " is quoted in Erikson, 1969, p. 447.

317   Nehru, "I am afraid . . . " is quoted in Brown, 1989, p. 270.

318   Gandhi, "I seek entirely . . . " is quoted in Fischer, 1983, pp. 188–189.

318   Gandhi, "Hitler . . . Mussolini . . . " is quoted in Fischer, 1983, p. 331.

318   On Gandhi's letter to Hitler, see Payne, 1990, p. 487.

319   Gandhi, "It is, I know, open to you . . . " is quoted in Bondurant, 1958, p. 93.

319   Gandhi, "For me there is no turning back . . . " is quoted in Payne, 1990, p. 391.

319   Fischer, "to pick up a pinch . . . " is from Fischer, 1950, p. 268.

319   Nehru, "It seemed as though . . . " is quoted in Bondurant, 1958, p. 94.

320   Gandhi, "I want world sympathy . . . " is quoted in Payne, 1990, p. 392.

320   Bondurant, "nothing less than to cause . . . " is from Bondurant, 1958, p. 89.

320   Miller, "Suddenly at a word . . . " is quoted in Mehta, 1976, p. 148; and Shirer, p. 98.

321   Gandhi, "to remind us of the famous . . . " is quoted in Shirer, 1979, p. 99.

321   For more on the resolution of the salt march, see Bondurant, 1958, p. 102.

321   Shirer, "The Mahatma kept breaking into the headlines . . . " is from Shirer, 1979, p. 165.

321 For more on Gandhi's burdensome schedule, see Payne, 1990, p. 412.

321 Gandhi, "I have come here . . . " is quoted in Shirer, 1979, p. 167.

322 Gandhi, "India has been held by the sword . . . " is quoted in Shirer, 1979, p. 169–170.

322 Gandhi, "Freedom is our birthright . . . " is quoted in Shirer, 1979, p. 190.

322 Gandhi, "By our internal squabbles . . . " is quoted in Shirer, 1979, p. 192.

322 Gandhi, "This has been . . . " is quoted in Shirer, 1979, p. 194.

322 Shirer, "I think it was this mistrust . . . " is from Shirer, 1979, p. 197.

323 On Gandhi seeking to introduce a balance, see Mehta, 1976, p. 155.

323 Nanda, "There was hardly an aspect . . . " is from Nanda, 1985, p. 126.

324 Gandhi, "Would it not be more appropriate . . . " is quoted in Mehta, 1976, p. 171.

325 Brown, "He went into the eye . . . " is from Brown, 1989, p. 360.

325 On the *Times* correspondent's view of Gandhi's achievements, see Nanda, 1985, p. 107.

325 Mehta, "Gandhi was Christ . . . " is from Mehta, 1976, p. xi.

326 Nehru, "a psychological change . . . " is quoted in Erikson, 1969, p. 265.

326 Martin Luther King Jr., "the only morally . . . " is quoted in Nanda, 1985, p. 34.

327 Einstein, "Gandhi had demonstrated . . . " is quoted in Fischer, 1950, p. 10.

327 Einstein, "Generations to come . . . " is quoted in Shirer, 1979, frontispiece.

INTERLUDE THREE

328 Gandhi, "Action is my domain . . . " is quoted in Mehta, 1976, p. 69.

CHAPTER 10. CREATIVITY ACROSS THE DOMAINS

341 Flaubert, "I love my work . . . " is quoted in Kakutani, 1992, Sec. C, p. 19.

343 On Freud's belief in a seven-year creative cycle, see E. Freud, 1960, p. 301.

344 For more on stages of development of creativity, see Wallas, 1926.

347 For more on Geertz's phrase "deep play," see Geertz, 1973.

354 For more on the notion of "escape from freedom," see Fromm, 1941.

EPILOGUE. THE MODERN ERA AND BEYOND

363 For a view of modernity as an earlier phenomenon, see Toulmin, 1990.

364 Mailer, "Look at who Picasso's . . . " is quoted in Spencer, 1991, p. 47.

364 For more on the modern era in the nineteenth century, see Johnson, 1991.

365 For doubts about the progressive view of the nineteenth century, see Le Rider, 1990; and Wohl, 1979.

366 Krauss, "A test lab for the end of the world" is quoted in Varnedoe, 1986, p. 18.

366 For portraits of different cities, see Lukacs, 1989, on Budapest; Eksteins, 1989, on major cities; Janik and Toulmin, 1973, on Vienna; Varnedoe, 1986, on Vienna; and Dangerfield, 1961.

366 Musil, "Out of the oil smooth spirit . . . " is from Musil, 1980, p. 59.

367 For more on historians' confirmation that the feeling of dramatic change was in the air, see Dangerfield, 1961; and Tuchman, 1967.

367 Woolf, "in or about December 1910 . . . " is quoted in Toulmin, 1990, p. 150.

367 On *Le sacre du printemps*, see Eksteins, 1989.

367 Hobhouse, "It turns out to be . . . " is quoted in Frankel, 1956, p. 39.

367 Pfaff, "it is now clear . . . " is from Pfaff, 1989, p. 21.

368 Baudelaire, "[Modernity] is the experience . . . " is quoted in Sennett, 1991, p. 6.

368 Baudelaire, "By modernity . . . " is quoted in Berman, 1988, p. 132.

368 For more on a "tradition of the new," see Rosenberg, 1959.

373 For Baudelaire's comment, see Donahue, 1991.

375 For more on the "disappearance of childhood," see Postman, 1982; and Winn, 1983.

BIBLIOGRAPHY

Ackroyd, P. *T. S. Eliot*. New York: Simon & Schuster, 1984.

Amabile, T. M. *The Social Psychology of Creativity*. New York: Springer-Verlag, 1983.

Apel, W., ed. *Harvard Dictionary of Music*. Cambridge, Mass.: Harvard University Press, 1972.

Apollinaire, G. *The Cubist Painters: Aesthetic Meditations*. Wienborn: Schultz, 1949; originally published in 1913.

Armitage, M. *Martha Graham: The Early Years*. New York: Da Capo Press, 1978.

Arnheim, R. *Picasso's* Guernica: *The Genesis of a Painting*. Berkeley: University of California Press, 1962.

Auerbach, E. *Mimesis*. New York: Doubleday/Anchor, 1953.

Baer, N. v.N. *The Art of Enchantment: Diaghilev's Ballets Russes, 1909–1929*. San Francisco: The Fine Arts Museums, 1988.

Bamberger, J. "Growing Up Prodigies: The Midlife Crisis." In D. H. Feldman, ed., *Developmental Approaches to Giftedness*. San Francisco: Jossey-Bass, 1982, pp. 61–78.

Barnes, C. Review of Martha Graham. *New York Times*, October 31, 1965.

Barnett, L. *The Universe and Dr. Einstein*. New York: Mentor, 1948.

Barr, A. H. *Picasso: Fifty Years of His Art*. New York: Museum of Modern Art, 1974; originally published in 1946.

———. *Cubism and Abstract Art*. Cambridge, Mass.: Harvard University Press, 1986.

Barron, F. *Creative Person and Creative Process*. New York: Holt, Rinehart, and Winston, 1969.

Bate, W. J. *John Keats*. Cambridge, Mass.: Harvard University Press, 1963.

Bedient, C. *He Do the Police in Different Voices:* The Wasteland *and Its Protagonist*. Chicago: University of Chicago Press, 1986.

Berger, J. *The Success and Failure of Picasso*. New York: Pantheon Books, 1965.

Berman, M. *All That Is Solid Melts into Air: The Experience of Modernity*. New York: Penguin, 1988.

Bernstein, J. "Profile of I. I. Rabi." In J. Bernstein, *Experiencing Science*. New York: Dutton, 1980.

———. "A Critic at Large: Besso." *New Yorker*, February 27, 1989, pp. 86–87.

Block, N., and G. Dworkin, eds. *The IQ Controversy*. New York: Pantheon, 1976.

Bloom, B., with L. Sosniak. *Developing Talent in Young Children.* New York: Ballantine Books, 1985.

Bloom, H., ed. *T. S. Eliot's* The Waste Land. New York: Chelsea House, 1986.

Blunt, A. *Picasso's* Guernica. London: Oxford University Press, 1969.

Boden, M. *The Creative Mind.* New York: Basic Books, 1990.

Boelich, W., ed. *The Letters of Sigmund Freud to Edward Silberstein.* Cambridge, Mass.: Harvard University Press, 1990.

Bondurant, J. *Conquest of Violence: The Gandhian Philosophy of Conflict.* Berkeley: University of California Press, 1958.

Boring, E. G. *A History of Experimental Psychology.* New York: Appleton-Century-Crofts, 1950.

Boucourechliev, A. *Stravinsky.* New York: Holmes and Meier, 1987.

Bresson, M. "Appraising African Art Through Western Eyes." *New York Times,* October 7, 1990, sec. 2, 37.

Breuer, J., and S. Freud. *Studies in Hysteria.* In J. Strachey, ed. *The Standard Edition of the Complete Psychological Works of Sigmund Freud, vol. 2.* London: Hogarth Press, 1966.

Brill, A. A., ed. *The Basic Writings of Sigmund Freud.* New York: Modern Library, 1938.

Brooker, J. S., and J. Bentley. *Reading* The Waste Land. Amherst: University of Massachusetts Press, 1990.

Brown, J. M. *Gandhi: Prisoner of Hope.* New Haven, Conn.: Yale University Press, 1989.

Burns, E., ed. *Gertrude Stein on Picasso.* New York: Liveright, 1980.

Cabanne, P. *Pablo Picasso: His Life and Times.* New York: Morrow, 1977.

Campbell, D. "Blind Variation and Selective Retention in Creative Thought as in Other Knowledge Processes." *Psychological Review* 67 (1960): 380–400.

Clampitt, Amy. "Remarks." Address to the Eliot Centennial, Harvard University, Cambridge, Mass., December 5, 1988.

Clark, R. W. *Einstein: The Life and Times.* New York: World, 1971.

———. *Freud: The Man and the Cause.* New York: Random House, 1980.

Clouzot, H. G. *Le mystère Picasso.* France: 1955. Film.

Cohen, S. J. *The Achievement of Martha Graham.* Chrysalis, n.d.

Cole, S. "Age and Scientific Performance." *American Journal of Sociology* 84 (1979): 859–977.

Conrad, B. "A Home for Picasso." *Horizon,* June 1986, pp. 11–16.

Conrad, P. "Review of Hugh Kenner, *A Sinking Island." Times Literary Supplement,* September 9, 1988, p. 981.

Coughlan, E. "Russian Folk-Wedding Music Said to Influence Stravinsky Ballet." *Chronicle of Higher Education,* June 13, 1990, p. A7.

Cox, C. B., and A. B. Hinchliffe. *T. S. Eliot:* The Waste Land: *A Casebook.* London: Macmillan, 1968.

Craft, R. *Stravinsky: Chronicle of a Friendship (1948–1971).* New York: Knopf, 1972.

———. *Stravinsky: Selected Correspondence, vol. 1.* New York: Knopf, 1982.

———. *Stravinsky: Selected Correspondence, vol. 2.* New York: Knopf, 1984.

————. *Stravinsky: Selected Correspondence, vol. 3*. New York: Knopf, 1985.

————. *Stravinsky: Glimpses of a Life*. New York: St. Martin's Press, 1992.

Croce, A. "Angel." *New Yorker*, October 15, 1990, pp. 124–133.

Csikszentmihalyi, M. "Motivation and Creativity: Towards a Synthesis of Structural and Energistic Approaches." *New Ideas in Psychology* 6 (1988a): 159–176.

————. "Society, Culture, and Person: A Systems View of Creativity." In R. J. Sternberg, ed., *The Nature of Creativity*. New York: Cambridge University Press, 1988b, pp. 325–339.

————. *Flow: The Psychology of Optimal Experience*. New York: Harper and Row, 1990.

Csikszentmihalyi, M., and I. Csikszentmihalyi, eds. *Optimal Experience*. New York: Cambridge University Press, 1988.

Csikszentmihalyi, M., and R. E. Robinson. "Culture, Time, and the Development of Talent." In R. Sternberg and J. E. Davidson, eds., *Conceptions of Giftedness*. New York: Cambridge University Press, 1986, pp. 263–284.

Dangerfield, G. *The Strange Death of Liberal England*. Reprint. New York: Capricorn Books, 1961.

Davie, D. "Anglican Eliot." In A. W. Litz, ed., *Eliot in His Time: Essays on the Occasion of the Fiftieth Anniversary of* The Waste Land. Princeton, N.J.: Princeton University Press, 1973.

De Mille, A. "Martha Graham." *Atlantic Monthly*, November 1950, pp. 25–31.

————. *Dance to the Piper*. Boston: Little, Brown/Atlantic Monthly Press, 1951.

————. "Measuring the Steps of a Giant." *New York Times*, April 7, 1991a, pp. 1, 22.

————. *Martha: The Life and Work of Martha Graham*. New York: Random House, 1991b.

Denby, E. *Looking at the Dance*. New York: Pelligrini and Cudahy, 1949.

Donahue, D. "The Poet of Modern Life [Baudelaire]." *New York Review of Books*, February 14, 1991, pp. 22–24.

Druskin, M. *Igor Stravinsky: His Life, Works, and Views*. New York: Cambridge University Press, 1983.

Dukas, H., and B. Hoffmann. *Albert Einstein: The Human Side*. Princeton, N.J.: Princeton University Press, 1979.

Duncker, K. "On Problem-Solving." *Psychological Monographs* 58 (1945): whole.

Dunning, J. "Martha Graham at 95 Does Something Different." *New York Times*, October 1, 1990, pp. C19–C20.

————. "Troupe Contemplates Life Without Graham." *New York Times*, April 3, 1991, sec. C, p. 11.

Einstein, A. *Relativity: The Special and General Theory*. New York: Holt, 1921.

Eksteins, M. *Rites of Spring: The Great War and the Birth of the Modern Age*. New York: Houghton Mifflin, 1989.

Eliot, T. S. *Selected Poems*. New York: Harcourt, Brace, and World, 1936.

————. *The Complete Poems and Plays*. New York: Harcourt Brace and World, 1952.

Eliot, V., ed. *T. S. Eliot:* The Waste Land: *A Facsimile and Transcript of the Original Drafts, Including the Annotations of Ezra Pound*. New York: Harcourt Brace Jovanovich, 1971.

———. *The Letters of T. S. Eliot, Vol. 1, 1898–1922*. New York: Harcourt Brace Jovanovich, 1988.

Ellenberger, H. F. *The Discovery of the Unconscious*. New York: Basic Books, 1970.

Ellmann, R. "The First *Waste Land*." In A. W. Litz, ed., *Eliot in His Time: Essays on the Occasion of the Fiftieth Anniversary of* The Waste Land. Princeton, N.J.: Princeton University Press, 1973.

Erikson, E. H. *Identity and the Life Cycle*. New York: International Universities Press, 1959.

———. *Gandhi's Truth*. New York: Norton, 1969.

Eysenck, H. J. "Measuring Individual Creativity." Paper presented to the Workshop on Creativity, the Achievement Project, Kent, England, December 13–15, 1991.

Feldman, D. H. *Beyond Universals in Cognitive Development*. Norwood, N.J.: Ablex, 1980.

———. *Nature's Gambit*. New York: Basic Books, 1986.

———. "Creativity: Dreams, Insights, and Transformations." In R. Sternberg, ed., *The Nature of Creativity*. New York: Cambridge University Press, 1988, pp. 271–297.

Findlay, S., and C. Lumsden. "The Creative Mind: Towards an Evolutionary Theory of Discovery and Innovation." *Journal of Social and Biological Structures* 11 (1988): 3–55.

Fischer, L. *The Life of Mahatma Gandhi*. New York: Harper and Brothers, 1950.

———, ed. *The Essential Gandhi*. New York: Vintage Books, 1983.

Flam, J. "Monet's Way." *New York Review of Books,* May 17, 1990, pp. 9–13.

Flavell, J. *The Developmental Psychology of Jean Piaget*. New York: Van Nostrand, 1963.

Foucault, M. *The Order of Things*. New York: Pantheon, 1970.

Frank, P. "Einstein's Philosophy of Science." *Review of Modern Physics*, July 1949, p. 21.

———. *Einstein: His Life and Times*. New York: Knopf, 1953.

Frankel, C. *The Case for Modern Man*. New York: Harper and Brothers, 1956.

Freud, E. L. ed. *Letters of Sigmund Freud*. New York: Basic Books, 1960.

Freud, S. *An Autobiographical Study*. New York: Norton, 1935.

———. *The Interpretation of Dreams*. In A. A. Brill, ed., *The Basic Writings of Sigmund Freud*. New York: Modern Library, 1938a, originally published 1900.

———. *Three Contributions to the Theory of Sex*. In A. A. Brill, ed., *The Basic Writings of Sigmund Freud*. New York: Modern Library, 1938b, originally published 1900.

———. *Collected Papers, vol. 1*, translated by Joan Riviere. London: Hogarth Press, 1949.

———. *The Origins of Psychoanalysis: Letters to Wilhelm Fliess, Drafts, and Notes 1887–1902*. New York: Basic Books, 1954.

———. *Totem and Taboo*. In J. Strachey, ed., *The Standard Edition of the Complete Psychological Works of Sigmund Freud, vol. 13*. London: Hogarth Press, 1955, pp. 1–161.

———. *Creativity and the Unconscious,* edited by B. Nelson. New York: Harper and Row, 1958.

————. "Creative Writers and Daydreaming." In J. Strachey, ed., *The Standard Edition of the Complete Psychological Works of Sigmund Freud*, vol. 9. London: Hogarth Press, 1959, pp. 143–144.

————. *Jokes and Their Relation to the Unconscious*. In J. Strachey, ed., *The Standard Edition of the Complete Psychological Works of Sigmund Freud*, vol. 8. London: Hogarth Press, 1960a.

————. *The Psychopathology of Every Day Life*. In J. Strachey, ed., *The Standard Edition of the Complete Psychological Works of Sigmund Freud*, vol. 6. London: Hogarth Press, 1960b.

————. *Civilization and Its Discontents*. In J. Strachey, ed., *The Standard Edition of the Complete Psychological Works of Sigmund Freud*, vol. 21. London: Hogarth Press, 1961a, pp. 59–145.

————. *Dostoevesky and Parricide*. In J. Strachey, ed., *The Standard Edition of the Complete Psychological Works of Sigmund Freud, vol. 21*. London: Hogarth Press, 1961b.

————. *Early Psychoanalytic Publications*. In J. Strachey, ed., *The Standard Edition of the Complete Psychological Works of Sigmund Freud, vol. 3*. London: Hogarth Press, 1962.

————. *A History of the Psychoanalytic Movement*. New York: Collier Books, 1963.

————. *Moses and Monotheism*. In J. Strachey, ed., *The Standard Edition of the Complete Psychological Works of Sigmund Freud, vol. 23*. London: Hogarth Press, 1964, pp. 3–137.

————. *Pre-Psychoanalytic Publications and Unpublished Drafts*. In J. Strachey, ed., *The Standard Edition of the Complete Psychological Works of Sigmund Freud, vol. 1*. London: Hogarth Press, 1966.

————. *Leonardo: A Study in Psychosexuality*. New York: Vintage, 1967.

Fromm, E. *Escape from Freedom*. New York: Rinehart, 1941.

Frye, N. *T. S. Eliot: An Introduction*. Chicago: University of Chicago Press, 1963.

Furbank, P. N. "Review of N. Braybrooke (Ed.), *Seeds in the Wind: Juvenilia from W. B. Yeats to Ted Hughes*." *Times Literary Supplement*, November 17, 1989, p. 1261.

Gablik, S. *Has Modernism Failed?* New York: Thames and Hudson, 1984.

Gandhi, M. *Non-Violent Resistance: Satyagraha*. New York: Schocken Books, 1951.

————. *Autobiography: The Story of My Experiments with Truth*. New York: Dover, 1963.

Garafola, L. *Diaghilev's Ballet Russes*. New York: Oxford University Press, 1989.

————. "A Lady and Her Legends." *Times Literary Supplement*, May 1, 1992, p. 16.

Gardner, Helen. "*The Waste Land*: Paris 1922." In A. W. Litz, ed., *Essays on the Occasion of the Fiftieth Anniversary of* The Waste Land. Princeton, N.J.: Princeton University Press, 1973.

Gardner, Howard. *Artful Scribbles*. New York: Basic Books, 1980.

————. *Frames of Mind: The Theory of Multiple Intelligences*. New York: Basic Books, 1983.

————. "Freud in Three Frames." *Daedalus* (Summer 1986): 105–134.

————. "Creative Lives and Creative Works: A Synthetic Scientific Approach." In R. Sternberg, ed., *The Nature of Creativity*. New York: Cambridge University Press, 1988a.

———. "Creativity: An Interdisciplinary Perspective." *Creativity Research Journal,* 1988b, 8–26.

———. *To Open Minds: Chinese Clues to the Dilemma of Contemporary Education.* New York: Basic Books, 1989.

Gardner, Howard, and Y. Dudai. "Biology and Giftedness." *Items* (Social Science Research Council), 35 (1985): 1–6.

Gardner, Howard, and R. Nemirovsky. "From Private Intuitions to Public Symbol Systems: An Examination of Creative Process in Georg Cantor and Sigmund Freud. *Creativity Research Journal* 4 (1991): 1–21.

Gardner, Howard, and C. Wolf. "The Fruits of Asynchrony: Creativity from a Psychological Point of View." *Adolescent Psychiatry* 15 (1988): 106–123.

Gay, P. *The Bourgeois Experience: Victoria to Freud.* New York: Oxford University Press, 1984.

———. *A Godless Jew: Freud, Atheism, and the Making of Psychoanalysis.* New Haven, Conn.: Yale University Press, 1987.

———. *Freud: A Life for Our Time.* New York: Doubleday/Anchor, 1988.

———. *Reading Freud: Explorations and Entertainments.* New Haven, Conn.: Yale University Press, 1990.

Gedo, J. *Portraits of the Artist.* New York: Guilford, 1983.

Gedo, M. "Art as Autobiography: Picasso's *Guernica.*" *Art Quarterly* (September 1979): 191–210.

———. *Art as Autobiography.* Chicago: University of Chicago Press, 1980.

———. "The Archaeology of a Painting: A Visit to the City of the Dead Beneath Picasso's *La vie.*" *Psychoanalytic Inquiry* 3 (1983): 371–430.

Geertz, C. *The Interpretation of Cultures.* New York: Basic Books, 1973.

Gergen, K. *The Saturated Self: Dilemmas of Identity in Contemporary Life.* New York: Basic Books, 1991.

Getzels, J., and P. Jackson. *Creativity and Intelligence.* New York: Wiley, 1962.

Ghiselin, B. *The Creative Process.* New York: Mentor, 1952.

Gilot, F. *Matisse and Picasso: A Friendship in Art.* New York: Doubleday, 1990.

Gilot, F., and C. Lake. *Life with Picasso.* New York: Avon Books, 1981.

Gish, N. The Wasteland: *A Poem of Memory and Desire.* Boston: Wayne, 1988.

Glaesemer, J. *Der junge Picasso. Fruhwerk und blaue Period.* Bern: Kunstmuseum, 1984.

Glimcher, A., and M. Glimcher. *Je suis le cahier: The Sketchbooks of Picasso.* Boston: Atlantic Monthly Press, 1986.

Goldberg, S. "The Early Response to Einstein's Special Theory of Relativity 1905–1911: A Case Study in National Differences." Ph.D. diss., Harvard University, 1968.

Golding, J. "Two Who Made a Revolution." *New York Review of Books,* May 31, 1990, pp. 8–11.

Goleman, D. "As a Therapist Freud Fell Short, Scholars Find." *New York Times,* March 6, 1990, pp. C1, C12.

Gombrich, E. H. "In Search of Cultural History." In E. H. Gombrich, *Ideals and Idols.* London: Oxford University Press, 1979.

———. "Styles of Art and Styles of Life." *The Reynolds Lecture*. London: Royal Academy of Arts, 1991.

Goodman, N. *Languages of Art*. Indianapolis: Hacket, 1976.

Gopnik, A. "High and Low: Caricature, Primitivism, and the Cubist Portrait." *Art Journal* (Winter 1983): 371–376.

Gordon, L. *Eliot's Early Years*. New York: Oxford University Press, 1977.

Graham, M. "How I Became a Dancer." *Saturday Review*, September 28, 1965, p. 54.

———. *The Notebooks of Martha Graham*. Edited by N. Wilson Ross. New York: Harcourt Brace, 1973.

———. "Martha Graham Reflects on Dance." *New York Times,* March 31, 1985, sec. 2, pp. 1, 8.

———. *Blood Memory: An Autobiography*. New York: Doubleday, 1991.

Gregory, F. "The Mysteries and Wonders of Natural Science: Bernstein's *Naturwissenschaftliche Volksbucher* and the Adolescent Einstein." Paper presented at the Workshop on Einstein's Early Life, Andover, Mass., October 6, 1990.

Groddeck, G. *The Book of the It*. New York: Vintage, 1961; originally published in 1923.

Gruber, H. E. *Darwin on Man*. 2d ed. Chicago: University of Chicago Press, 1982.

Gruber, H. E., and S. N. Davis. "Inching Our Way up Mount Olympus: The Evolving Systems Approach to Creative Thinking." In R. J. Sternberg, ed., *The Nature of Creativity*. New York: Cambridge University Press, 1988, pp. 24370.

Gruenbaum, A. *Foundations of Psychoanalysis: A Philosophical Critique*. Berkeley: University of California Press, 1984.

Guilford, J. P. "Creativity." *American Psychologist* 5 (1950): 444–454.

Gyongyi, E. and Z. Jobbagyi. *A Golden Age: Art and Society in Hungary, 1896–1914*, translated by Z. Beres and P. Doherty. Miami: Center for the Arts, 1989.

Hall, C., and G. Lindzey. *Theories of Personality*. New York: Wiley, 1957.

Hall, D. "Interview with Ezra Pound." *Writers at Work: The Paris Review Interviews, vol. 2.* Harmondsworth, England: Penguin, 1963a.

———. "Interview with T. S. Eliot." *Writers at Work: The Paris Review Interviews, vol. 2.* Harmondsworth, England: Penguin, 1963b.

Harding, D. W. "What the Thunder Said." In A. D. Moody, ed., The Waste Land *in Different Voices*. London: Arnold, 1974.

Harris, W. H, and J. S. Levy, eds. *The New Columbia Encyclopedia*. New York: Columbia University Press, 1975.

Hayes, J. R. *The Complete Problem Solver*. Philadelphia: Franklin Institute Press, 1981.

Henahan, D. "Creator vs. Creator: Who Wins?" *New York Times*, August 4, 1980, sec. 1, p. 23.

Hestenes, D. "Secrets of Genius: Review of A. Miller, *Imagery in Scientific Thought*." *New Directions in Psychology* 8 (1990): 231–246.

Hilton, T. *Picasso*. London: Thames and Hudson, 1975.

———. "The Genesis of Painting." *Times Literary Supplement*, September 19, 1986, p. 1034.

Ho, W-C. *Yani: The Brush of Innocence*. New York: Hudson Hills Press, 1989.

Hoffmann, B. *Einstein*. St. Albans, England: Paladin, 1975.

Hoffman, B., and H. Dukas. *Albert Einstein: The Human Side*. Princeton, N.J.: Princeton University Press, 1979.

Holmes, F. L. *Lavoisier and the Chemistry of Life*. Madison: University of Wisconsin Press, 1985.

Holton, G. *Thematic Origins of Scientific Thought*. 2d ed. Cambridge, Mass.: Harvard University Press, 1988.

Holton, G., and Y. Elkana, eds. *Albert Einstein: Historical and Cultural Perspectives*. Princeton, N.J.: Princeton University Press, 1982.

Horan, R. "The Recent Theater of Martha Graham." *Dance Index*.

Horgan, P. *Encounters with Stravinsky*. Middletown, Conn.: Wesleyan University Press, 1989.

Hudson, L., and B. Jacot. *The Way Men Think*. New Haven, Conn.: Yale University Press, 1991.

Huffington, A. S. *Picasso: Creator and Destroyer*. New York: Simon & Schuster, 1988.

Hughes, H. S. *Consciousness and Society*. New York: Knopf, 1958.

Infeld, L. *Albert Einstein: His Work and Its Influence on Our World*. New York: Scribner's, 1950.

Inkeies, A. *Exploring Individual Modernity*. New York: Columbia University Press, 1983.

Jack, H. A. *The Gandhi Reader*. Bloomington: Indiana University Press, 1956.

Jameson, F. *Postmodernism, or the Cultural Logic of Late Capitalism*. Durham, N.C.: Duke University Press, 1991.

Janik, A., and S. Toulmin. *Wittgenstein's Vienna*. New York: Touchstone Press, 1973.

Johnson, P. *The Birth of the Modern*. New York: HarperCollins, 1991.

Johnson-Laird, P. N. "Freedom and Constraints in Creativity." In R. J. Sternberg, ed., *The Nature of Creativity*. New York: Cambridge University Press, 1988, pp. 202–219.

John-Steiner, V. *Notebooks of the Mind*. Albuquerque: University of New Mexico Press, 1985.

Jones, E. *The Life and Work of Sigmund Freud*. Edited and abridged by Lionel Trilling and Steven Marcus. New York: Basic Books, 1961.

Jowitt, D. *Time and the Dancing Image*. New York: Morrow, 1988.

Kakutani, M. "Review of A. N. Wilson, *Eminent Victorians*," *New York Times*. n.d.

———. "Henri Troyat and His Life of Flaubert." *New York Times*, December 8, 1992, p. C19.

Kaye, E. "I See You as a Goddess." *Mirabella,* July 1991, pp. 42–46.

Kenner, H., ed. *T. S. Eliot: A Collection of Critical Essays*. Englewood-Cliffs, N.J.: Prentice-Hall, 1972.

———. "The Urban Apocalypse." In A. W. Litz, ed., *Eliot in His Time: Essays on the Occasion of the Fiftieth Anniversary of* The Waste Land. Princeton, N.J.: Princeton University Press, 1973.

Kernan, A. "Radical Literal Criticism May Represent the Last Phases of an Older Order Collapsing." *Chronicle of Higher Education*, September 19, 1990, pp. B1-B3.

Kimmelman, M. "Modern to Show New Picasso Tomorrow." *New York Times*, February 10, 1992, p. C13.

Kisselgoff, A. "A Graham Family Reunion." *New York Times*, June 10, 1977.

———. "Martha Graham." *New York Times Magazine*, February 19, 1984, pp. 44–55.

———. "Martha Graham Dies at 96: A Revolutionary in Dance." *New York Times*, April 2, 1991, pp. A1, B7.

Koehler, W. *The Task of Gestalt Psychology*. Princeton, N.J.: Princeton University Press, 1969.

Kozinn, A. "Raising Questions in 9 All-Stravinsky Concerts." *New York Times*, July 6, 1990, p. C5.

Kramer, H. "Yet Another Surprise from Picasso." *Insight*, December 22, 1986, pp. 69–70.

———. "The Man Who Held the Cubists Together: Review of P. Assouline, *An Artful Life: A Biography of D. H. Kahnweiler*." *New York Times Book Review*, September 2, 1990, pp. 8–9.

Kroeber, A. *Configurations of Cultural Growth*. Berkeley: University of California Press, 1944.

Kuhn, T. *The Structure of Scientific Revolutions*. 2d ed. Chicago: University of Chicago Press, 1970.

Langbaum, R., "Modes of Characterization in *The Waste Land*." In A. W. Litz, ed., *Eliot in His Time: Essays on the Occasion of the Fiftieth Anniversary of* The Waste Land. Princeton, N.J.: Princeton University Press, 1973.

Langley, P., H. Simon, G. L. Bradshaw, and J. M. Zytkow. *Scientific Discovery*. Cambridge, Mass.: MIT Press, 1987.

Laporte, P. M. "Cubism and Relativity with a Letter of Albert Einstein." *Leonardo* 21 (1988): 313–315.

Lasch, C. *The True and Only Heaven*. New York: Norton, 1991.

Leavis, F. R. "*The Waste Land*." In H. Kenner, ed., *T. S. Eliot: A Collection of Critical Essays*. Englewood Cliffs, N.J.: Prentice-Hall, 1972.

Lehman, H. C. *Age and Achievement*. Princeton, N.J.: Princeton University Press, 1953.

Leighten, P. *Re-Ordering the Universe: Picasso and Anarchism, 1897–1914*. Princeton, N.J.: Princeton University Press, 1989.

Lenin, V. I. *Essential Works of Lenin*. New York: Dover, 1916.

Le Rider, J. *Modernité viennoise et crises de l'identité*. Paris: Presses Universitaires de France, 1990.

Lesure, F. *Igor Stravinsky: Le sacre du printemps. Dossier de Press*. Geneva: Minkoff, 1980.

Lewis, W. "Early London Environment." In H. Kenner, ed., *T. S. Eliot: A Collection of Critical Essays*. Englewood Cliffs, N.J.: Prentice-Hall, 1972.

Libman, L. *And Music at the Close*. New York: Norton, 1972.

Litz, A. W. "*The Waste Land* Fifty Years After." In A. W. Litz, ed., *Eliot in His Time: Essays on the Occasion of the Fiftieth Anniversary of* The Waste Land. Princeton, N.J.: Princeton University Press, 1973.

Lukacs, J. *Budapest 1900: A Historical Portrait of a City and Its Culture*. London: Weidenfeld and Nicholson, 1989.

———. "The Short Century. It's Over." *New York Times*, February 17, 1991, sec. 4, p. 13.

Lutz, T. *American Nervousness, 1903*. Ithaca, N.Y.: Cornell University Press, 1991.

MacKinnon, D. W. "Personality Correlates of Creativity: A Study of American Architects." In G. S. Neilsen, ed., *Proceedings of the Fourteenth International Congress of Applied Psychology, vol. 2*. Copenhagen: Munksgaard, 1962, pp. 11–39.

Malraux, A. *The Voices of Silence*. Garden City, N.Y.: Doubleday, 1963.

Martin, J. Dance reviews. *New York Times*, October 22, 1929; January 5, 1930; and February 8, 1931.

———. *America Dancing*. New York: Dodge, 1936.

Martin, M. *A Half Century of Eliot Criticism: An Annotated Bibliography of Books and Articles in English, 1916–1965*. Lewisburg, Pa.: Bucknell University Press, 1975.

Martindale, C. *The Clockwork Muse*. New York: Basic Books, 1990.

Masson, J. M. *The Complete Letters of Sigmund Freud to Wilhelm Fliess, 1887–1904*. Cambridge, Mass.: Harvard University Press, 1985.

Matthews, G. *Philosophy and the Young Child*. Cambridge, Mass.: Harvard University Press, 1980.

Mazo, J. H. *Prime Movers: The Makers of Modern Dance in America*. New York: Morrow, 1977.

McCosh, C. *Martha Graham: An American Original*. Film. n.d.

McDonough, D. *Martha Graham: A Biography*. New York: Praeger, 1973.

McLuhan, M. *Understanding Media: The Extensions of Man*. New York: McGraw-Hill, 1964.

Medawar, P. *Induction and Intuition*. Philadelphia: American Philosophical Society, 1969.

Medcalf, S. "The Shaman's Secret Heart: T. S. Eliot as Visionary, Critic, and Humorist." *Times Literary Supplement*, October 2, 1992, pp. 10–12.

Mehta, V. *Mahatma Gandhi and His Apostles*. New York: Viking Press, 1976.

Merton, R. K. "Singletons and Multiples in Scientific Discovery. A Chapter in the Sociology of Science." *Proceedings of the American Philosophical Society* 105 (1961): 470–486.

———. "The Matthew Effect in Science." *Science* 159 (1968): 56–63.

Miller, A. I. *Frontiers of Physics: 1900–1911: Selected Essays*. Boston: Birkhauser, 1986a.

———. *Imagery in Scientific Thought*. Cambridge, Mass.: MIT Press, 1986b.

———. "Scientific Creativity: A Comparative Study of Henri Poincaré and Albert Einstein." *Creative Research Journal*, in press.

Miller, J. E. *T. S. Eliot's Personal Waste Land: Exorcism of the Demons*. University Park: Pennsylvania State University Press, 1977.

Morgan, B. *Martha Graham: Sixteen Dances in Photographs*. Dobbs Ferry, N.Y.: Morgan Press, 1941.

Morris, G. L. K. "Marie, Marie, Hold On Tight." In H. Kenner, ed., *T. S. Eliot: A Collection of Critical Essays*. Englewood Cliffs, N.J.: Prentice-Hall, 1972.

Moss, H. "Masterpieces: A Review of Rainer Maria Rilke's *Letters on Cezanne*." *New Yorker*, July 7, 1986, pp. 80–82.

Murray, H. A. *Endeavors in Personality*. New York: Harper and Row, 1981.

Museum of Modern Art. *Picasso and Braque: Pioneering Cubism*. New York: Museum of Modern Art, 1989.

Musil, R. *Man Without Qualities*. New York: Perigree, 1980.

Nambodiripod, E. M. S. *The Mahatma and the Isms.* Calcutta: National Book Agency, 1981.

Nanda, B. R. *Gandhi and His Critics.* Delhi: Oxford University Press, 1985.

Nayar, R. "Crises of Nation and Creed." *Times Literary Supplement,* June 8–14, p. 603.

Nelson, B., ed. *Freud on Creativity and the Unconscious.* New York: Harper and Row, 1958.

Newell, A., and H. Simon. *Human Problem Solving.* Englewood Cliffs, N.J.: Prentice-Hall, 1972.

Nordmann, C. *Einstein and the Universe: A Popular Exposition of the Famous Theory.* London: Unwin, 1922.

Norman, C. *Ezra Pound,* rev. ed. London: Minerva Press, 1969.

Norris, C. *What's Wrong with Post Modernism.* Hemel Hempstead, England: Harvester Wheatsheaf, 1991.

Nunberg, H., ed. *Minutes of the Vienna Psychoanalytic Society, vol. 1.* New York: International Universities Press, 1962.

Ozick, C. "T. S. Eliot at 101." *New Yorker,* November 20, 1989, pp. 119–154.

Pais, A. *Subtle Is the Lord. The Science and the Life of Albert Einstein.* New York: Oxford University Press, 1982.

Payne R. *The Life and Death of Mahatma Gandhi.* New York: Dutton, 1990.

Penrose, R. *Picasso: His Life and Works,* 3d ed. Berkeley: University of California Press, 1981.

Perkins, D. N. *The Mind's Best Work.* Cambridge, Mass.: Harvard University Press, 1981.

———. "Creativity: Beyond the Darwinian Paradigm." Paper prepared for the Achievement Project Symposium, Kent, England, December 13–15, 1991.

Perloff, M. *Post-Modern Genres.* Norman: University of Oklahoma Press, 1989.

Pfaff, W. "Fallen Hero. Review of T. E. Lawrence." *New Yorker,* May 8, 1989, pp. 105–115.

Piaget, J. *The Child's Conception of the World.* Totowa, N.J.: Littlefield, Adams, 1965.

———. *The Child's Conception of Movement and Speed.* London: Routledge and Kegan Paul, 1970.

Pierpont, C. R. "Maenads," *New Yorker,* August 20, 1990, pp. 82–91.

Polak, H. S., H. N. Brailsford, and L. Pethick-Lawrence. *Mahatma Gandhi.* London: Odheims Press, 1949.

Polanyi, M. *Personal Knowledge.* Chicago: University of Chicago Press, 1958.

Popper, K. *The Poverty of Historicism.* New York: Harper and Row, 1964.

———. *Unended Quest.* London: Fontana/Collins, 1976.

Postman, N. *The Disappearance of Childhood.* New York: Delacorte Press, 1982.

Pound, E. "Mr. Eliot's Solid Merit." In H. Kenner, ed., *T. S. Eliot: A Collection of Critical Essays.* Englewood Cliffs, N.J.: Prentice-Hall, 1972.

Pribram, K., and M. M. Gill. *Freud's Project Reassessed.* New York: Basic Books, 1976.

Pyenson, L. *The Young Einstein.* Bristol, England: Hilger, 1985.

Raymond, C. "Study of Patient Histories Suggests Freud Expressed or Distorted Facts That Contradicted His Theories." *Chronicle of Higher Education,* May 19, pp. A4–A6.

Rhodes, R. *The Making of the Atomic Bomb.* New York: Simon & Schuster, 1986.

Richardson, J. *A Life of Picasso: Volume 1, 1881–1906.* New York: Random House, 1991.

Riding, A. "Contrite Paris Hails Nijinsky's 'Sacre.'" *New York Times,* September 29, 1990, p. 17.

Rieff, P. *Freud: The Mind of the Moralist.* New York: Viking Press, 1959.

Rogosin, E. *The Dance Makers.* New York: Walker, 1980.

Rolland, R. *Mahatma Gandhi.* New York: Century, 1924.

Rosenberg, H. *The Tradition of the New.* New York: Horizon, 1959.

Rotenstreich, N. "Relativity and Relativism." In G. Holton and Y. Elkana, eds., *Albert Einstein: Historical and Cultural Perspecti*ves. Princeton, N.J.: Princeton University Press, 1982, pp. 175–204.

Russell, F. D. *Picasso's* Guernica. London: Thames and Hudson, 1980.

Russell, J. "Picasso's Sketchbooks Show the Prolific Talent of a Genius." *New York Times,* April 27, 1986, pp. 1, 39.

Saal, H. "Goddess in the Wings." *Newsweek,* May 14, 1973, p. 87.

Schank, R. C. "Creativity as a Mechanical Process." In R. J. Sternberg, ed., *The Nature of Creativity.* New York: Cambridge University Press, 1988, pp. 220–242.

Schiff, G. *Picasso in Perspec*tive. New York: Prentice Hall, 1976.

Schilpp, P. A. *Albert Einstein: Philosopher-Scientist.* Evanston, Ill.: Library of Living Philosophers, 1949.

Schonberg, H. "It All Came Too Easily for Camille Saint-Saëns." *New York Times,* January 12, 1969, sec. 2, p. 17.

Schorske, C. *Fin-de-Siècle Vienna: Politics and Culture.* New York: Knopf, 1979.

Scott, Spencer. "Being Different." *New York Times Magazine,* September 22, 1991, p. 47.

Sencourt, R. *T. S. Eliot: A Memoir.* New York: Dodd Mead, 1979.

Sennett, R. "Fragments Against the Ruin: A Review of A. Giddens, *The Consequences of Modernity.*" *Times Literary Supplement,* February 8, 1991, p. 6.

Shattuck, R. *The Banquet Years.* London: Cope, 1969.

Sheean, V. *Kindly Light.* New York: Random House, 1949.

Shelton, S. *Divine Dancer: A Biography of Ruth St. Denis.* New York: Doubleday, 1981.

Shirer, W. L. *Gandhi: A Memoir.* New York: Simon & Schuster, 1979.

Showalter, E. *Sexual Anarchy: Gender and Culture at the Fin-de-Siècle.* New York: Viking, 1990.

Siegel, M. B. "A Visit to the Lighthouse That Is Martha Graham." *Boston Globe,* April 29, 1973, sec. B, p. 21.

———. *The Stages of Design: Images of American Dance.* Boston: Houghton Mifflin, 1979.

Simonton, D. K. *Genius, Creativity, and Leadership.* Cambridge, Mass.: Harvard University Press, 1984.

———. "Creativity, Leadership, and Chance." In R. J. Sternberg, ed., *The Nature of Creativity.* New York: Cambridge University Press, 1988, pp. 386–436.

———. *Scientific Genius.* New York: Cambridge University Press, 1989.

———. *Psychology, Science, and History.* New Haven, Conn.: Yale University Press, 1990.

Skinner, B. F. *The Science of Behavior.* New York: Macmillan, 1953.

Snell, R. "A Dialogue with Tradition: Review of *On Classic Ground: Picasso, Leger, de Chirico and the New Classicism.*" *Times Literary Supplement,* June 22–28, 1990, p. 669.

Snow, C. P. *The Two Cultures and the Scientific Revolution.* New York: Cambridge University Press, 1959.

Spencer, S. "Being Different." *New York Times Magazine,* September 22, 1991.

Spender, S. *T. S. Eliot.* New York: Viking Press, 1975.

Spurr, D. "The Inner Xanadu of *The Waste Land.*" In D. Spurr, ed., *Conflicts in Consciousness: T. S. Eliot's Poetry and Criticism.* Urbana: University of Illinois Press, 1984.

Stachel, J., ed. *The Collected Papers of Albert Einstein, Volume 1: The Early Years, 1879–1902.* Princeton, N.J.: Princeton University Press, 1987.

――――. *The Collected Papers of Albert Einstein, Volume 2: The Swiss Years: Writings 1900–1909.* Princeton, N.J.: Princeton University Press, 1989.

――――. Presentation on Einstein and Judaism at Workshop on the Young Einstein, N. Andover, Mass., October 1990.

Staller, N. "Early Picasso and the Origins of Cubism." *Arts Magazine* 61 (1986): 80–90.

Steegmüller, F. *Your Isadora.* New York: Random House, 1974.

Stein, G. *Gertrude Stein on Picasso.* Edited by E. Burns. New York: Liveright, 1970.

Steinberg, M. "Graham: Sometimes Maddening but Exciting." *Boston Globe,* November 19, 1973.

Sternberg, R. J. *Beyond IQ.* New York: Cambridge University Press, 1985.

――――, ed. *The Nature of Creativity.* New York: Cambridge University Press, 1988.

Stevens, M. "Low and Behold." *New Republic,* December 24, 1990, pp. 27–33.

Stodelle, E. *Deep Song: The Dance Story of Martha Graham.* New York: Schimer, 1984.

Stoppard, T. *Travesties.* New York: Grove, 1975.

Strachey, L. *Eminent Victorians.* London: Bloomsbury, 1988; originally published, 1918.

Stravinsky, I. *An Autobiography.* New York: Simon & Schuster, 1936.

――――. *An Autobiography.* New York: Norton, 1962.

――――. *The Rite of Spring, Sketches, 1911–1913.* London: Boosey and Hawkes, 1969.

――――. *The Poetics of Music.* Cambridge, Mass.: Harvard University Press, 1970.

――――. *The Rite of Spring in Full Score.* New York: Dover, 1989.

Stravinsky, I., and R. Craft. *Expositions and Developments.* London: Faber and Faber, 1962.

Sulloway, F. J. *Freud: Biologist of the Mind.* New York: Basic Books, 1983.

Svarny, E. *"The Men of 1914": T. S. Eliot and Early Modernism.* Philadelphia: Open University Press, 1988.

Swenson, L. S. *Genesis of Relativity: Einstein in Context.* New York: Burt and Franklin, 1979.

Tansman, I. *Igor Stravinsky: The Man and His Music.* New York: Putnam's, 1949.

Taylor, C. *Sources of the Self: The Making of the Modern Identity.* Cambridge, Mass.: Harvard University Press, 1989.

Taylor, R. "Picasso: Sketchbooks a Dazzling Revelation of Artist's Imagination." *Boston Globe,* July 13, 1986a, pp. A1, A13.

———. "Picasso: Fragments of Genius." *Boston Sunday Globe*, October 19, 1986b, pp. 105–106.

Terry, W. "The Legacy of Isadora Duncan and Ruth St. Denis." *Brooklyn* 1960, number 5.

———. *Frontiers of Dance: The Life of Martha Graham*. New York: Crowell, 1975.

———. *Ted Shawn: Father of American Dance*. New York: Dial, 1976.

———. *I Was There: Selected Dance Reviews and Articles 1936–1976*. Dekker: Audience Arts, 1978.

Teuber, M. *Kubismus: Kuenstler, Themes, Werke, 1907–1920*. Cologne: Josef-Haubrich-Kunsthalle, 1982.

Tobias, T. "A Conversation with Martha Graham." *Dance Magazine*, March 1984, p. 64.

Torrance, E. P. *Guiding Creative Talent*. Englewood Cliffs, N.J.: Prentice Hall, 1962.

———. "The Nature of Creativity as Manifest in Its Testing." In R. J. Sternberg, ed., *The Nature of Creativity*. New York: Cambridge University Press, 1988, pp. 43–75.

Toulmin, S. *Cosmopolis: The Hidden Agenda of Modernity*. New York: Free Press, 1990.

Tuchman, B. *The Proud Tower*. New York: Bantam Books, 1967.

Van den Toorn, P. C. *The Music of Igor Stravinsky*. New Haven, Conn.: Yale University Press, 1983.

———. *Stravinsky and* The Rite of Spring: *The Beginnings of a Musical Language*. Berkeley: University of California Press, 1987.

Varnedoe, K. *Vienna 1900: Art, Architecture, and Design*. Boston: Little, Brown, 1986.

Vasari, G. *Lives of the Artists*. New York: Penguin, 1987.

Vernon, P., ed. *Creativity*. London: Penguin, 1970.

Vlad, R. *Stravinsky*. 2d ed. London: Oxford University Press, 1967.

Waldholz, M. "Doubted and Resisted, Freud's Daring Map of the Mind Endures." *Wall Street Journal*, December 2, 1991, sec. A, pp. 1, 8.

Wallace, D., and H. E. Gruber, eds. *Creative People at Work*. New York: Oxford University Press, 1990.

Wallach, M. *The Intelligence-Creativity Distinction*. Morristown, N.J.: General Learning Corporation, 1971.

Wallas, G. *The Art of Thought*. New York: Harcourt Brace and World, 1926.

Walters, J., and H. Gardner. "The Crystallizing Experience." In R. J. Sternberg and J. Davidson, eds., *Conceptions of Giftedness*. New York: Cambridge University Press, 1986, pp. 306–331.

Ward, N. "Fourmillante Cité: Baudelaire and *The Waste Land*." In A. D. Moody, ed., The Waste Land *in Different Voices*. London: Arnold, 1974.

Watkins, M. Dance reviews. *New York Herald Tribune*, January 25, 1931, and February 22, 1931.

Weisberg, R. *Creativity, Genius, and Other Myths*. New York: Freeman, 1986.

Wertheimer, M. *Productive Thinking*. New York: Harper and Row, 1959.

White, E. W. *Stravinsky: A Critical Survey*. London: Lehmann, 1947.

White, E. W., and J. Noble. "Igor Stravinsky." In *The New Grove Dictionary of Music*. London: Macmillan, 1980, pp. 240–265.

Whitehead, A. N. *The Aims of Education and Other Essays*. New York: Macmillan, 1926.

Whitney, C. "Two More T. S. Eliot Poems Found." *New York Times*, November 2, 1991, p. 13.

Whyte, L. L. *The Unconscious Before Freud*. New York: St. Martin's Press, 1978.

Williamson, G. *A Reader's Guide to T. S. Eliot: A Poem by Poem Analysis*. New York: Farrar, Straus and Cudahy, 1953.

Wilford, J. N. "Letter to Supporter Records Einstein's Search for Proof." *New York Times*, March 24, 1992, p. C 1.

Wilson, A. N. *Eminent Victorians*. New York: Norton, 1990.

Wilson, E. *Axel's Castle*. New York: Fontana, 1959.

———. *To the Finland Station*. London: Fontana, 1962.

———. *The Twenties*. Edited by Leon Edel. New York: Farrar, Straus and Giroux, 1975.

Winn, M. *Children Without Childhood*. New York: Pantheon Books, 1983.

Wittels, F. *Sigmund Freud: His Personality, His Teaching and His School*. London: George Allen and Unwin, 1924.

Wohl, R. *The Generation of 1914*. Cambridge, Mass.: Harvard University Press, 1979.

Wolpert, S. *Tilak and Gokhale: Revolution and Reform in the Making of Modern India*. Berkeley: University of California Press, 1962.

Zuckerman, H., and R. K. Merton. "Age, Aging, and Age Structure in Science." In M. W. Riley, ed., *Aging and Society*. New York: Russell Sage Foundation, 1972, pp. 292–356.

NAME INDEX

SUBJECT INDEX